COMPREHENSIVE BIOCHEMISTRY

ELSEVIER/NORTH-HOLLAND BIOMEDICAL PRESS

335 Jan van Galenstraat, P.O. Box 211, Amsterdam

ELSEVIER/NORTH-HOLLAND, INC.

52, Vanderbilt Avenue, New York, N.Y. 10017

With 45 plates and 63 figures

PRINTED IN THE NETHERLANDS

COMPREHENSIVE BIOCHEMISTRY

COMPREHENSIVE
BIOCHEMISTRY

SECTION I (VOLUMES 1—4)

PHYSICO-CHEMICAL AND ORGANIC ASPECTS
OF BIOCHEMISTRY

SECTION II (VOLUMES 5—11)

CHEMISTRY OF BIOLOGICAL COMPOUNDS

SECTION III (VOLUMES 12—16)

BIOCHEMICAL REACTION MECHANISMS

SECTION IV (VOLUMES 17—21)

METABOLISM

SECTION V (VOLUMES 22—29)

CHEMICAL BIOLOGY

SECTION VI (VOLUMES 30—34)

A HISTORY OF BIOCHEMISTRY

COMPREHENSIVE BIOCHEMISTRY

EDITED BY

MARCEL FLORKIN

Emeritus Professor of Biochemistry, University of Liège (Belgium)

AND

ELMER H. STOTZ

Professor of Biochemistry, University of Rochester, School of Medicine and Dentistry, Rochester, N.Y. (U.S.A.)

VOLUME 33A

A HISTORY OF BIOCHEMISTRY

PART V. THE UNRAVELLING OF
BIOSYNTHETIC PATHWAYS

by

MARCEL FLORKIN

ELSEVIER SCIENTIFIC PUBLISHING COMPANY

AMSTERDAM · OXFORD · NEW YORK

1979

GENERAL PREFACE

The Editors are keenly aware that the literature of Biochemistry is already very large, in fact so widespread that it is increasingly difficult to assemble the most pertinent material in a given area. Beyond the ordinary textbook the subject matter of the rapidly expanding knowledge of biochemistry is spread among innumerable journals, monographs, and series of reviews. The Editors believe that there is a real place for an advanced treatise in biochemistry which assembles the principal areas of the subject in a single set of books.

It would be ideal if an individual or a small group of biochemists could produce such an advanced treatise, and within the time to keep reasonably abreast of rapid advances, but this is at least difficult if not impossible. Instead, the Editors with the advice of the Advisory Board, have assembled what they consider the best possible sequence of chapters written by competent authors; they must take the responsibility for inevitable gaps of subject matter and duplication which may result from this procedure.

Most evident to the modern biochemist, apart from the body of knowledge of the chemistry and metabolism of biological substances, is the extent to which we must draw from recent concepts of physical and organic chemistry, and in turn project into the vast field of biology. Thus in the organization of Comprehensive Biochemistry, sections II, III and IV, Chemistry of Biological Compounds, Biochemical Reaction Mechanisms, and Metabolism may be considered classical biochemistry, while the first and fifth sections provide selected material on the origins and projections of the subject.

It is hoped that sub-division of the sections into bound volumes will not only be convenient, but will find favour among students concerned with specialized areas, and will permit easier future revisions of the individual volumes. Towards the latter end particularly, the Editors will welcome all comments in their effort to produce a useful and efficient source of biochemical knowledge.

M. Florkin

Liège/Rochester

E.H. Stotz

PREFACE TO SECTION VI

(Volumes 30—34)

In the many chapters of previous sections of *Comprehensive Biochemistry* covering organic and physicochemical concepts (Section I), chemistry of the major constituents of living material (Section II), enzymology (Section III), metabolism (Section IV), and the molecular basis of biological concepts (Section V), authors have been necessarily restricted to the more recent developments of their topics. Any historical aspects were confined to recognition of events required for interpretation of the present status of their subjects. These latest developments are only insertions in a science which has had a prolonged history of development.

Section VI is intended to retrace the long process of evolution of the science of Biochemistry, framed in a conceptual background and in a manner not recorded in recent treatises. Part I of this section deals with Proto-biochemistry or with the discourses imagined concerning matter-of-life and forces-of-life before molecular aspects of life could be investigated. Part II concerns the transition between Proto-biochemistry and Biochemistry and retraces its main landmarks. In Part III the history of the identification of the sources of free energy in organisms is depicted. Part IV is devoted to early studies in Biosynthesis and Part V to the unravelling of biosynthetic pathways. While these latter parts are concerned with the molecular level of integration, Part VI is more specifically directed towards the history of molecular interpretations of physiological and biological concepts, and of the origins of the concept of life as the expression of a molecular order.

The *History* narrated in Section VI thus leads to the thresholds of the individual histories in the recent developments recorded by the authors of Section I—V of *Comprehensive Biochemistry*.

M. Florkin

Liège/Rochester
E.H. Stotz

CONTENTS

VOLUME 33A

A HISTORY OF BIOCHEMISTRY

Part V. The Unravelling of Biosynthetic Pathways

General Preface i
Preface to section VI viii
List of Plates..................................... xvii
Errata (Volume 32) xx

Introduction 1

Chapter 53. The energetics of biosynthesis

1. Introduction 29
2. ATP and biosynthesis 31
3. Discovery by Nachmansohn of the utilization of the energy of ATP
   in the biosynthesis of acetylcholine 33
4. Sulfanilamide acetylation 35
5. Coenzyme A 37
6. Mechanism of acetate activation 38

References 43

Chapter 54. Peptide bonds

(A) Replacement reactions in proteins and transpeptidations 45
1. The theory of incorporation by substitution 45

2. Replacement reactions in protein synthesis (peptide transfers) 49
3. Template hypothesis and peptide hypothesis 51
(B) Peptide bond synthesis 54
4. Demise of peptide bond formation by reversal of Mass Action 54
5. Inhibition of peptide synthesis by inhibition of oxidation 54
6. Peptide bond synthesis in homogenates 55
7. Activated form of benzoic acid in hippuric acid synthesis 55

References 59

Chapter 55. The pentose phosphate cycle

(A) Oxidative irreversible sequence from glucose-6-phosphate to ribu-
lose-5-phosphate 61
1. From glucose-6-phosphate to D-gluconate-6-phosphate 61
2. Enzymes of the oxidative pathway 64
    (a) Glucose-6-phosphate dehydrogenase 64
    (b) Lactonase 65
    (c) Phosphogluconate dehydrogenase 65
(B) Non-oxidative, reversible transfer from pentose phosphate to
hexose phosphate 67
3. Introduction 67
4. Phosphoribose isomerase and phosphoribose epimerase 67
5. Transketolase 69
6. Mechanism of transketolase action 70
7. Transaldolase 71
(C) Pentose phosphate cycle 72

References 78

Chapter 56. The photosynthetic cycle of carbon reduction

1. First experiments with ^{11}C. Demise of the formaldehyde theory.
    Discovery of CO_2 fixation in the dark 81
2. Identification of an ATP-dependent carboxylative phase 85
    (a) Discovery of ^{14}C and recognition of phosphoglyceric acid as an
        early intermediate 85
    (b) Introduction of paper chromatography 91
    (c) Ribulose and sedoheptulose as intermediates 92
    (d) Phosphoglyceric acid formation with cell-free enzyme prepara-
        tions 94
    (e) Enzymatic formation of phosphoglyceric acid from ribulose
        diphosphate and carbon dioxide 96

3. Identification of an ATP and NADPH-dependent reductive phase . . 97
   (a) Kinetic studies 97
   (b) Enzymes of the reductive phase 100
4. Regenerative phase 101
5. Synthesis of hexose phosphate from CO_2 in the dark in a model
   multi-enzyme system 102
6. The reductive carbohydrate cycle 103

References 106

Chapter 57. CO_2 fixation in heterotrophs. Gluconeogenesis

 1. Introduction 109
 2. Fixation of CO_2 in heterotrophic bacteria 111
 3. Carbon dioxide fixation in pigeon liver preparations 117
 4. Evans and Slotin's experiments demonstrating the participation of
    CO_2 in the biosynthesis of α-oxoglutarate 121
 5. Ruben and Kamen's experiments 122
 6. Direct demonstration of CO_2 fixation in mammalian tissues 124
 7. Phosphoenolpyruvate carboxykinase 129
 8. Pyruvate carboxylase 131
 9. Fructose diphosphatase 134
10. From glucose-6-phosphate to glucose 135
11. Recapitulation of intermediate stages of gluconeogenesis 135

References 138

Chapter 58. Biosynthesis of complex saccharides from monosaccharides

 1. Introduction 141
 2. The concept of transglycosylation 141
 3. The phosphorylase theory of glycogen and starch biosynthesis
    (1939—1957) 144
 4. The unravelling of glucan structures 148
 5. Phosphorylase theory of sucrose biosynthesis 150
 6. Reactions catalysed by glucosyltransferase 150
 7. Introduction of the concept of a branching enzyme 151
 8. Sugar nucleotides and their function in the interconversions of
    sugars 153
 9. The role of sugar nucleotides in transglycosylation 156
    (a) The formation of glycosidic bonds by transglycosylation 156
    (b) UDPG as glucose donor in the synthesis of trehalose phosphate 157

(c) UDPG as glucose donor in the synthesis of sucrose and of sucrose phosphate 157
(d) UDPG as glucose donor in the synthesis of cellulose in microorganisms 158
(e) UDPG as glucose donor in the biosynthesis of glycogen 159
(f) UDPG as glucose donor in the biosynthesis of starch 160
(g) UDP-galactose as a galactose donor to glucose in lactose biosynthesis 161
(h) Glycosyl transfer from UDPAG in chitin biosynthesis 164
10. The enzymes of the biosynthesis of complex saccharides from sugar nucleotides 164

References 166

Chapter 59. Biosynthesis of fatty acids and glycerides

1. Fatty acid biosynthesis considered as the reversal of β-oxidation .. 171
2. Attempts at biosynthesis of fatty acids in the presence of β-oxidation enzymes and coenzymes 173
3. Biosynthesis of fatty acids in cell-free preparations 173
4. Absolute requirement for bicarbonate. Malonyl-CoA as intermediate 175
5. Acetyl-CoA carboxylase 178
6. Fatty acid synthase 179
7. Discovery of acyl carrier protein 180
8. Transacylases 181
9. Acyl-malonyl-ACP condensing enzyme 181
10. β-Ketoacyl-ACP reductase 182
11. Enoyl-ACP hydrase 182
12. Enoyl-ACP reductase 183
13. Elongation of fatty acids in animal tissues 183
    (a) Mitochondrial elongation system 183
    (b) Microsomal elongation system 183
14. Desaturation of fatty acids 185
15. Biosynthesis of triglycerides 186

References 190

Chapter 60. Biosynthesis of tetrapyrroles and of corrinoids

1. Introduction 193
2. Metabolic sources of the atoms of porphyrins 196
3. Relationship of the tricarboxylic acid cycle and porphyrin formation 207

4. Recognition of the role of δ-aminolaevulinic acid and of porphobilinogen 213
5. Biosynthesis of δ-aminolaevulinic acid 217
6. From δ-aminolaevulinic acid to porphobilinogen 220
7. From porphobilinogen to tetrapyrroles 221
8. Metalloporphyrins 227
   (a) The iron branch 227
   (b) The magnesium branch 229
9. Biosynthesis of corrinoids 231
10. Prodigiosin 233

References 235

Chapter 61. The isprenoid pathway (terpenes, carotenoids, side chain of tocopherols, ubiquinones, dolichols, rubber, sterols and steroids, steroid hormones, bile acids and bile alcohols, arthropod hormones, insect pheromones)

1. Introduction 239
2. "Acetate" as starting point of the isoprenoid pathway 240
3. Acetoacetate as an intermediate in the biosynthesis of cholesterol 243
4. Distribution of acetate carbons in cholesterol 243
5. Squalene confirmed as cholesterol precursor 247
6. Biosynthesis of cholesterol by cell-free preparations of rat liver and
   by microsomes 249
7. Role of coenzyme A 250
8. Cyclisation of squalene 250
9. Metabolic precursors of the isoprene units of squalene 257
10. From mevalonic acid to isopentenyl pyrophosphate and dimethyl-
    allylpyrophosphate 260
11. Identification of the "biological isoprene units" and their conver-
    sion to farnesyl pyrophosphate 263
12. Condensation of farnesyl pyrophosphate. Presqualene 265
13. Conversion of lanosterol to cholesterol 273
14. Biosynthesis of mevalonic acid from acetic acid 273
15. From mevalonic acid to rubber 275
16. Biosynthesis of plant terpenes 276
17. Nature and biosynthesis of the C_{40} carotenoid precursor
    (phytoene) 277
18. Biochemical sequence to C_{40} conversion 282
19. From mevalonic acid to isoprenoid alkaloids 284
20. From mevalonic acid to isoprenoid quinones, chromanes and
    chromers 284
21. From cholesterol to steroid hormones 287

22. Biosynthesis from pregnenolone in the adrenal cortex 289
23. Biosynthesis from pregnenolone in the testis 294
24. Biosynthesis from pregnenolone in the ovary 296
25. Biosynthesis from pregnenolone in the corpus luteum 297
26. Biosynthesis from pregnenolone in the placenta 297
27. Enzymes of the biosynthesis of steroid hormones 300
28. Biosynthesis of bile acids and bile alcohols 300
29. Conjugation of bile acids and alcohols 311
30. Insect and crustacean ecdysones 311
31. Isoprenoids in insects 313

References 316

Chapter 62. Biosynthesis of the purine nucleus

1. Isotopic experiments on whole animals 325
2. Hypoxanthine synthesis by pigeon liver homogenates 330
3. Inosinic acid and purine biosynthesis 330
4. Experiments with 5-amino-4-imidazole carboxamide 333
5. Enzymatic synthesis and utilization of PRPP 333
6. Origin of PRPP 334
7. Identification of two ribotide derivatives of glycinamide as intermediates in the biosynthesis of inosinic acid 335
8. From PRPP to FGAR 336
9. Conversion of FGAR to FGAM and then to AIRP, with formation of the imidazole ring 338
10. From AIRP to AICRP 341
11. From AICRP to IMP 343
12. Interconversions among purine nucleotides 345
13. Formation of adenosine monophosphate 345
14. Formation of xanthosine and guanosine phosphates 345
15. From IMP to uric acid 348
16. Enzymes for the further degradation of uric acid 350
17. Persistence of the theory of Wiener in the case of molluscs and the demise of this theory 350
18. The case of uricotelic animals 352

References 353

Chapter 63. Extensions on the pathway of purine biosynthesis: riboflavin, pterin, pteridine. Tetrahydrofolic acid as coenzyme of C_1 transfer in biosynthesis

1. Pterins and pteridines 357
2. Riboflavin 357

 3. Biosynthesis of riboflavin 358
 4. Biosynthesis of pteridines 359
 5. Folic acid 362
    (a) Introduction 362
    (b) Biosynthesis 363
    (c) Enzymic aspects of folic acid biosynthesis 368
 6. Concept of transfer of C_1 units as catalysed by pteroin proteins
    containing, as coenzymes, derivatives of tetrahydrofolic acid 368
 7. Enzymatic production of tetrahydrofolate 371
 8. Formylated natural derivatives related to folic acid and their inter-
    conversions 372
    (a) Rhizopterin 372
    (b) Teropterin 373
    (c) N^5-Formyltetrahydrofolic acid (folinic acid, citrovorum factor,
        leucovorin) 373
    (d) N^{10}-Formyltetrahydrofolic acid 374
 9. "Active formate" and "active formaldehyde" 378
10. Polyglutamates of pteroic acid 381
11. Recognition of specific participation of folate factors in C_1 trans-
    fers 381

References 385

Subject Index 389
Name Index 419

Section VI

A HISTORY OF BIOCHEMISTRY

Vol. 30.   A History of Biochemistry
            Part I. Proto-Biochemistry
            Part II. From Proto-Biochemistry to Biochemistry
Vol. 31.   A History of Biochemistry
            Part III. History of the identification of the sources of
            free energy in organisms
Vol. 32.   A History of Biochemistry
            Part IV. Early studies on biosynthesis
Vol. 33.   A History of Biochemistry
            Part V. The unravelling of biosynthetic pathways
Vol. 34.   A History of Biochemistry
            Part VI. History of molecular interpretations of physio-
            logical and biological concepts, and the origins of the
            conception of life as the expression of a molecular order

LIST OF PLATES

(Unless otherwise stated, the portraits belong to the author's personal collection)

Plate 178. Hans Thacker Clarke. By courtesy of *Annual Reviews of Biochemistry*, 27 (1958) 1.
Plate 179. Rudolph Schoenheimer.
Plate 180. David Rittenberg.
Plate 181. David Nachmansohn, photograph by Lotte Jacobi, New York.
Plate 182. Paul Berg.
Plate 183. Max Bergmann.
Plate 184. Joseph Stewart Fruton.
Plate 185. Hubert Chantrenne.
Plate 186. Frank Dickens.
Plate 187. Bernard Leonard Horecker.
Plate 188. Efraim Racker. By courtesy of Dr. A.G. Katki, Hanover, New Hampshire.
Plate 189. Martin David Kamen (1913).
Plate 190. Sam Ruben (left) and William Zev Hassid (right). By courtesy of Martin D. Kamen.
Plate 191. Melvin Calvin.
Plate 192. A.A. Benson.
Plate 193. Chester Hamlin Werkman. By courtesy of Russel W. Brown, Biographical Memoirs, 44 (1974) 329.
Plate 194. Harland Goft Wood.
Plate 195. Earl A. Evans Jr.
Plate 196. Albert Baird Hastings.
Plate 197. Merton F. Utter.
Plate 198. William Zev Hassid.
Plate 199. Luis Federico Leloir.
Plate 200. Samuel Gurin.
Plate 201. Salih J. Wakil.
Plate 202. Richard Willstätter. By courtesy of Prof. Kretovich, Moscow.
Plate 203. David Shemin.
Plate 204. Albert Neuberger.
Plate 205. Sam Granick, by courtesy of Rockefeller University Archives.
Plate 206. L. Bogorad.
Plate 207. Konrad Bloch.

Plate 208. Sir John Cornforth and Lady R. Cornforth.
Plate 209. George Joseph Popjàk.
Plate 210. Harold John Channon.
Plate 211. Leopold Ruzicka. By courtesy of *Annu. Rev. Biochem.*, 42 (1973) 1.
Plate 212. Karl Folkers.
Plate 213. T.W. Goodwin.
Plate 214. Leo T. Samuels.
Plate 215. Henry Danielsson.
Plate 216. G.A.D. Haslewood.
Plate 217. Peter Karlson.
Plate 218. John Machlin Buchanan.
Plate 219. Donald Devereux Woods. By courtesy of Sir Hans Krebs.
Plate 220. Gene M. Brown.
Plate 221. T. Shiota.
Plate 222. Lothar Jaenicke.
Plate 223. F.M. Huennekens.

Volume 32

Errata and Corrigenda

p. 4,   line 32, instead of *materialist,* read *mecanicist*

p. 27, line 9, instead of *1749,* read *1730.*

p. 58, caption, read *Plate 134. Sergei Nikolaevich Winogradsky*

p. 77, line 15, instead of *Listar,* read *Lister*

p. 78, last line, instead of *structure,* read *structureless*

p. 96, line 18, instead of *(Chapter III—V),* read *(Chapters III—V)*

p. 107, line 25, instead of *do not read,* read *do not react*

p. 153, line 3, instead of *Chapter I—IV,* read *Chapters I—IV*

p. 159, line 19, instead of *Ritthausen was voiding,* read *Ritthausen was voicing*

p. 175, line 18, instead of *on the formation,* read *on the information*

p. 208, reference 35, instead of *J, Needham,* read *J. Needham*

p. 220, caption, read *Plate 162. Frank C. Mann*

p. 226, reference 20, instead of *cited by Bollman et al.,* read *cited by Bollman et al. [16]*

p. 283, line 6, instead of *Schoenmeier,* read *Schoenheimer*

p. 290, at the bottom of the page, the structure formula of creatinine should read

$$\begin{array}{ccc} & N(CH_3) \cdot CH_2 & \\ & | & | \\ NH-C & & \\ | & & \\ NH & \!\!\!\!-\!\!\!\!- & CO \end{array}$$

p. 294, line 5, from below, instead of *the same data,* read *the same date*

p. 307, line 11, instead of *maltase,* read *maltose*

p. 335, column 1, line 8, instead of *aerial mutation,* read *aerial nutrition*

p. 335, column 1, line 9, instead of *phlogistotheory,* read *phlogiston theory.*

p. 342, column 1, line 15, instead of *Person*, read *Persoz*.

p. 343, column 2, line 32, instead of *Isalis*, read *Isatis*.

p. 344, column 2, line 1, instead of *Laserta*, read *Lacerta*

p. 344, column 2, line 14, instead of *Menta*, read *Mentha*

p. 348, column 2, line 45, instead of *Salaskow*, read *Salaskin*

p. 349, column 1, line 17, instead of *Rana esculanta*, read *Rana esculenta*

p. 349, column 2, line 11, instead of *salon sperm heads*, read *salmon sperm heads*

p. 350, column 2, line 33, instead of *Tropicone*, read *Tropinone*

p. 352, column 2, line 3, instead of *U. Bernard*, read *C. Bernard*

p. 353, column 2, line 6, read *Berthelot, D., 148*

p. 353, column 2, line 8, read *Bethelot, M. 177—179, 180, 181.*

p. 354, column 1, line 4, delete *Clausius, 118*

p. 354, column 1, line 6, instead of *Bordeau*, read *Bordeu*

p. 354, column 2, line 13, instead of *Clausius, 118*, read *Clausius, R.E., 118*

p. 354, column 2, line 20, instead of *Compte, A., 11*, read *Comte, A., 11*

Introduction

1. Foreword

The period of early studies on biosynthesis reviewed in Part IV of this *History*, spans from Joseph Black's studies on "fixed air" to the introduction of deuterium in metabolic studies by Rudolph Schoenheimer in 1935. During this period, we have studied the interplay between chemistry and biology in the field of anabolic * studies. This interplay, during the period considered, suffered of a number of roadblocks.

Though aimed at grasping the phenomena of biosynthesis in their natural settings inside organisms, the physiologists were often warped in their chronic sin, organicism. Many of them, including the most justly famous, believed in the obscure properties of a biological entity, "protoplasm", within which, as stated by M. Stephenson [1],

"occurred pseudo-chemical events, mysterious and undecipherable"

against which Hopkins has fought with rational methodology, illustrating the dictum of Bachelard:

"The history of sciences is the history of the defeats of irrationalism".

Given the keys to unlock, the biosynthetic theories developed

---

* As we have noted before (Vol. 30, p. 9) the terms *anabolism* and *catabolism* were coined by the physiologist Gaskell in 1886.

after 1935 are of internalist nature. They belong to research programs in Chemistry. *

While other parts of this *History* have been concerned with outworn and discredited theories and with the reasons for their introduction and for their dismissal, we already have, in Part III, studied an aspect of contemporary history of science, the identification of the sources of free energy in organisms, also of internalist nature. In Part IV (Vols. 33A and 33B), we shall again consider a topic of contemporary history, the unravelling of biosynthetic pathways. In both cases, the presentation deals with the historical development of present knowledge, e.g. it retraces the methodology of progress in modern science and provides the basic justification of the theories accepted today. It may be noticed that the textbooks of biochemistry, while providing a heavy burden of memorizable material to the brains of unfortunate students, do not provide this kind of information. The history of biochemistry thus becomes a part not only of history but also of biochemistry.

On the other hand, during the period covered in Part IV, the chemists became more and more able to accomplish in vitro all forms of possible biosyntheses of a given biomolecule, but they remained unable to grasp which one was adopted by "the chemist in the body", lacking, as they were, the ability to identify in situ the biological pathways. Altogether, organic chemistry had proven a poor guide in the study of biosynthesis. But the chemists paid no attention to the "mysterious and undecipherable" properties of "protoplasm". Their methodology had originated in the rational approach of Dumas, based on the properties of material units defined according to criteria and conceptual systems elaborated by physical and chemical sciences. Their dominant principle was that there was, in Nature, only one chemistry. The first key which allowed to unlock the biosynthetic secrets was of entirely experimental nature and was provided by Schoenheimer when he introduced the tracer isotopes. Other keys were introduced with the recourse to exacting mutants of microorganisms and with the accomplishment of enzyme purification, permitting to identify single steps in metabolic chains. These methodological acquisitions

---

* This does not apply to the theories of the integration and regulation of metabolic pathways, which will be considered at a later stage of this *History*.

belong to an episode of horizontal history flowing from several sources. They must be considered in the context of the psychology of the scientist and of the sociology of science, as we shall do in this Introduction.

In the perspective adopted, as stated by Dijksterhuis [2],

"The history of Sciences forms not only the memory of science, but also its epistemological * laboratory".

2. The use of isotopes

a. Heavy isotopes

It has been stated that if, from the rather confused state described in Part IV of this *History*, biochemistry began around the middle of the 1930s to move towards the gratifying knowledge of the pathways of biosynthesis which we have now, we owe it to Rudolph Schoenheimer, who introduced the method of tagging organic molecules with stable isotopes. This statement needs qualification, as the development of Schoenheimer's methodology flows from several sources.

In the period of early approaches to biosynthesis described in Part IV, it was very difficult to identify in their biological setting the chemical reactions taking place in organisms.

We have, in the *Introduction* to Part III, underlined the fact that the discovery of deuterium by Urey in 1931 was the basis of a new period of utilization of tracer studies. Heavy water was used, for instance, in studies on water metabolism (see Vol. 31, p. 12). When Urey, at Columbia, had prepared enough heavy water to release some of it for work in laboratories other than his own, he envisaged its application to a "heavy water biology" and, for such a project (which at the end revealed itself as frustrating) he applied to the Rockefeller Foundation for funds [3].

In 1932, Warren Weaver, a mathematical physicist, had been

---

* Epistemology, as opposed to ontology, consists in the critical study of science in order to determine the logical origin and the value of conceptualist systems.

References p. 26

appointed Director of the natural sciences at the Rockefeller
Foundation [4]. In full agreement with Max Mason, President of
the Rockefeller Foundation and his colleague at Wisconsin, Weaver
suggested to the trustees that the science program of the Foundation be shifted from its previous preoccupations with the physical
sciences to an interest in stimulating and aiding the application of
the methods so effectively developed in physical sciences, to basic
biological problems. The trustees accepted Weaver's proposal. The
progress in the programme was so promising as to allow Weaver,
when he wrote a report for 1938, to entitle the corresponding section "*Molecular Biology*",

"a relatively new field in which delicate modern techniques are being used to
investigate ever more minute details of certain life processes".

Weaver was convinced that biology and medicine, if they became
quantitative by a recourse to exact methods of mathematics, physics and chemistry, could become truly scientific disciplines. In the
field of his duties he directed the shift which led from public
health and medical education problems to a concern with the
quantitative aspects of biomedical sciences. Among Weaver's reali-
zations along this line is the cooperative programme started in
1934 at Copenhagen. It included the cooperation of A. Krogh and
G. Hevesy in the field of application of heavy water to fluid transport studies.
    Paradoxically, Weaver's prospects were favoured by the eco-
nomic recession. In such crisis periods, the highly specialized scientists find it difficult to get jobs and they are happy enough to be
included in interdisciplinary teams.
    Along the lines suggested by Urey for the biological application
of deuterium studies, a number of departments headed by his colleagues of Columbia University became involved. One of these was
the Department of Biological Chemistry. The head of this department, Hans T. Clarke [3], had studied organic chemistry in London under Sir W. Ramsay and J. Norman Collie *. After stays in
Fischer's laboratory in Belfast and in University College, London,
Clarke became head of a section for organic chemical research at

---

* Collie first proposed that two carbon units were important in biosynthesis.

Eastman Kodak Co. at Rochester, New York. In 1928 he accepted an invitation from Columbia University, to direct the Department of Biological Chemistry at the College of Physicians and Surgeons. He was asked to introduce new viewpoints, derived from his expertness in organic chemistry, in biological and medical research. This was in harmony with the traditions of Columbia, one of the first medical schools to establish a department of biochemistry in 1898.

The choice of Clarke as head of the Department of Biological Chemistry of Columbia had been a symptom of the increasing relevance of organic chemistry in the way biochemistry was conceived as a science. Upon arriving at Columbia, Clarke launched a programme of hormone research, and recruited several experts in this field.

In the Columbia heavy water project, Clarke had agreed to study the optical properties of deuterated organic molecules. Not only did Urey provide the heavy water, but he selected four of his students to assist the biologists of Columbia. Among them was David Rittenberg, who had just finished his Ph.D. He had no prospect of a job in physical chemistry. He was assigned to Clarke's group in order to help its members to study the optical activity of deutero compounds.

Clarke asked Rittenberg to talk with the other members of the Department about possible applications of heavy water to their own work. It was in the course of this talk with Rittenberg that Schoenheimer, who had arrived in Clarke's laboratory in 1933, proposed to use deuterium as a molecular tracer to follow cholesterol metabolism in its natural settings.

Rudolph Schoenheimer — born in Berlin on May 10, 1898 — whose father and maternal grandfather were members of the medical profession obtained his M.D. in Berlin in 1922. He became interested in atherosclerosis, a disease which could be induced in rabbits by a diet of fats and cholesterol. He became a resident pathologist at the Moabit Hospital in Berlin where he stayed one year, devoting his research work to the physiology of deposition of cholesterol in the arteries. Anxious to increase his chemical knowledge, Schoenheimer obtained a Rockefeller Fellowship and went to Leipzig to follow the two years' chemical training course organized for M.D.s by Karl Thomas with the help of the Rockefeller Foundation. In 1926, Schoenheimer became an assistant to

Plate 178. Hans Thacker Clarke.

Ludwig Aschoff at Freiburg. Aschoff was an expert in the field of atherosclerosis and was interested in the biochemical aspects of the disorders of cholesterol metabolism, a viewpoint which influenced Schoenheimer. At the same time, in Freiburg, Ad. Windaus was active in the elucidation of sterol structure, an area which increased the interest of Schoenheimer in cholesterol studies. It was known that cholesterol was synthesized in the animal organisms, with the exception of insects (see Chapter 50), and it was known that squalene was a precursor of cholesterol (Channon, 1926), but the metabolic origin and the intermediates of the isoprenoid pathway remained unknown.

In 1932, Schoenheimer married Salome Glucksohn, an embryologist. Having been compelled to resign, because of his Jewish origin, from his job in Aschoff's department, Schoenheimer left Germany with his wife, in 1933, for New York where he obtained a post at Columbia.

The details of this aspect of Schoenheimer's curriculum have been reviewed by Kohler [5] (on Schoenheimer, see also Peyer [6]). When the Nazi pressure became felt in Germany, Schoenheimer's chief Aschoff in Freiburg urged him to emigrate to the U.S.A. where he already had contacts having spent the years 1930—1931 at the University of Chicago as a Douglas-Smith Fellow in the Department of Surgery. The Josiah Macy Foundation had begun operating in 1930 with a special interest in degenerative diseases and its president L. Kast had asked Aschoff to assume the preparation of a statistical survey of atherosclerosis in America. Aschoff declined and suggested Schoenheimer, who would be in America anyway. Schoenheimer was not chosen but the Macy Foundation, in 1931, began to support his work on atherosclerosis and cholesterol. When, in 1933, Aschoff urged Schoenheimer to move to the States, it was with Kast that he exchanged views concerning possible jobs. Kast approached Cornell and Columbia. At Columbia, Kast approached the professor of pathology, J.W. Jobbling, who was interested in atherosclerosis studies and who put Kast in touch with Clarke. Schoenheimer was in 1934 given a post as a research assistant in Clarke's department, the department of Biological Chemistry in Columbia School of Physicians and Surgeons, with salary and support provided by the Macy Foundation.

As we have stated above, Rittenberg had come to Clarke's

Plate 179. Rudolph Schoenheimer.

laboratory to study a physicochemical topic, the optical activity of deutero compounds. It is typical of Clarke's openmindedness that he agreed instead to allow Rittenberg and Schoenheimer to proceed with a study intending to trace the metabolic fate of fats labelled with heavy hydrogen. Tracer studies accomplished before were not concerned with the use of deuterium as a molecular tracer to follow biochemical reactions. This was Schoenheimer's idea. He conceived it in the context of his interests in atherosclerosis, a disease which played here the part gout had played in stimulating the biochemical studies on nucleic acid metabolism. Kohler [5] has acutely analysed the origins of this significant innovation, due to Schoenheimer, in following the changes within molecules. Schoenheimer had become familiar with the isotopic method in a general way when he had collaborated with Hevesy in studies on the partition of lead in organisms [7]. Kohler [5] has underlined the importance of the biochemical tradition with respect to this development. We have described in Chapter 54 how Knoop succeeded in following the β-oxidation of fatty acids by attaching benzene rings or chlorine atoms to an end of these molecules. From the configuration of the marked degradation products he deduced the process of transformation of the molecules. But the pressing subject for Schoenheimer was to define how cholesterol was formed and destroyed in the body, in order to explain the metabolic conditions of its concentration in biological media.

To substantiate this interpretation, Kohler [5] quotes from a grant proposal (1938), which is in the Archives of the Rockefeller Foundation and is obviously written by Schoenheimer. In this document, the following sentence is found:

"our first experiments with deuterium have already shown that the use of isotopes as a label for physiological substances is not only applicable to the study of cholesterol metabolism, *for which it was originally devised*, but has a very wide scope and can most probably be used for the investigation of all types of physiological compounds".

According to Kohler [5], the transformation of biochemistry as a scientific discipline in the period preceding World War II was the result of the expansion of Schoenheimer's work from atherosclerosis to studies on cholesterol metabolism, and of the expansion from cholesterol metabolism to a study of all types of biomolecules.

Plate 180. David Rittenberg.

It must be noted that the first experiments of Schoenheimer et al. [8] on the isoprenoid pathway were rather frustrating. For instance, in 1936, they fed deutero-coprostane to dogs and found no deuterium in bile acids.

In contrast to these frustrating results on the isoprenoid pathway, exploratory experiments with fatty acids were unexpectedly highly promising. Schoenheimer and Rittenberg [9], feeding deuterated linseed oil to mice with the purpose of recovering it as excretion products, found that nearly half of the deuterium was inserted in the fat depots.

Kohler [5] has epitomized the profound change introduced by deuterium studies in the conceptual system of biochemistry, particularly concerning biosynthesis.

"The pace of conceptual change accelerated sharply. Schoenheimer's demonstration that the turnover time of fatty acids in mice was only about six to eight days radically revised previous ideas of the stability of depot fats [10]. Experiments in which deuterated saturated fatty acids were converted by mice to deuterated unsaturated fatty acids proved the existence of hydrogenation and hydrogenation cycles in the tissues [11]. J.B. Leathes had suggested in 1909 that desaturation was the first step in the degradation of fatty acids; Schoenheimer's experiments were the first clear evidence for this belief. Similarly, isotope experiments provided the first unambiguous evidence for a theory, proposed thirty years earlier by Knoop, that fatty acids were synthesized and degraded by two carbon units. Mice fed deutero-stearic acid were shown to excrete deutero-palmitic acid, two carbons shorter [12]. However, experiments with deuterated butyric and hexanoic acids showed that these were not direct intermediates in the synthesis of the higher fatty acids; contrary to Knoop's scheme and the common belief [13]. Finally, by isolating deuterated cholesterol from mice fed on heavy water and on fat and sterol free diet, he showed that cholesterol was synthesized in the tissues by the coupling of small molecules, just as the fatty acids were [14]. By contrast it was widely believed, on grounds of structural similarities, that sterols were synthesized by the cyclisation of long chain fatty acids.

Again and again, this pattern was repeated: studies with isotopically labeled compounds confirmed, modified or disproved traditional biochemical ideas originally inferred indirectly from feeding experiments or from a similarity between biological processes and known chemical reactions. The isotope method promised solutions for problems to which specialists had devoted entire careers. It was their ability to bring new evidence to a whole range of traditional ideas that gave biochemists in the mid-1930s a sense of a sharp break with the past, a sense that what had been done before was mere speculation. The suddenness with which traditional ideas were outdated (or put on sounder footing) helps explain why, for a whole generation of biochemists, biochemistry seemed to begin in the 1930s. Reading Schoen-

References p. 26

heimer's papers, one senses the atmosphere of excitement and expectation in which biochemistry became an intellectually dynamic and coherent field in the late 1930s".

If the great hopes which had been invested in the heavy water biology had been frustrating, the isotopic tracer methods soon appeared as opening up broad possibilities for metabolic studies. Clarke and Urey, as Kohler [5] has concluded from a study of unpublished documents, suggested that the Columbia group should contemplate the development of an extended programme of biological and medical applications of all the isotopes of biological interest that Urey was busy to concentrate (^{2}H, ^{13}C, ^{15}N, ^{18}O). Urey succeeded in early 1937 in concentrating sufficient ^{15}N to provide Schoenheimer with a supply of this heavy isotope.

A large supply of ^{15}N having become available in the spring of 1938, Schoenheimer, helped by the mass spectrometer of Rittenberg and the ability of Sarah Ratner, an organic chemist who had completed her Ph.D. with Clarke, fed rats on [^{15}N]tyrosine. To their surprise, the authors found that one half of the ingested isotope was retained in the tissues [15]. All the non-essential amino acids contained ^{15}N. That the $-NH_2$ groups of the amino acids were in a state of constant and rapid metabolic activity was confirmed by experiments with ^{15}N, [^{2}H]leucine [16]. This confirmed the view proposed in 1935 by Borsook, and Keighley (see Chapter 11), who had seen that in an animal in nitrogenous balance at a high level of nitrogen intake, about half the nitrogen in the 24-h urine came from tissue protein breakdown. This implied that a corresponding amount was resynthesized. The data of Borsook, confirmed by Schoenheimer, gave the fatal blow to the theory of endogenous and exogenous metabolism of Folin according to which the animal was pictured as a machine with a frictional wear and tear loss (the endogenous metabolism) the rest of the daily food intake being burned as fuel.

By 1939, when the fission of uranium was announced, Urey turned his attention towards the separation of the uranium isotopes. Another development which took place in the late 1930s was an increasing competition between the schools interested respectively in radioactive isotopes and in heavy isotopes. Ernest Lawrence, in Berkeley, looking for sources of financing for his radiation laboratory, became interested in the biological and medical applications of radio-isotopes. The inconvenience was that the radio-isotopes of light elements were short-lived. In september 1939, Lawrence, returning from scientific meetings on the East Coast instructed his collaborators Kamen and Ruben to devote both Berkeley cyclotrons to a search for long-lived radio-isotopes of hydrogen, carbon, nitrogen and oxygen. In February 1940 Kamen and Ruben discovered ^{14}C. Kamen [17] has retraced the history of this major discovery which changed the face of metabolic research. The Columbia School continued to work on intermediary metabolic studies with isotopes, but Schoenheimer had committed suicide on 15 September 1941. His insight had been a determinant factor in the development of the most potent tool of the isotopic method. He had the acuteness of abandoning the direct consideration of primary evidence in the disease which interested him to shift towards a situation of biosynthesis at the chemical level. His contribution consists in situating the problem of biosynthesis on a chemical level, a problem he was able to pursue owing to his insight in the discovery of a method for the recognition of intramolecular changes until then inaccessible.

The spectacular biochemical development of biochemistry in the 1930s is an American phenomenon. But it appears that the emigration from Europe of a host of eminent people played an important role. In the 1930s, the concepts introduced by Schoenheimer, for instance, in the field of the turnover of biomolecules, revolutionized the whole thinking. Radio-isotopes became indeed available in quantity to American biochemists before they did to European researchers. But it would be unfair to disregard the highly powerful influence of the support of private or federal agencies. When Urey isolated heavy isotopes, their application to biochemical problems was permitted by the support of the Rockefeller Foundation as we have seen. Later on the support of the National Institute of Health was the fountainhead of the development of biochemical research in the U.S.A.

3. Isolation of enzymes

One of the important factors which allowed recognition of chains of reactions and to break them up in the complete sequence of their separate links was the isolation of pure enzymes. This went hand in hand with the progress of protein chemistry since the enzymes were recognized as proteins (see Chapter 13). A large amount of knowledge accumulated in the field of enzymology permitted the systematization of the methods for the isolation and purification of the enzymes. A large body of empirical data has been accumulated (Reviews [18—21]).

One of the difficulties consists in the possible differences in several tissues, between enzymes catalysing the same reaction. For instance, as shown by Shack [22], dehydropeptidase from liver is soluble and that from kidney is not. The present author, as President of the International Union of Biochemistry, has founded, in 1956, an Enzyme Commission [23] in order to establish a nomenclature of enzymes. The enzymes were to be classified according to the reactions catalysed. This still is the state of affairs presently and it should be kept in mind by historians of biochemistry that this state of affairs will not last for ever.

Experience has been obtained, when a proper form of starting material was selected, concerning the precautions necessary to avoid denaturation in the process of getting the enzyme in solution. The application to such a solution or to the natural fluid if chosen as starting material of methods, has led to the availability of a number of procedures (precipitation with salts or with cold organic solvents; removal of water under distillation, lyophilisation, etc.).

In Chapter 17 we have recalled the importance of cell breakage in the history of glycolysis and in the isolation of the components of Büchner's zymase. Many methods for breaking the cells of plants, animals and microorganisms have been devised (see Schwimmer and Pardee [24]), as well as a variety of methods for the extraction of enzymes by means of solvents.

In this issue, the use of methods of assay of activity are of paramount importance. These methods not only give information on the amount of enzyme in preparations corresponding to several degrees of purification, but they provide a way of distinguishing between enzymes and inert proteins which constitute the main impurities to eliminate.

In order to determine the degree of purification, enzyme activity and protein determination are combined to give the specific activity (units of enzyme activity for mg of protein N). Enrichment (ratio of specific activity after a given step to the specific activity before) and yield (ratio of total enzyme activity after a procedure to total activity before) are independent of one another. The yield indicates losses of enzyme. Enrichment values give information as how far progress has been made toward isolation. Purity checks lead to selecting further processes towards purification. Methods of appraisal of purity are numerous (see Li [25]).

4. The use of exacting mutants in biosynthetic studies

As we have retraced in the Introduction to Part III, microbial mutants were introduced in 1941 by Beadle in the analysis of metabolic pathways by taking advantage in a mutant, of a block of one of the enzymatic steps of the pathway.

A first successful application of exacting mutants methodology to biosynthetic studies was recorded in 1943 by David Bonner and his colleagues [26] who obtained mutants of *Neurospora crassa* requiring both isoleucine and valine, an observation which was of importance in the course of unravelling the biosynthesis of branched-chain amino acids (isoleucine, leucine, valine) (see Chapter 67). We shall record in relevant chapters the contribution of exacting mutant studies to the knowledge of several biosynthetic pathways.

5. Dominant importance of chemistry in the unravelling of biosynthetic pathways

We have, in Part IV, described early studies on biosynthesis. This single-handed, clumsy approach dominated by animal chemists and plant chemists aimed at the target of situating the chemical aspects of life in their proper biological setting. But it remained jeopardized by the lack of specificity of analytical procedures, by the complexity of the mixtures, by the lack of sufficient chemical knowledge. The current ideas were indirectly inferred from nutrition experiments or from known chemical reactions. In the dark sky of the period described in Part IV, the ornithine cycle appeared as a *stella nova*. In this firmament many new constellations were to appear during the 1930s. From the historical viewpoint, a particular interest is attached to those transitions at which

References p. 26

long periods of sterility and of approaches getting nowhere are replaced by fertile processes and high rewards.

In no other chapter on what Fruton [38] has so adequately called "the interplay of chemistry and biology" has this interplay been more successful.

To focus the reasons for this success on the introduction of the isotopic method would not give a complete picture, as of even greater importance were the progress of chemical knowledge and the impact of the introduction of this chemical knowledge into the activity of biochemical laboratories, by competent scientists. The freedom of invention in forming hypotheses concerning the sequence of chemical steps in a biosynthetic pathway is subordinate to the limitations imposed by the configurations of the chemical compounds involved. The biochemist, in this domain, invents only within the framework established by the chemist. This often unspoken reality, but always kept in sight, is the background of his conceptual systematization. Biochemists were not always, at every step of their inquiry, able to rely on a proper chemical body of knowledge. Organic chemistry has developed along the lines of its own scientific methodology and practical applications, and it is easy to understand that the majority of organic chemists have preferably pursued their own scientific ideal, as well as material interests. It is generally along the lines of their own scientific methodology that organic chemists synthesized biological compounds but they have occasionally done it with the recognized purpose of contributing to biochemical progress.

This was for instance the case with Hermann O.L. Fischer, who had heard the appeal to a new alliance formulated by his father, the great Emil Fischer (see ref. 27).

As stated by Hartman and Buchanan [28]:

"Although frequently organic reactions have no direct counterpart in biochemistry, there are many instances ... where they may serve as excellent guides for the formulation of preliminary ideas" (p. 81).

For instance, when mevalonic acid was experimentally recognized as an intermediate in the biosynthesis of squalene (see Chapter 50), reasoning by analogy suggested that the production of a derivative reactive enough to undergo condensation reactions could result from an oxidation of the C-5 of mevalonic acid to an aldehyde, from an oxidation of this carbon to a carboxyl, or from the formation of double bonds between C-2 and C-3 and between C-4 and C-5. This chemical notion led to experimentation, for instance, with deuterium-labelled mevalonic acid in position 5. It was found that at least three quarters of the hydrogens of the C-5 of mevalonic acid were retained in squalene. As an oxidation of the alcoholic group or of the acid group would result in the loss of half or all of the hydrogens on C-5, both these possibilities were eliminated.

The dominating epistemological impact of chemical knowledge on the interpretations of biosynthetic pathways recurs ever and anon in the "Discussion" section of scientific papers concerned.

The methodology of anabolic studies differs from the one of catabolic research whose free energy-producing aspects have been dealt with in Part III. The studies on catabolic aspects were first mainly based on the methods of trapping intermediates, of reducing the size and complexity of the samples involved in experimentation, while introducing artefacts which were sometimes difficult to detect (see Part III). Even more than in catabolic studies, the unravelling of biosynthetic pathways was led by methodological approaches, such as the use of isotopes and of paper chromatography, with reference to epistemological background of chemical knowledge. It is an example, in modern science, of investigative (experimental) processes combined with conceptual chemical aspects. This Berthelot would have called a queer mixture of chemistry and natural history. To this we cannot agree as we recognize that transformations of matter in organisms, as well as in crucibles, are ruled by quantum chemistry (Lavoisier has said that there is only one chemistry), what Berthelot would not have recognized when dealing with aspects of a higher level than that of the proximate principles.

But the originality of the successful biosynthesis studies was that their subject matter was formulated as a chemical problem: to unravel the sequence of chemical reactions leading to the formation of a definite kind of biomolecule and the assembly line governed by enzymes. While the theory of intercalation of components of the food had governed the ideas relating to development and maintenance of organisms, the change of concepts related to the possibilities of synthesis in animals (Chapter 42) had modified the views, and Claude Bernard had insisted on the indirect nature of nutrition, the blood deprived of metabolic activity providing the cells with precursors for biosynthesis, these precursors being derived from metabolic processes and added to the circulating blood. It was common belief at the time that the constituents of the food were decomposed into their chemical elements, these elements being associated into the constituents of blood from which the constituents of the cells were derived by "vital creation".

While the problem the biochemists who studied catabolic pathways and their involvement in the formation of ATP were facing was to follow a kind of known biomolecule in a series of metabolic changes leading to unknown products, the unravelling of biosynthetic pathways supposed the identification of the precursors utilized by the cells and until then totally unsuspected. The isotopic tracer method, which was so useful in catabolic studies was even more so in experimental biosynthetic studies. It was one of the major concepts introduced during the 1930s that the precursors utilized in the biosynthetic pathways, while not reduced to the nature of elements, were small building units.

As stated by H.G. Wood [29] in a Harvey Lecture delivered in 1950:

"The plan of the body machinery appears to be to tear down the foodstuffs and rebuild with the pieces much as does the contractor when he builds from materials recovered from a razed structure" (p. 127).

6. The impact of the ornithine cycle

During the period of early studies described in Part IV, the identification of the ornithine cycle occurred as a pioneer work in a biochemically almost uncharted area. Although single reactions had been proposed before as explanatory, the ornithine cycle was the first success in sequencing a biosynthetic pathway into a chain of enzymatic separate links. The ornithine cycle, revealing a new pattern of metabolic process, emerged out of context, an isolated archipelago lost in a sea of empiricism and half-failure (see Chapter 48). The concept of the ornithine cycle was favourably received and enthusiastically accepted by the majority of biochemists.

As we have stated in Chapter 48, when Krebs conceived the "ornithine cycle", there was in fact a hint that ornithine might play a role in ureogenesis. For 20 years, the parallel between the occurrence of arginase and ureogenesis had been repeatedly stated. But nobody conceived that high arginase activity might have anything to do with ureogenesis.

When Krebs submitted the paper for publication in the *Zeitschrift für physiologische Chemie*, Knoop, the editor of the periodical,

"commented in a very complimentary way and added that he felt really stupid that it never occurred to him before that arginase might play a role in the synthesis of urea" * (Krebs [30], p. 8).

To quote Krebs further concerning the reception of the ornithine cycle,

"But there were adverse criticisms. In 1934, a Russian physiologist of Leningrad, E.S. London [31], argued that ornithine had no effect in his experiments on the isolated perfused dog liver. In 1942, Trowell [32], of the Cambridge Physiological laboratory, reported that the perfused rat liver did not respond to ornithine.

At the same time Bach and Williamson [33] of the Cambridge Biochemical Laboratory, claimed to have shown that dog liver can form urea from ammonia even where arginase is completely inhibited by high concentrations of ornithine, and they concluded that liver can synthesize urea without participation of arginase. As late as 1956 Bronk and Fischer [34] reported — from my own department at Oxford — that under certain conditions citrulline is less effective than ornithine in promoting urea formation from ammonia, and they concluded that citrulline cannot be an intermediate. The results of London and Trowell were due to inadequate perfusion technique. The conclusions of Bach and Williamson were based on the wrong assumption that arginine and ornithine readily penetrate liver slices. In fact the rates of penetration are relatively slow. Although the observations of Bronk and Fischer [34] were correct, their interpretation was mistaken. They expected the kinetics of an homogeneous solution in the highly compartmented system of

---

* Genuine reasons for this lack of recognition are found by Krebs in a relative lack of consideration of evolutionary concepts, particularly in biochemistry.

"Now we know that non-functional characters do not survive in the course of evolution. Nowadays if one were faced with an exceptionally high enzyme activity in one particular tissue (as is true for arginase in liver) biologically orientated biochemists would at once ask themselves why is there so much activity? Apparently nobody has asked this question. Of course occasionally this kind of question cannot be answered but it ought to be searchingly asked. Sometimes the failure to answer the question is due to methodological limitations. In the case of arginase and urea synthesis the tissue slice technique overcame this limitation" (From a letter of Sir Hans Krebs to the author, 5 April, 1978).

References p. 26

living cells. Permeability barriers, often unpredictable, can interfere and may cause deviations, usually of a minor quantitative kind, from a postulated kinetic behavior. At the time, in 1956, the importance of permeability barriers at the plasma membrane and mitochondrial membrane was not yet sufficiently appreciated.

Experience shows that to postulate, as Bronk and Fischer [34] did, a second mechanism on the basis of some kinetic discrepancies is far more rash and fanciful than to accept kinetic abnormalities on account of permeability barriers and other complications when dealing with very complex systems".

In a review article of 1943, Borsook and Dubnoff [35] started a discussion of urea formation with the words:

"There is an increasing expression of doubt regarding the ornithine cycle".

As in the case of the tricarboxylic acid cycle, the objections here mentioned were due to misinterpretation of results in the application of the isotopic technique. We have retraced in Chapter 33 how Ogston clarified the problem in the case of the tricarboxylic acid cycle. As underlined by Grisolia [36], a number of opposing findings remained present until around 1946, and were interpreted by several biochemists as indications of the existence of a mechanism other than the ornithine cycle. Between 1932 and 1946, the experiments were accomplished with tissue slices and the main findings in support of the ornithine cycle were the incorporation of CO_2 and the catalytic role of ornithine and citrulline (see Chapter 48). The main opposing findings were the failure to isolate citrulline from liver, the poor or low enhancement of urea synthesis from citrulline and the lack of incorporation of citrulline into urea, while glutamine was rapidly incorporated (see Grisolia [36]).

The introduction of homogenates by Cohen and Hayano [37] in 1946 allowed, during the period of 1946–1948 (as will be retraced in Chapter 65), the separation of the steps ornithine → citrulline and citrulline → urea, the recognition of aspartic acid as a specific and direct amino donor, the demonstration of equal specific activity for $^{14}CO_2$ → [^{14}C]citrulline → [^{14}C]urea, and the recognition of the catalytic role of carbamoylglutamate (see Grisolia [36]). As pointed out by Grisolia this development illustrates the heuristic effect of apparently contradictory findings. For instance, the participation of an amide donor, considered as a contradictory finding finally led to the clarification of the urea cycle (see Chapter 65).

7. Order of chapters

We shall start the sequence of chapters with the energetic sources of biosynthesis (Chapter 53) and the formation of peptide bonds (Chapter 54), an aspect which was the subject of early studies in the field of the energetics of biosynthesis *.

When a radioactive isotope of carbon, ^{11}C, became available, the biochemists turned to the fundamental problem of the conversion of CO_2 and H_2O into carbohydrates by illuminated green plants. The early history of the approaches to this subject has been retraced in Chapters 40 and 45, and the studies on photosynthetic phosphorylation, generator of ATP, has been reviewed in Chapter 38. The interpretation of the reductive carbohydrate cycle (Chapter 56) was facilitated by the unravelling of the pentose phosphate cycle (Chapter 55) essential for the formation of specific structural and functional components such as ribose and erythrose-4-P. The pentose phosphate cycle also provides NADPH, used in many biosynthetic reactions.

An important step in biosynthetic studies was the discovery that CO_2 utilization is not limited to autotrophs, but can be recognized in heterotrophs (Chapter 57).

As stated by Fruton [38],

"Not only did it represent one of the first indications of the power of the isotope method to reveal the operation of unsuspected biochemical processes, but it also drew the attention of biochemists concerned with problems of animal metabolism to the general importance of the extensive studies that had been conducted before 1935 on the metabolism of a large variety of microorganisms (Van Niel, 1949; Woods, 1953). In particular, Marjory Stephenson's valuable book *Bacterial Metabolism* (first edition, 1930; third edition, 1949) clearly indicated the contribution microbiology had made to the study of biochemical dynamics. In 1935, CO_2 was considered to be a

---

* To this aspect we shall, in Part V, limit the historical approach to protein synthesis, which will find a more proper place in the history of the transition from chemical genetics to molecular genetics in Part VI.

References p. 26

completely inert end product of cellular respiration in heterotroph organisms, i.e. those which require some organic compounds as nutrients. When Wood and Werkman (1936) found that the formation of succinic acid by propionic acid bacteria is related to the uptake of CO_2, little attention was given to their discovery, but, when the use of carbon isotopes showed CO_2 fixation to be effected under conditions where no *net* uptake of CO_2 was evident its significance came to be widely appreciated. An early example was the finding in 1941, that the carbon of $^{11}CO_2$ is incorporated by animals in their liver glycogen" (p. 468).

The chapters on photosynthesis and on gluconeogenesis are followed by a chapter on the biosynthesis of complex saccharides from monosaccharides (Chapter 58).

We have reviewed in Part III the discovery by Lipmann of the structure of coenzyme A and of its role in the formation of acetyl-CoA, the "active acetate" which is obtained from acetate, from sugars through pyruvate, as well as from fatty acids. It was also indicated in Part III that acetyl-CoA can derive in catabolism from a number of amino acids (leucine, isoleucine, threonine, tryptophan). First, according to the theory of reversible zymo-hydrolysis, it was conceived that fatty acid biosynthesis was the reversal of β-oxidation, starting from acetyl-CoA. The isotope tracer method was considered as supporting this view. It was, as will be narrated in Chapter 59, in 1957 that fatty acid biosynthesis was accomplished from acetyl-CoA in the absence of any of the β-oxidation enzymes and that the theory of reversible zymo-hydrolysis was discarded from the field of fatty acid biosynthesis. An absolute requirement for bicarbonate, discovered in 1958, led to the recognition of malonyl-CoA as an intermediate, as described in Chapter 59 in which the history of the biosynthesis of triglycerides is also retraced.

It was at the end of the 1940s that Neuberger in London and Shemin in New York simultaneously launched an approach to the biosynthesis of porphyrins, which they pursued shoulder to shoulder and which led to the developments reviewed in Chapter 60, also dealing with the biosynthesis of metalloporphyrins of the iron branch and of the magnesium branch. The enzymology of this domain became an active subject of investigation in the 1960s. Corrinoid biosynthesis was recognized as an extension on the pathway of porphyrin biosynthesis. It is also treated in Chapter 60, in which the biosynthesis of prodigiosin, a pyrrolic compound not derived from δ-aminolevulinic acid is considered too.

The first recognition of the participation of deutero acetate in the formation of a member of what we now call the isoprenoid pathway dates from 1937, but the recognition in 1953 of the participation of acetyl-CoA was an important step, followed by the unravelling of the isoprenoid pathway retraced in Chapter 61.

Another success of the isotope tracer technique took place during the 1950s with the unravelling of the biosynthetic pathway of the purines and its extensions (Chapters 62 and 63).

The utilization and interconversion of C_1 units other than CO_2 (methyl, hydroxymethyl and formyl groups) was also cleared up by the method of isotope tracers. It was inferred that a central role is played in the utilization of these C_1 units by the cofactor tetrahydrofolic acid. We shall deal with this subject when treating the extensions of the biosynthetic pathway of the purines, the biosynthesis of tetrahydrofolic acid being one of these extensions (Chapter 63).

The selection of primary biosynthetic pathways and of their terminal or lateral extensions, treated in the present volume will leave for another volume the biosynthetic aspects concerning amino acids, a field in which, besides isotope tracers and pure enzymes, the recourse to auxotroph mutants of microorganisms has played a more important role. A series of general considerations concerning the concept of biosynthetic pathway will also find place in that volume (Vol. 33B. *The unravelling of biosynthetic pathways. Continued: The biosynthesis of amino acids and its extensions*).

In references to common names and systematic names of enzymes, the author has adopted the suggestion formulated by the editors of this Treatise in the Preface to the third edition of Volume 13. This suggestion reads as follows:

"A plea may be made, as the texts of the three documents (Report of the Enzyme Commission, 1961; Enzyme Nomenclature, 1964 and Enzyme Nomenclature, 1972) are widely different, to insure correct quotation of each of them. While keeping the symbol EC for the Report of the Enzyme Commission (1961), EN 1964 and EN 1972 could be used respectively to designate the two successive editions of Enzyme Nomenclature, which have kept the character of recommendation, a feature the following editions should maintain".

Unless otherwise stated, EN in this volume corresponds to EN 1972. The bookshelf of the history of molecular aspects of life is more and more frequently enriched. The book of R. Olby, *The Path to the Double Helix* [39], published in 1974, has supplied a rich source of information on the origins of "molecular biology". In Olby's book, a section involving four chapters is devoted to the field of bacterial transformation, in which an important impact is attributed to the work of Oswald T. Avery. One of Avery's former collaborators, René J. Dubos, has since written a detailed biography of Avery, entitled *The Professor, the Institute and DNA* [40]. We shall pay due attention to these scholarly researches in Part VI of this *History*.

In 1975 there appeared a revised edition of Nachmansohn's monograph *Chemical and Molecular Basis of Nerve Activity* [41], first published in 1959. To this revised version has been added a supplement by Nachmansohn (Properties and function of the proteins of the acetylcholine cycle in excitable membranes) and a second supplement by E. Neumann (Towards a molecular model of bioelectricity). The whole volume has incontestable historical value of particular importance in the contemporary history of the knowledge — beyond electrophysiology — of the molecular aspects of the nervous system.

In the same line, emphasizing the increasing interplay between physiology and chemistry, those interested in the history of Biochemistry will peruse with pleasure and profit a series of "informal essays on the history of physiology", "written as part of the celebrations of the Centenary of the Physiological Society in 1976" and published under the title *The Pursuit of Nature* [42]. It is a collection of brilliant essays by such experts as A.L. Hodgkin, A.F. Huxley, W. Feldberg, W.A.H. Rushton, R.A. Gregory and R.A. McCance, on such important topics as nerve, muscle, neuromuscular transmission, vision, gastrointestinal hormones, and perinatal physiology, including many molecular correlates of "functions".

Like the monograph of Nachmansohn mentioned above, Feldberg's essay in *The Pursuit of Nature* is centered on aspects of acetylcholine functions. This is also the case for the main topic treated in Z.M. Bacq's little book, *Chemical Transmission of Nerve Impulses* [43], which was presented to the participants of the Sir Henry Dale Centennial Symposium in Cambridge on 17—19 September 1975.

The recognition of the history of contemporary biochemistry as its "epistemological laboratory" is confirmed by the introduction of historical presentations in such scientific symposia as *The Urea Cycle* [44], edited by S. Grisolia, R. Baguena and F. Mayor (1976) or *Die aktivierte Essigsäure und ihre Folgen, Autobiographische Beiträge* [45], edited by G.R. Hartmann (1976).

To celebrate Severo Ochoa's 70th birthday, his students and colleagues have edited a collection of essays reflecting on the development of a subject, of a concept or of an approach to Biochemistry. This collection of scholarly and inspiring essays has appeared under the title *Reflections on Biochemistry. In Honour of Severo Ochoa*, edited by A. Kornberg, B.L. Horecker, L. Cornudella and J. Oro [46].

The author, here again, expresses his debt of gratitude towards colleagues and friends who helped him in endeavouring to make this narration accurate, and in allowing him to quote from their writings. Such thanks are particularly due to H. Borsook, J.M. Buchanan, H. Chantrenne, F. Dickens, T.W. Goodwin, S. Grisolia, G.A.D. Haslewood, A.B. Hastings, B.L. Horecker, L. Jaenicke, M. Kamen, R.E. Kohler, H.A. Krebs, L.F. Leloir, D. Nachmansohn, A. Neuberger, G. Popjàk, D. Shemin, E.H. Stotz, G.H. Tait, M.F. Utter, S.J. Wakil, and H.G. Wood.

June 1978   Marcel Florkin

REFERENCES

1   M. Stephenson, in J. Needham and E. Baldwin (Eds.), Hopkins and Bio-
     chemistry 1861—1947, Cambridge, 1949, p. 1.
2   E.J. Dijksterhuis, in M. Clagett (Ed.), Critical Problems in the History
     of Science, 1959, 2nd ed., Madison, 1962, p. 3.
3   H.T. Clarke, Annu. Rev. Biochem., 27 (1958) 1.
4   W. Weaver, Science, 170 (1970) 581.
5   R.E. Kohler, Historical Studies in the Physical Sciences, Vol. 8, 1977,
     in press.
6   U. Peyer, Rudolph Schoenheimer (1898—1941), Zürich, 1972.
7   G. Hevesy, Cold Spring Harbor Symp., 13 (1948) 129.
8   R. Schoenheimer, D. Rittenberg, B.N. Berg and L. Rousselot, J. Biol.
     Chem., 115 (1936) 635.
9   R. Schoenheimer and D. Rittenberg, J. Biol. Chem., 111 (1935) 175.
10   R. Schoenheimer and D. Rittenberg, J. Biol. Chem., 114 (1936) 381.
11   R. Schoenheimer and D. Rittenberg, J. Biol. Chem., 113 (1936) 505;
     114 (1936) 381; 117 (1937) 485.
12   R. Schoenheimer and D. Rittenberg, J. Biol. Chem., 120 (1937) 155.
13   D. Rittenberg, R. Schoenheimer and E.A. Evans Jr., J. Biol. Chem., 120
     (1937) 503.
14   D. Rittenberg and R. Schoenheimer, J. Biol. Chem., 121 (1937) 235.
15   R. Schoenheimer, S. Ratner and D. Rittenberg, J. Biol. Chem., 127
     (1939) 333.
16   R. Schoenheimer, S. Ratner and D. Rittenberg, J. Biol. Chem., 130
     (1939) 703.
17   M.D. Kamen, Science, 140 (1963) 584.
18   E. Bamann and K. Myrbäck (Ed.), Methoden der Fermentforschung,
     Leipzig, 1941.
19   F.F. Nord and R. Weidenhagen (Ed.), Handbuch der Enzymologie,
     Leipzig, 1940.
20   J.B. Summer and K. Myrbäck (Ed.), The Enzymes, New York, 1950—
     1952.
21   J.H. Northrop, M. Kunitz and R.M. Herriott, Crystalline Enzymes, New
     York, 1948.
22   J. Shack, J. Biol. Chem., 180 (1949) 411.
23   O. Hoffmann-Ostenhof and R.H.S. Thompson, Nature (London), 181
     (1958) 452.
24   S. Schwimmer and A.B. Pardee, Adv. Enzymol., 14 (1953) 375.
25   Choh Hao Li, in D.M. Greenberg (Ed.), Amino Acids and Proteins,
     Springfield, 1951.
26   D. Bonner, E.L. Tatum and G.W. Beadle, Arch. Biochem., 3 (1944) 71.
27   H.O.L. Fischer, Annu. Rev. Biochem., 29 (1960) 1.
28   S.C. Hartman and J.M. Buchanan, Ergebn. Physiol., 500 (1959) 75.
29   H.G. Wood, Harvey Lect., 45 (1950) 127.
30   H.A. Krebs, in S. Grisolia, R. Baguena and F. Mayor (Eds.), The Urea
     Cycle, New York, 1976, ref. 44.

31   E.S. London, A.K. Alexandry and S.W. Nedswedski, Z. Physiol. Chem., 227 (1934) 233.
32   A.O. Trowell, J. Physiol., 100 (1942) 432.
33   S.J. Bach and S. Williamson, Nature (London), 150 (1942) 575.
34   J.R. Bronk and R.B. Fischer, Biochem. J., 64 (1956) 118.
35   H. Borsook and J.W. Dubnoff, Annu. Rev. Biochem., 12 (1943) 183.
36   S. Grisolia, in S. Grisolia, R. Baguena and F. Mayor (Eds.), The Urea Cycle, New York, 1976, p. 13.
37   P.P. Cohen and M. Hayano, J. Biol. Chem., 166 (1946) 239.
38   J.S. Fruton, Molecules and Life. Historical Essays on the Interplay of Chemistry and Biology, New York, 1972.
39   R. Olby, The Path to the Double Helix, London, 1974.
40   R.J. Dubos, The Professor, the Institute and DNA, New York, 1976.
41   D. Nachmansohn, Chemical and Molecular Basis of Nerve Activity, New York, 1975.
42   A.L. Hodgkin, A.F. Huxley, W. Feldberg, W.A.H. Rushton, R.A. Gregory, R.A. McCance, The Pursuit of Nature — Informal Essays on the History of Physiology, London, 1977.
43   Z.M. Bacq, Chemical Transmission of Nerve Impulse, Oxford, 1975 (English translation of Z.M. Bacq, Les Transmissions Chimiques de l'Influx Nerveux, Paris, 1974).
44   S. Grisolia, R. Baguena and F. Mayor (Eds.), The Urea Cycle, New York, 1976.
45   G.R. Hartmann (Ed.), Die aktivierte Essigsäure und ihre Folgen. Autobiographische Beiträge, Berlin, 1976.
46   A. Kornberg, B.L. Horecker, L. Cornudella and J. Oro (Eds.), Reflections on Biochemistry. In honour of Severo Ochoa, Oxford, 1976.

Chapter 53

The energetics of biosynthesis

1. Introduction

"The terms "life" and "living systems" are difficult to define with any degree of precision; for our purposes, it is sufficient to consider living systems as those in which the replicative process is dependent upon energy derived from the breakdown of metabolites. In addition, most living systems require energy for maintenance during non-replicative phases. Specifically, energy is utilized for chemical purposes such as the synthesis of metabolites or structural materials, or it may be used for mechanical work, as represented by the transport of materials or the movement of structural proteins. All types of energy-requiring reactions ultimately depend upon a rather small and highly specialized group of "energy-rich" chemical compounds to supply the driving force" (Huennekens and Whiteley [1].

As we have indicated in Chapter 42, Section 2, it was Dumas who first introduced the concept that food shared with solar light the quality of a source in the flux of energy through organisms.

The concept of "energy-rich" chemical compound was, as we have retraced in Chapters 25 and 32, introduced by Meyerhof and his school and it has since become generally recognized, as they first showed, that the hydrolysis of ATP is the source of energy of muscular contraction.

However, how is the energy of food used in the body for the functions of living cells? This question was first raised by Meyerhof [1a] in a lecture on the energetics of the living cells given in 1913 at the Philosophical Society in Kiel. In his penetrating analysis he postulated that the use of thermodynamics was essential for finding clues that may provide the answer. It was necessary to establish the sequence of energy transformations by which the food would produce, during the metabolic degradation, the specific compounds required for the various functions of cells. The lecture marks the beginning of the notion of the paramount importance of bioenergetics for biochemistry. He stressed in his lecture the difference between free and total energy of a compound and their relative significance in the analysis of biological systems. The momentous importance of these notions was widely recognized. Jacques Loeb asked Meyerhof for a translation of the lecture into English and its publication in his monograph series. Due to the interruption by the war it appeared in 1925 [1b]. The experimental use of these thermodynamic notions in the analysis of metabolic pathways was decisively enhanced when Meyerhof and his associates tested in the years 1927—1934 a series of newly discovered phosphate derivatives, among them phosphocreatine and ATP, as to the energy released during their hydrolysis and found that two types of phosphate derivatives exist, one which releases large, the other only small amounts of energy on hydrolysis. One of the main general conclusions of studies on biosynthetic pathways is that the high energy phosphate derivatives such as adenylpyrophosphate, phosphocreatine etc. are the general source of energy in endergonic reactions. As was shown in Volume 31 of this Treatise, the energy released by the hydrolysis of ATP has been recognized as the universal collecting locus of free energy which is used in organisms. It has emerged from the large body of research that adenylpyrophosphate acts in biosynthetic phenomena as a universal condensing agent between the small organic units involved. We have in Part III of this *History* related the progressive elucidation of the methods by which ATP is obtained in organisms from inorganic phosphate by energy coupling (redox potential, conversion in group potential).

Besides muscle contraction, a variety of physiological processes have been identified as using metabolic free energy via ATP: nerve conduction and electric discharge, light production, active transport, etc. (Vol. II of *Comparative Biochemistry* [1] is devoted to these aspects.)

In his review article of 1941, Lipmann [2] tackles the possibility of the participation of ATP in biosynthetic reactions. At that time, the concept of active acetate had not yet been identified with acetyl-CoA, as was shown in Chapter 33.

2. ATP and biosynthesis

We have repeatedly, in Part III of this *History*, devoted to the history of the identification of the sources of free energy in organisms, emphasized the importance of the review article published by Lipmann [2] in 1941, and in which he formulated the general notion of the "phosphate cycle", introduced the concept of "high group potential" and recognized that a variety of energy-rich groups could be derived from the energy of certain metabolic processes and transferred to certain metabolic situations. This review article introduced the concept of a widespread utilization of energy-rich phosphate (and other) derivatives * as energy carriers in different phenomena. As stated by Fruton [3] (p. 375),

"although . . . some of the specific mechanisms envisaged by Lipmann were not found to be operative in biological systems, his general approach influenced a generation of biochemists after World War II".

On the basis of his work on *Lactobacillus delbrueckii*, Lipmann considered acetylphosphate as accomplishing a double function, that of acetyl donor (for instance in the biosynthesis of acetylcholine), and that of a phosphoryldonor, for instance to ADP, and conversely from ATP to acetylphosphate. At that time (before the discovery of CoA), Lipmann considers acetylphosphate as the "active acetate" involved in acetylations. Lipmann starts from an

---

* We shall use the expression "energy-rich phosphate derivative" rather than "energy-rich phosphate bond". In the early 1940s, physical chemists, soon after its formulation by Lipmann [2], rejected the notion of energy-rich phosphate "bonds", but it took several years until biochemists recognized that the high energy released on hydrolysis is not due to the energy of phosphate bonds, but to the higher resonance stabilization of the hydrolytic product (and in the case of ATP in addition to the electrostatic repulsion of the negatively charged groups). In a short article in TIBS (September 1976) Pauling [2a] recalled how Coryell brought this to the attention of Kalckar [2b], who then used the appropriate formulation in his review article in 1942. Fruton and Simmonds [2c], in General Biochemistry, were the first to stress the proper definition. In spite of this, the distinction of the two types of symbols, $\sim$P and —P may still be useful as long as one does not associate the squiggle $\sim$ with the energy of the phosphate bond, as Lipmann [2] had originally assumed.

analogy with the use of acetylchloride in organic chemistry.
To quote Lipmann,

"For such purpose a constantly high acetyl group potential might be maintained in the cell, as would be utilized for the purpose of acetylation in
organic chemistry in the form of acetylchloride. Apparently the analogous
acetylphosphate is utilized in cell metabolism in an identical manner. In this
case, then, the purpose with the generation of an acylphosphate it not to
utilize the potential imposed upon the phosphate, but that likewise imposed
upon the acyl part, the acetyl group potential generated metabolically by
anhydrization with phosphoric acid.

$$\text{acetyl} \rightarrow \text{acetyl} \sim + \text{phosphate}$$
$$\downarrow$$
acetate + phosphoryl $\sim$ $\leftarrow$ phosphate   choline

In such a case, anhydrization with phosphoric acid would be utilized to
force the organic partner into a desired linkage" (p. 153).

Lipmann, in his 1941 review article, extends the mechanism of
acylphosphate formation hypothetically to the biosynthesis of fats
and proteins.

"The major part of the constituents of protoplasma are compounds which
contain the ester or the peptide linkage. In the routine procedure of organic
chemistry for synthesis of compounds of this type the acylchloride of the
acid part is first prepared and then brought into reaction with the hydroxyl
or amino group of the other part. In an analogous procedure the cell might
first prepare the acylphosphate with adenylpyrophosphate as the source of
energy-rich phosphate groups. The acylphosphate of fatty acids might then
condense with glycerol to form the fats. The acylphosphate of amino acids
likewise might condense with the amino groups of other amino acids to form
the proteins.
    The phosphoryl groups, continuously generated in metabolizing cells,
might thus find application for protoplasmic synthesis" (pp. 153—154).

In 1941, Lipmann thus considers the role of ATP in biosynthesis
as accomplishing the formation of acylphosphates: acetylphosphate from acetate, and the corresponding acylphosphates from
amino acids and fatty acids.

3. Discovery by Nachmansohn of the utilization of the energy of ATP in the biosynthesis of acetylcholine

In the 1930s, the physiologists were highly interested in the role of acetylcholine in nerve activity.

The acetylation of choline, an important utilization of "active acetate" in biosynthesis, had been shown by Mann and Quastel [4] to be dependent on pyruvic acid oxidation.

It was known that decomposition and reconstitution of acetylcholine took place in nervous tissues [5,6] and that the process was dependent on the oxidative metabolism [7].

Nachmansohn found in 1937 a high acetylcholinesterase activity in the electric tissue of the eel *Torpedo marmorata*. It appeared as essential to learn about the formation and hydrolysis of acetylcholine, and to know how these aspects are integrated in the sequence of reactions associated with electrical activity.

To quote Nachmansohn [8]:

"But how was the energy spent during activity restored; what reactions provided the energy for the resynthesis of acetylcholine? I had worked in Meyerhof's laboratory for several years on the role of phosphocreatine in muscle contraction. I was surprised to find that its concentration in electric tissue of the eel, in spite of the extremely low protein content of this tissue, is as high or higher than in striated muscle. ATP is nearly but not quite as high as in muscle. In view of the essential role of these two compounds in the energy transformations in muscular contraction their conspicuously high concentration in the highly specialized electric tissue suggested to me the possibility that they may also be essential in providing the energy requirement for genesis of bioelectricity although, for many reasons, through different mechanisms.

A strong breakdown of phosphocreatine was indeed found to coincide roughly with electrical activity [9].

It appeared safe to assume that phosphocreatine was used for the resynthesis of ATP used during activity. Even if a large part of the energy of ATP hydrolysis were used for the restoration of the ionic concentration gradient, as was later shown to be the case, the data suggested that part of the energy of ATP hydrolysis may be used for the acetylation of choline.

When ATP was added to electric tissue extracts, acetylation of choline was disappointedly small. This could be due to rapid hydrolysis of ATP by ATPase. Indeed, when ATP was added in the presence of sodium fluoride, a strong ACCh formation was obtained [10]. This was the first enzymatic acetylation obtained in a soluble system in which the free energy of ATP hydrolysis was used" (pp. 15—16).

Plate 181. David Nachmansohn.

The publication of the paper of Nachmansohn and Machado [10] in the *Journal of Neurophysiology* resulted from the intermission of John F. Fulton. The paper had previously been successively refused by *Science*, by the *Journal of Biological Chemistry* and by the *Proceedings of the Society of Experimental Biology and Medicine*. At that time (and for 2 years afterwards) the argument of authority imposed the concept of acetylphosphate as the general "active acetate" used in acetylations.

In their paper of 1943, Nachmansohn and Machado [10] described the extraction, from brain, of an enzyme (choline acetylase) which, in a cell free solution, acetylated choline on the addition of ATP. This discovery came as a shock at a time when it was a widespread belief that acetylation was accomplished by acetylphosphate and that the function of ATP was to produce acetylphosphate (or in general produce acylphosphates).

Meyerhof [11] has concisely qualified the importance of Nachmansohn's discovery, which became in fact the root of our whole theory of biosynthesis:

"Acetate in the presence of ATP and the choline acetylating enzyme systems forms acetylcholine. In this way enzymatic acetylation was shown for the first time" (p. 5).

Nachmansohn's reaction was indeed the first enzymatic acetylation obtained in a soluble system. It was also the first of the many endergonic reactions, as will appear in the course of this volume, which have been demonstrated to utilize the energy liberated by the splitting of ATP. This must be recognized as a highlight in the history of biosynthesis.

4. Sulphanilamide acetylation

At the end of 1944, Lipmann who had been an opponent of the concept of the utilization of the energy of ATP in endergonic biosynthetic reactions, and particularly in acetylation, convinced himself, during a stay in Nachmansohn's laboratory [8] of the difference between the animal system of acetylation discovered by Nachmansohn, and the system at work in the bacteria he had himself studied. Lipmann and Nachmansohn are friends. They both belong to the brilliant school of Meyerhof's disciples.

As was stated in Chapter 33 (p. 272) Lipmann [12] turned to a study of the enzymatic acetylation of sulphanilamide in pigeon liver homogenates and extracts. The conclusion of the introduction of his paper records his recognition that

"the coupling with energy-yielding reactions could be traced to de dependence on the supply of energy-rich phosphate bonds through adenyl pyrophosphate" (p. 173).

In the conditions of his experiments, Lipmann recognized acetylphosphate as "a poor acetyl donor".

Glutathione, glutamate and citrate had been observed by Nachmansohn, John and Waelsch [13] to activate the enzymatic acetylation of choline. These compounds were unsuccessfully tried by Lipmann [12] who adds

"In other respects the enzymatic mechanism of sulphanilamide acetylation appears strikingly similar to choline acetylation in which Nachmansohn and Machado likewise observed a linking between acetylation and phosphorylation" (p. 186).

"The poor reactivity of acetylphosphate makes a separate reaction between adenylpyrophosphate which exergonically breaks up into acetylsulphanilamide, adenylic acid, and inorganic phosphate.

$$CH_3COO^- + NH_2C_6H_4SO_2NH_2 + ad^- \sim HO.PO_3^-$$
$$\rightarrow CH_3CO.NHC_6H_4SO_2NH_2 + ad^- + HO.PO_3^-$$

$$\Delta Fo = +3 - 12 = -9 \text{ calories}$$

The liberation of inorganic phosphate is overshadowed in the present conditions by a large phosphate liberation through adenylpyrophosphate. A stoichiometric participation of adenylpyrophosphate, however, in the process of acetylation is indicated strongly by the quantities required for acetylation. Furthermore, the endergonic nature of the condensation makes a coupling with an exergonic process necessary and excludes a mere catalysis" (p. 188).

This work of Lipmann on the acetylation of sulphanilamide by liver extracts confirmed, as first recognized by Nachmansohn in the case of choline acetylation, that ATP provides the energy of acetylation.

5. Coenzyme A

In his paper on the acetylation of sulphanilamide (1945) Lipmann [12] records the presence of a heat-resistant component to be a part of the enzyme system and he describes reversible inactivation through dialysis or autolysis.

"The heat-stable principle is present in larger amounts in liver; kidney and muscle contain less, and results with yeast have been negative so far. In the experiments described in Table X a reversible activation was brought about by keeping extracts overnight at 7°, with or without dialysis. Likewise, partial inactivation occurred on 2 h incubation at 37°. Frequently the reactivation by boiled liver preparation was incomplete, indicating an additional inactivation of the residual compound. Rather concentrated preparations of the boiled organs had to be used for reactivation. The best results were obtained with undiluted boiled juices, prepared by heating in boiling water the mashed tissues without added fluid" (p. 186).

Almost simultaneously three other groups of researchers suggested the participation of a coenzyme in ATP activated acetylation reaction first described by Nachmansohn. Shortly after the discovery of choline acetylase, Nachmansohn and Machado observed that this enzyme rapidly lost its activity on dialysis, an observation which was confirmed by Nachmansohn, John and Waelsch [13]. Nachmansohn and Berman [14] obtained a purified preparation from *Kochsaft* of liver, heart and brain of a coenzyme which reactivated completely a dialyzed choline acetylase solution prepared from acetone-dried powder of rat brain.

In Barron's laboratory, Lipton [15] also independently described a coenzyme for choline acetylase. Feldberg and Mann [16] also observed the need of an activator in choline acetylation. It soon appeared that the coenzyme is used in acylations in general and the name of "coenzyme of acetylation" (CoA) was coined for it by Lipmann. In Chapter 33 we have retraced the brilliant work of Lipmann which led to the recognition of coenzyme A as containing pantothenic acid in combination with adenylic acid.

We shall here again quote an excerpt of the Harvey Lecture of Lipmann [17], a pregnant, concise and correct statement of the history of the discovery of coenzyme A and of its nature:

"On dialysis as well as on ageing the enzyme solution lost the ability to acetylate, which was regained on addition of boiled extracts. None of the known coenzymes could replace this factor and the isolation of this apparently new coenzyme was therefore attempted. On purification it appeared that the new coenzyme, coenzyme A, was a pantothenic acid derivative. The same coenzyme was found to activate the acetylation of choline [18]. Concurrently, Nachmansohn and Berman [14] and Feldberg and Mann [16] observed the need for an activator in choline acetylation. We find the activators of Nachmansohn and Feldberg and Mann to be identical with coenzyme A''.

6. Mechanism of acetate activation

The mechanism of acetylation, after the discovery of CoA, was elaborated along two pathways. The first, based on acylphosphate as acetylating agent as first proposed by Lipmann, is initiated by a phosphorylation of acetate with ATP in the presence of aceto-kinase (Lipmann [19]; Stern and Ochoa [20]; Rose, Grunberg-Manago, Korey and Ochoa [21]). This is followed by the transfer of the acetyl group of coenzyme A in the presence of phospho-transacetylase (Stadtman and Barker [22]; Stadtman, Novelli and Lipmann [23]; Stadtman [24]).

ATP + acetate $\rightleftharpoons$ acetylphosphate + ADP

acetylphosphate + CoA $\rightleftharpoons$ acetyl-CoA + phosphate

That this pathway is found only in certain microorganisms was recognized by Lipmann [17] and by Barker [25].

Another pathway involving a reaction of ATP, acetate and CoA involves the formation of acetyl-CoA resulting from a split of ATP. We shall call it Nachmansohn's reaction. It has also been termed the aceto-CoA-kinase reaction [26] from the name given to the enzyme involved

ATP + acetate + CoA $\rightleftharpoons$ acetyl-CoA + A5P + PP

Nachmansohn's reaction was, as stated above, discovered in brain extracts. Its presence in animal tissues was confirmed (Lipmann [12]; Chou and Lipmann [27]; Beinert, Green, Hele, Hift, Von Korpp and Ramakrishnan [28]). It has also been demonstrated in yeast (Lipmann, Jones, Black and Flynn [26]; Jones, Black, Flynn and Lipmann [29]) in plants [30] and in *Rhodospirillum rubrum* [31]. It appears clearly, therefore, that the general mechanism of acetate activation is Nachmansohn's reaction, e.g. an activation of acetate by ATP (with as shown later, a formation of adenylacetate) while the activation through acetylphosphate described by Lipmann in his review article of 1941 is an exceptional pathway limited to certain bacteria. When the paper of Nachmansohn and Machado was successively refused by three biochemical periodicals, the referees' reason was uniformly that there was no mechanism to explain the activation of acetate by ATP [8].

Jones, Lipmann, Hilz and Lynen [32] reported that a partially purified enzyme from yeast catalysed an exchange of PP and ATP in the absence of acetate and CoA. They also observed that [^{14}C]acetate exchanged with the acetyl group of acetyl-CoA in the absence of A5P and PP. The authors proposed a mechanism involving enzyme-bound substrates:

(1) ATP + enzyme ⇌ enzyme — A5P + PP

(2) Enzyme — A5P + CoA ⇌ enzyme — CoA + A5P

(3) Enzyme — CoA + acetate ⇌ acetyl-CoA + enzyme

This was derived from isotope exchange experiments. Paul Berg [33] reinvestigated the hypothesis and found that, with a more highly purified enzyme, the exchange of PP and ATP does not occur unless acetate is added. In the presence of CoA alone, no exchange of A5P and ATP was observed.

Berg [33] proposed a mechanism involving a primary reaction of ATP and acetate, as obtaining in Nachmansohn's reaction, and forming a hitherto undescribed compound, adenylacetate.

Adenylacetate

This compound has been chemically synthesized by Berg [34] by the reaction of acetic anhydride and A5P in pyridine and from acetylchloride and silver adenylate. That adenylacetate is the phos-

Plate 182. Paul Berg.

phoacetyl derivative of A5P was indicated by studies with A5P deaminase and 5'-nucleotidase.

Synthetic adenylacetate was demonstrated by Berg to react enzymatically as follows

ATP + acetate $\rightleftharpoons$ adenylacetate + PP

(an example of pyrophosphate cleavage of ATP in contrast to the more common orthophosphate cleavage)

Adenylacetate + CoA $\rightleftharpoons$ acetyl-CoA + A5P

That adenyl acetate acts as an intermediary in acetyl-CoA formation by yeast aceto-CoA-kinase was based by Berg [33] on the following observations:

"(1) The exchange of PP and ATP requires acetate;
(2) Synthetic adenylacetate is readily converted to ATP in the presence of PP and to acetyl-CoA with CoA;
(3) the CoA-independent accumulation of acetohydroxamic acid, PP, and A5P in the presence of hydroxylamine;
(4) the requirement of acetate and CoA for the exchange of A5P with ATP;
(5) and the requirement of A5P and PP for the exchange of acetate with acetyl-CoA" (p. 1009).

Later [18]O experiments of Boyer, Koeppe, Luchsinger and Falcone [35] have shown that [^{18}O]carboxyl-labelled acetate gave rise to excess [18]O in the phosphate group of A5P, which gives further support to the mechanism proposed by Berg.

As we have shown above, the process of acetylcholine biosynthesis, later extended to acylations in general was cleared by the successive contributions of Nachmansohn, Lipmann and Berg.

When it appeared that at least two enzymatic steps are operative in the acetylation process, choline acetylase was redefined as catalysing the transfer of the acetyl group from acetyl-CoA to choline (Korey, Braganza and Nachmansohn [36]).

Acetylation was the first process of biosynthesis in which ATP was shown to provide the energy of an endergonic reaction by the splitting of phosphate bonds. This process has been recognized as providing the energy of endergonic joining of precursors in the biosynthesis of different biomolecules as will be shown in the following chapters. ATP thus carries to the endergonic reactions of

References p. 43

anabolism the energy-rich phosphate derivatives formed in catabolism as was shown Part III of this *History*.

As Lipmann has adequately stated, the small chemical units
from which the material of organisms is built

"are glued together by an enormously versatile condensing reagent, ATP"
(p. 212).

As it will appear in following chapters, ATP, during its utilization
in biosynthetic reactions may undergo loss of either a pyrophosphate or an orthophosphate group.

The free energy required by endergonic reactions of biosynthesis provided by the phosphate bond energy of ATP is not only
generated along the pathways described in Part III of this *History*,
but also by the electrons provided by the reduced coenzymes
NADH and NADPH. The steps of the discovery of the generation
of NAD in metabolism were described in Chapter 36 and the
history of the unravelling of the biosynthetic pathway of its formation will be retraced in Chapter 71, devoted to the metabolic
pathways extending from tryptophan biosynthesis, in Volume 33B.

As will be shown in Chapter 55, NADP has been recognized as
formed in the pentose phosphate cycle.

REFERENCES

1   F.M. Huennekens and H.R. Whiteley, in M. Florkin and H.S. Mason (Eds.), Comparative Biochemistry, A Comprehensive Treatise, Vol. I, New York and London, 1960, p. 107—108.
1a   O. Meyerhof, Zu Energetik der Zellvorgänge, Göttingen, 1913.
1b   O. Meyerhof, Chemical Dynamics of Life Phenomena, Philadelphia and London, 1925.
2   F. Lipmann, Advan. Enzymol., 1 (1941) 99.
2a   L. Pauling, TIBS, September 1976, p. 214.
2b   H.M. Kalckar, Chem. Rev., 28 (1941) 71.
2c   J.S. Fruton and S. Simmonds, General Biochemistry, 2nd ed., New York, 1953, p. 377.
3   J.S. Fruton, Molecules and Life. Historical Essays on the Interplay of Chemistry and Biology, New York, 1972.
4   P.J.G. Mann and J.H. Quastel, Nature (London), 145 (1940) 856.
5   H.H. Dale and H.S. Gasser, J. Pharmacol., 29 (1929) 53.
6   D. Nachmansohn, Yale J. Biol. Med., 12 (1940) 565.
7   P.J.G. Mann, M. Tennenbaum and J.H. Quastel, Biochem. J., 32 (1938) 243.
8   D. Nachmansohn, Annu. Rev. Biochem., 41 (1972) 1.
8a   D. Nachmansohn, Chemical and Molecular Basis of Nerve Activity, New York, 1975.
9   D. Nachmansohn, R.T. Cox, C.W. Coates and A.L. Machado, J. Neurophysiol., 6 (1943) 383.
10   D. Nachmansohn and A.L. Machado, J. Neurophysiol., 6 (1943) 397.
11   O. Meyerhof, in W.D. McElroy and B. Glass (Eds.), Phosphorus Metabolism, Vol. 1, Baltimore, 1951, 3.
12   F. Lipmann, J. Biol. Chem., 160 (1945) 173.
13   D. Nachmansohn, H.M. John and H. Waelsch, J. Biol. Chem., 150 (1943) 485.
14   D. Nachmansohn and M. Berman, J. Biol. Chem., 165 (1946) 551.
15   M.A. Lipton, Fed. Proc., 5 (1946) 145.
16   W. Feldberg and T. Mann, J. Physiol., 104 (1946) 411.
17   F. Lipmann, Harvey Lectures, 44 (1950) 99.
18   F. Lipmann and N.O. Kaplan, J. Biol. Chem., 162 (1946) 743.
19   F. Lipmann, J. Biol. Chem., 155 (1944) 55.
20   J.R. Stern and S. Ochoa, J. Biol. Chem., 191 (1951) 161.
21   I.A. Rose, M. Grunberg-Manago, S.R. Korey and S. Ochoa, J. Biol. Chem., 211 (1954) 737.
22   E.R. Stadtman and H.A. Barker, J. Biol. Chem., 184 (1950) 769.
23   E.R. Stadtman, G.D. Novelli and F. Lipmann, J. Biol. Chem., 191 (1951) 365.
24   E.R. Stadtman, J. Biol. Chem., 196 (1952) 527—535.
25   H.A. Barker, in W.D. McElroy and B. Glass (Eds.), Phosphorus Metabolism, Vol. 1, Baltimore, 1951, p. 204.
26   F. Lipmann, M.E. Jones, S. Black and R.M. Flynn, J. Am. Chem. Soc., 74 (1952) 2384.

27   T.C. Chou and F. Lipmann, J. Biol. Chem., 196 (1952) 89.
28   H. Beinert, D. Green, P. Hele, H. Hift, R.W. Von Korpp and C.V. Rama-
      krishnan, J. Biol. Chem., 203 (1953) 35.
29   M.E. Jones, S. Black, R.M. Flynn and F. Lipmann, Biochim. Biophys.
      Acta, 12 (1953) 141.
30   A. Millerd and J. Bonner, Arch. Biochem. Biophys., 49 (1954) 343.
31   M.A. Eisenberg, Biochim. Biophys. Acta, 16 (1955) 58.
32   M.E. Jones, F. Lipmann, H. Hilz and F. Lynen, J. Am. Chem. Soc., 75
      (1953) 3285.
33   P. Berg, J. Biol. Chem., 222 (1956) 991.
34   P. Berg, J. Biol. Chem., 222 (1956) 1015.
35   P.D. Boyer, O.J. Koeppe, W.W. Luchsinger and A.B. Falcone, Fed.
      Proc., 14 (1955) 185.
36   S.R. Korey, B. de Braganza and D. Nachmansohn, J. Biol. Chem., 189
      (1951) 705.
37   F. Lipmann, Wanderings of a Biochemist, New York, 1971.

Chapter 54

Peptide bonds

(A) REPLACEMENT REACTIONS IN PROTEINS AND
TRANSPEPTIDATIONS

1. The theory of incorporation by substitution

Even when, as was described in Chapter 15, proteins were recognized as definite chemical entities, amenable to physical and chemical precise investigations, experimental research on protein biosynthesis long remained limited to global aspects, such as changes in the total amount of proteins in growth under different conditions. The advent of isotopes changed the situation and brought to light a host of peripheral information. Every tissue tested was found to incorporate every common amino acid of the L-form. It was recognized that the rate of incorporation varies with the tissue and with the amino acid considered (from 0.1 to 10 μmoles per gram of protein per hour). Evidence was provided to show that the amino acids were incorporated by peptide linkage. It was also demonstrated that the incorporation of a single amino acid is largely independent of the presence of others, except for specific accelerating effects of a few amino acids. In the majority of the cases studied it was found that inhibitors of respiration and phosphorylation inhibit amino acid incorporation and protein synthesis (for the large literature on these peripheral aspects, see Borsook [1]).

We have retraced in Chapter 10 the origins of the concept of the dynamic state of body constituents which introduced the notion of turnover and we have referred in that chapter to one of the particular aspects of this major acquisition, the concept, introduced by Schoenheimer, of a chemical exchange between free amino acids and amino acids bound in a protein and involving breakage of two strong covalent bonds.

Returning to the paper of Schoenheimer, Ratner and Rittenberg [2], we see that the authors, having demonstrated the rapid incorporation of amino acids of the diet into the body proteins of their rats, wrote (p. 726):

"There are two general reactions possible which might lead to amino acid replacement: (1) Complete breakdown of the protein into its units, followed by resynthesis, or (2) only partial replacement of units. Metabolic studies with isotopes indicate only end-results but no intermediate steps of a reaction. We have no indication as to what has happened to the protein molecule in the animals. Both reactions are conceivable".

This was a neutral position to which the authors did not adhere in the following part of the paper. They write (p. 730)

"peptide linkages open, the amino acids liberated mix with other of the same species of whatever source, diet or tissue".

and (p. 731)

"The process requires the continuous opening and closing of peptide linkages".

Edsall [3] wonders why

"having earlier stated the alternative so clearly, they later decided to back what eventually turned out to be the wrong horse".

We may try to retrace the reasons which led Schoenheimer to the conclusions he adopted. In the experiments of Schoenheimer, Ratner and Rittenberg [2] we see the rapid incorporation of doubly labelled leucine (deuterium in the carbon chain; ^{15}N — which had become available in 1937 — in the amino group). Most of the amino nitrogen appeared in the body proteins, not in the urine. A large proportion of the carbon skeleton appeared in the leucine of the protein. Thus the rapid incorporation of amino acid in the diet into body protein was strikingly demonstrated. This rapid rate was astonishing, at least to the biochemists of the time who, as Edsall recalls had been "brought up on the doctrine of separate exogenous and endogenous metabolism. They must have startled Schoenheimer himself".

We have retraced in Chapter 10 the origins of Folin's theory of separate endogenous ("wear-and-tear") and exogenous protein metabolism formulated in 1905, which were still widely accepted at the time of Schoenheimer's experiments. The correct interpretation that protein molecules were being constantly resynthesized from amino acids while others were degraded, so as to maintain a steady state, was difficult to reconcile with the current theory of endogenous and exogenous metabolism. Edsall feels that a constant opening and closing of peptide links was perhaps easier to imagine, though

"it could also have raised the question: why did not the peptide chain fall apart, if an anterior amino acid was sliced out from it. Schoenheimer, I think, never discussed this point, though it is not difficult to imagine mechanisms whereby an enzyme could hold the two pieces in place, during the process of excision and replacement. At any rate, Schoenheimer apparently held to the conception of the continuous opening and closing of peptide linkages, during the tragically brief remainder of his life".

Nevertheless, observed facts were presented against the concept of constant opening and closing of peptide linkages. For instance, Humphrey and McFarlane [4] injected radioactive lysine and phenylalanine into rabbits hyperimmunized against pneumococcus type III. From these rabbits they obtained radioactive antibody and they transfused them into normal rabbits and immune rabbits. Concentration of passively transfused antibody falls progressively, but the specific activity of the antibody remaining at any time remained unchanged in the course of the experiment, an observation which does not support the opening and closing of peptide linkages. This theory nevertheless persisted for some time and it is still for instance, mentioned by Gale and Folkes [5] in 1955 in a study on the incorporation of labelled amino acids into disrupted staphylococcal cells. The authors put the query concerning the incorporation of a single labelled amino acid "as a result of protein synthesis or an exchange reaction" (p. 673). In the case of the incorporation of a single amino acid such as glutamic acid they conclude that in their experimental conditions, the incorporation

Plate 183. Max Bergmann.

"arises from substitution of corresponding residues in preformed proteins by molecules of that amino acid, labelled or otherwise, in the medium".

Gale and Folkes [5] note in their paper that the incorporation requires a source of energy (ATP or HDP). The process was inhibited if nucleic acids were removed from the experimental mixture, and restored by adding back the nucleic acids.

2. Replacement reactions in protein synthesis (peptide transfers)

Could whole segments of peptide chain from a protein undergoing breakdown, be taken over bodily and incorporated as such into another protein molecule that was being synthesized? As remarked by Edsall [3], it would be an economical arrangement, saving considerable energy, if proteins could be done that way rather than starting from amino acids and it was a possibility in the frame of the knowledge available around 1950.

Bergmann and Fraenkel-Conrat [6] discovered a first example of peptide synthesis in an exchange reaction during hydrolysis.

$C_6H_5CO-NHCH_2COOH + NH_2C_6H_5$
benzoylglycine   aniline
(hippuric acid)

$\rightarrow C_6H_5CO-NHCH_2CO-NHC_6H_5$
benzoylglycinanilide

This was the first evidence produced in favour of the important concept of transpeptidation. Bergmann and Fraenkel-Conrat [6] found that the rate of formation of benzoylglycinanilide from benzoylglycinamide and aniline in the presence of cysteine-activated papain was more rapid than the synthesis of the anilide from hippuric acid and aniline. On this basis they developed the theory that proteolytic enzymes such as papain catalyse transpeptidation reactions. The term transpeptidation, or transamidation was coined by analogy with transglycosidation, designating the enzyme-catalysed replacement of one participant in a glycosidic linkage by another.

Studies on transpeptidation by Fruton and his collaborators have greatly extended the knowledge of the subject (see Fruton [7,8]; Borsook [1]).

References p. 59

Plate 184. Joseph Stewart Fruton.

Transpeptidation reactions have assumed a new actuality in the interpretation of the process of polypeptide chain elongation in protein synthesis (Lipmann [9]) as we shall see in a later section of this *History*. Transpeptidation remains as a positive contribution to protein synthesis which found the justification of its importance at a later stage.

3. Template hypothesis and peptide hypothesis

The study of amino acid incorporation in proteins rendered possible by the isotopic method has led to two main hypotheses of protein synthesis: the template hypothesis, according to which many amino acids met on a template and reacted at once with a formation of protein, and the peptide hypothesis according to which proteins were progressively formed by means of peptide intermediates (literature in Borsook [1,10—12]).

Anfinsen and Steinberg [13—15] considered the biosynthesis of ovalbumin by the hen's oviduct. Minced oviduct was incubated with $^{14}CO_2$ and the ovalbumin produced was digested with a bacterial enzyme hydrolysing the protein to plakalbumin and three peptides: (1) a hexapeptide (3 alanine, 1 aspartic acid, 1 glycine and 1 valine); (2) a tetrapeptide (1 alanine, 1 aspartic acid, 1 glycine and 1 valine) and (3) a dipeptide (alanylalanine).

The authors isolated the aspartic acid from the plakalbumin and from the hexapeptide. They found the specific activity always greater in the hexapeptide than in the plakalbumin. Labelled albumin was found to be incorporated unequally into the different fragments obtained from the protein.

It was considered that, in the case of biosynthesis directly from the amino acid pool by a template mechanism, any one of the labelled amino acids would have the same specific activity throughout the protein molecule. As this was found not to be the case, the authors excluded a complete template mechanism.

From the whole of their studies, Steinberg and Anfinsen [15] envisaged three possible causes for the differences in specific activities of the intermediates. To quote these authors:

"First, the labelled amino acid entering different sized pools of preformed, unlabelled peptides could undergo different degrees of addition. The specific activity of that amino acid would then differ from peptide pool to peptide pool. When these peptides combined to form the protein, the different segments corresponding to these peptides could contain the labelled residue at different specific activities.

"Second, if the possibility of true dynamic equilibrium between free amino acids and peptides is considered, differences in the rate constants would permit differences in the rate of isotope equilibration even at the steady state.

"Finally, if peptide fragments released by catabolic reactions from a preformed protein were utilized directly in the synthesis of another protein, these fragments would be expected to contain residues of considerably lower specific activity than those in the remainder of the molecule".

Borsook ([16] p. 213) considers the relative probabilities of the three possibilities presented by Steinberg and Anfinsen. He suggests deriving the relative order of probabilities from the quantitative aspect of amino acid incorporation and to turn, for this purpose, to the system of the rabbit reticulocyte, in which 90% or more of the protein is haemoglobin. Muir, Neuberger and Perrone [17] injected labelled glycine and valine in rats and determined the specific activities of the terminal and non-terminal valine in the circulating haemoglobin 12, 24 and 72 h and 1 and 2 weeks after the injection. The valine had the same specific activity in the two sets of loci, and the ratio of incorporated glycine to valine remained the same. From this, Muir, Neuberger and Perrone [17] conclude that

"The findings are compatible with the hypothesis that globin synthesis consists either of a simultaneous condensation of amino acids or of a rapid successive formation of peptide bonds without any appreciable accumulation of intermediates".

Experimentation with bacteria led to some data in support of the theory of a complete template mode of protein synthesis, while other data supported protein synthesis through intermediates (see Borsook [16]).

As noted by Raacke ([18] pp. 346—349) three lines of evidence have been produced in favour of the theory according to which proteins were built up slowly, by means of peptide intermediates. The first line was based on observations that partial protein hydrolysates afford better growth of plants (literature in Raacke).

Other authors tried to compare the effectiveness with which a labelled free amino acid can compete with larger protein fragments in the formation of new proteins. This line of evidence was discarded when Campbell and Stone [19] found no evidence for a utilization of breakdown products other than free amino acids.

A third line of evidence for peptide intermediates has been derived from kinetic studies. Raacke [18] has for instance studied the kinetics of formation and disappearance of different nitrogenous fractions in ripening pea seeds. She found

"that toward the end of the ripening period there was a fall in the concentration of free amino acids while the amount of peptide material continued to increase, and that the decline of this latter fraction was accompanied by a sharp decline in protein" (p. 348).

One of the arguments against the peptide hypothesis has been based on the notion that peptides could not be found in appreciable amounts in cells. Since many peptides were identified, the argument has lost its weight. When Snoke, Yanari and Bloch [20] had demonstrated the formation of glutathione from γ-glutamyl-cysteine and glycine by the enzyme glutathione synthetase, glutathione was for some time a subject of experimentation.

It was recognized that in protein synthesis particles were required while glutathione synthesis occurs in a solution. Since it was also recognized that while requiring ATP, glutathione biosynthesis did not involve the splitting of ATP in the same manner, the interest in glutathione synthesis dropped. Against the peptide intermediate theory and in favour of the template theory several arguments played their role and led to a discredit of the former. We may illustrate for instance with an observation cited by Glass [21] who describes a mutant strain of *E. coli* forming an adaptive enzyme (β-galactosidase) only when supplied with a particular amino acid even if it has previously synthesized polypeptides containing this amino acid.

The template hypothesis got a new stimulating development when it was suggested that the activated amino acids align themselves along specific sites on a nucleic acid. This aspect will be dealt with in a later section of this *History*.

(B) PEPTIDE BOND SYNTHESIS

4. Demise of peptide bond formation by reversal of Mass Action

The period covered by this part of our *History* extends between the recognition of the peptide linkage of the amino acids identified as protein constituents and the beginning of the interplay of biochemistry, not only with chemistry, but with molecular genetics. During the period considered, it was generally admitted that the reactions of hydrolysis, biosynthesis and substitution were constantly and simultaneously taking place within cells, the nature of available substrates and the specificity of available proteolytic enzymes governing the relative proportions of the three categories of actions. The concept of the biosynthesis of a protein in a cell was in fact defined by the lack of hydrolysis or peptide transfer undergone by the protein, which depended upon the nature of the proteolytic enzymes involved (Bergmann and Fraenkel-Conrat [6]). Concerning the biosynthesis of peptide bonds itself, it was generally considered, until the end of the 1930s as accomplished through the influence of proteolytic enzymes. As we have retraced in Chapter 51, contrary to these beliefs, Borsook showed by thermodynamic data that protein synthesis could not proceed to any significant extent by reversal of Mass Action. An input of 2500—3000 calories per peptide bond biosynthesized was required.

5. Inhibition of peptide synthesis by inhibition of oxidation

Recognizing that the synthesis of hippuric acid from benzoic acid and glycine resembles in several aspects the synthesis of a peptide bond, Borsook and Dubnoff [22] turned their attention towards hippuric acid synthesis as a model. It had long been known that animals form hippuric acid from benzoic acid and glycine (Bunge and Schmiedeberg [23]). That hippuric acid synthesis resembles in many aspects the synthesis of a peptide bond was emphasized by Borsook and Dubnoff. The group which is formed, the CONH group, is the same, and is in the same α position to a carboxyl group. As we have stated in Chapter 51, Borsook and Dubnoff showed that the synthesis of hippuric acid was attended by a gain in free energy. The tendency of the reaction, if allowed to proceed spontaneously at 25°C to 38°C is not towards synthesis but towards practically complete hydrolysis of hippuric acid. Yet, when benzoic acid is fed, hippuric acid is rapidly synthesized. The authors experimented on tissue liver slices suspended in Ringer's solutions containing low concentrations of benzoic acid and glycine. More than half of the benzoic acid was converted to hippuric acid. The free energy of formation is of the same order of magnitude as in the case of peptide bonds. As the authors observed, the synthesis of hippuric acid by liver slices is completely inhibited by 0.001 M KCN. They concluded, therefore, that cell respiration was essential for the synthesis.

6. Peptide bond synthesis in homogenates

Cohen and McGilvery [24] showed that p-aminohippuric acid is formed from p-aminobenzoic acid and glycine, not only by liver slices [24], but also by homogenates [25,26], under aerobic conditions or, anaerobically, with ATP as energy source.

Borsook and Dubnoff [27] also reported the formation of hippuric acid in homogenized guinea pig liver suspended in a phosphate-saline solution. They observed that to obtain more than traces of synthesis, it was necessary to homogenize the liver in the presence of the substrate, i.e. benzoic acid and glycine. The yield of hippuric acid was nearly doubled when adenylic acid and α-ketoglutaric acid were added. As stated by the authors,

"Evidently the oxidation of the latter substance provided ATP which in turn furnished the free energy for the synthesis" (p. 397).

7. Activated form of benzoic acid in hippuric acid synthesis

As we have retraced in Chapter 33, Lipmann showed that the acetylation of sulphanilamide in liver required ATP. A coenzyme was involved, which he identified as coenzyme A. Chantrenne [28], hoping to identify the activated form of benzoic acid which, after the experiments of Borsook and Dubnoff, appeared to be involved in hippuric acid formation, synthesized monobenzoylphosphate

Plate 185. Hubert Chantrenne.

and dibenzoylphosphate. He found that the dibenzoylphosphate reacted more rapidly with amino acids. This also applies to phenyl-benzoylphosphate. Chantrenne suggested that diacylphosphates or substituted acylphosphates rather than mono-acylphosphates could be the intermediary activated states of the carboxyl group in the peptide bond synthesis.

While a guest in P.P. Cohen's laboratory at Madison, Wisconsin, Chantrenne [29] tried to determine whether the synthesis of hippuric acid requires the presence of a coenzyme as well as the acetylation of amines. This research was favoured by the fact that unpublished work of McGilvery in P.P. Cohen's laboratory had shown that the synthesis of p-aminohippuric acid occurs in extracts of acetone powder of rat liver. Chantrenne found that the same extracts also catalyse hippuric acid formation from benzoic acid plus glycine in the presence of ATP.

He showed that benzoylphosphate was not the high-energy intermediate in hippuric acid synthesis, and further that a coenzyme is required for the condensation of benzoic acid with glycine. Chantrenne first believed that he had identified a new coenzyme, active in benzoylations. But it was made possible, by Lipmann, to test a variety of preparations of the coenzyme of acetylations, CoA. Chantrenne recognized that the content of each of these preparations in "benzoylation coenzyme" corresponded to their content of coenzyme A, which he recognized as required for hippuric acid synthesis. He showed that benzoylphosphate was unable to replace benzoic acid plus ATP, thus excluding benzoylphosphate as an intermediate. The enzyme system is inactivated when treated by Dowex-1; the inactivated enzyme is reactivated by coenzyme A. As well as in the acetylation of amines, coenzyme A is involved in the peptide bond formation of hippuric acid. When, as in liver, the energy for the peptide bond formation comes from ATP, in the case of hippuric acid, the immediate high-energy intermediate is benzoyl-CoA and the action of ATP from which energy is derived is indirect, in fact two steps removed. It must be noted that Braunstein and Yefimochkina [30] had observed an effect of pantothenic acid deficiency upon hippuric acid synthesis in rats.

As shown by Schachter and Taggart [31,32], an enzyme-bound adenylbenzoate is formed. It transfers the benzoyl radical to coenzyme A.

ATP + benzoate = adenylbenzoate + PP

adenylbenzoate + CoA = benzoyl-CoA + AMP

benzoyl-CoA + glycine = hippuric acid + CoA

Synthetic adenylbenzoate can replace benzoate + ATP, as was shown by Moldave and Meister [33].

REFERENCES

1 H. Borsook, Adv. Prot. Chem., 8 (1953) 128.
2 R. Schoenheimer, S. Ratner and D. Rittenberg, J. Biol. Chem., 130 (1939) 703.
3 J. Edsall, Mol. Cell. Biochem., 5 (1974) 103.
4 J.H. Humphrey and A.S. McFarlane, Biochem. J., 57 (1954) 195.
5 E.F. Gale and J.P. Folkes, Biochem. J., 59 (1955) 661.
6 M. Bergmann and H. Fraenkel-Conrat, J. Biol. Chem., 119 (1937) 707.
7 J.S. Fruton, Yale J. Biol. Med., 22 (1950) 263.
8 J.S. Fruton, Harvey Lect., 51 (1957) 64.
9 F.L. Lipmann, Science, 164 (1969) 1024.
10 H. Borsook, Physiol. Rev., 30 (1950) 206.
11 H. Borsook, Fortschr. Chem. Org. Naturst., (1952) 292.
12 H. Borsook, C.L. Deasy, A.J. Haagen-Smit, G. Keighley and P.H. Lowy, Fed. Proc., 8 (1949) 589.
13 C.B. Anfinsen and D. Steinberg, J. Biol. Chem., 189 (1951) 739.
14 C.B. Anfinsen and D. Steinberg, Fed. Proc., 10 (1951) 156.
15 D. Steinberg and C.B. Anfinsen, J. Biol. Chem., 199 (1952) 25.
16 H. Borsook, in D.M. Greenberg (Ed.), Chemical Pathways of Metabolism, Vol. II, New York, 1954, p. 173.
17 H.M. Muir, A. Neuberger and J.C. Perrone, Biochem. J., 49 (1951) LV.
18 I.D. Raacke, in D.M. Greenberg (Ed.), Metabolic Pathways (2nd ed. of Chemical Pathways of Metabolism), Vol. II, New York and London, 1961, p. 263.
19 P.N. Campbell and N.E. Stone, Biochem. J., 66 (1957) 669.
20 J.E. Snoke, S. Yanari and K. Bloch, J. Biol. Chem., 201 (1953) 573.
21 B. Glass, in W.D. McElroy and B. Glass (Eds.), Phosphorus Metabolism, Vol. 2, Baltimore, 1952, p. 826.
22 H. Borsook and J.W. Dubnoff, J. Biol. Chem., 132 (1940) 307.
23 G. Bunge and O. Schmiedeberg, Arch. Exp. Pathol. Pharmacol., 6 (1876) 233.
24 P.P. Cohen and R.W. McGilvery, J. Biol. Chem., 166 (1946) 261.
25 P.P. Cohen and R.W. McGilvery, J. Biol. Chem., 169 (1947) 119.
26 P.P. Cohen and R.W. McGilvery, J. Biol. Chem., 171 (1947) 121.
27 H. Borsook and J.W. Dubnoff, J. Biol. Chem., 168 (1947) 397.
28 H. Chantrenne, Nature (London), 160 (1947) 603.
29 H. Chantrenne, J. Biol. Chem., 189 (1951) 227.
30 A.E. Braunstein and E.F. Yefimochkina, Dokl. Akad. Nauk SSSR, 71 (1950) 347 (in Russian).
31 D. Schachter and J.V. Taggart, J. Biol. Chem., 208 (1954) 263.
32 D. Schachter and J.V. Taggart, J. Biol. Chem., 211 (1954) 271.
33 K. Moldave and A. Meister, Biochim. Biophys. Acta, 24 (1957) 654.

Chapter 55

The pentose phosphate cycle

The pentose phosphate cycle can be considered as essential for the formation of specific structural and functional components: ribose and erythrose-4-phosphate. It is also highly significant in biosynthesis by virtue of its role as a generator of NADPH for many synthetic reactions. In no reaction of the pentose phosphate cycle is ATP produced directly. It is therefore justifiable to deal with this important metabolic pathway in the section on biosynthesis.

(A) OXIDATIVE IRREVERSIBLE SEQUENCE FROM
GLUCOSE-6-PHOSPHATE TO RIBULOSE-5-PHOSPHATE

1. From glucose-6-phosphate to D-gluconate-6-phosphate

The background of the series of enzyme and coenzyme discoveries which led to the formulation of this pathway is to be found in Warburg's theory of oxygen activation by "Fermenthemin". He asked himself why the "Sauerstoffübertragendes Fermenthemin" oxidizes the substrates of respiration inside the cells, while hemin does not react with them in vitro. As was shown by Warburg, Kubowitz and Christian [1] the red cells exhibit a respiration in the presence of methaemoglobin or methylene blue, which generates methaemoglobin in these cells. This is not observed with haemolysed blood cells, when centrifuged. The clear solution does not show any respiration after the addition of glucose, but if the glucose is replaced by glucose-6-phosphate (G-6-P), a respiration (oxygen consumption) appears (Warburg and Christian [2]). This fact led to the discovery of G-6-P dehydrogenase first identified in red cells by Warburg and Christian. They first called this dehy-

Plate 186. Frank Dickens.

drogenase Zwischenferment [3]. It was isolated by Julian, Wolfe and Reithel [4] in the crystalline form from bovine mammary gland, and from red cells by Yoshida [5].

Warburg and Christian, in publications of 1936 and 1937 [6,7], noted that D-gluconate-6-P formed from G-6-P by G-6-P dehydrogenase can, in the presence of NADP and another enzyme of Lebedev's juice of yeast, and with reduction of NADP and production of CO_2, be oxidized to a product of undetermined nature. This product was the subject of further studies. Lipmann [8] proposed that the product resulting from the oxidative decarboxylation of 6-phosphogluconate upon addition of bromacetate was a pentose phosphate, arabinose-5-phosphate. He observed the formation of 1 mole of CO_2 per atom of oxygen utilized.

After incubation of gluconate-6-P with a fraction of Lebedev's juice of yeast, Dickens [9] obtained a barium precipitate. The elementary analysis of this precipitate led him to consider the occurrence of a mixture of ketophosphohexonate and phosphopentonate. With a less purified preparation, Dickens obtained a lead salt of a compound, the elementary composition of which corresponded to a phosphotetronic acid. Dickens suggested a hypothetical pathway: G-6-P → gluconate-6-P → 2 keto-6-phosphogluconate → pentose-5-phosphate → 5-phosphopentonate → 2 keto-5-phosphopentonate → tetrose-4-phosphate → 4-phosphotetronate.

In spite of the fact that it was eventually discarded, this hypothetical path had the merit of leading Dickens to experiment with Lebedev's juice and with haemolysates of horse blood, and to reach the conclusion that arabinose-5-P was not involved, but rather ribose-5-P, which he was the first to test. These researches prior to 1938 became the background for the generalization of the concept, based on the observation of a production of CO_2, according to which there was a general pathway of direct oxidation of glucose, as an alternative for glycolysis.

Cohen and Scott [10] recognized the production of an unknown orcinol positive substance and of pentose enediol. The product of the oxidation of gluconate-6-P was recognized by Horecker and his school [11–13] as D-ribulose-5-P which was isolated by column chromatography and separated from D-ribose-5-P, both of which were definitely identified.

It is of interest to note that Horecker first identified ribose-5-P (R-5-P) by incubating yeast (or liver 6-phosphogluconate dehydrogenase) with gluconate-6-P dehydrogenase and NADP, but he observed that the primary product (short incubation time) was a ketopentose phosphoric acid ester. At equilibrium this ester rearranged into R-5-P.

When Scott and Cohen [14] allowed a purified dehydrogenase preparation from *E. coli* to act on gluconate-6-P they also obtained ribulose-5-P as the first product.

At the time, an oxidative pathway leading from glucose to ribulose-5-P (Ru-5-P) was formulated as follows

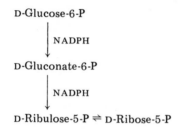

The overall reaction corresponds to

$$6 \text{ G-6-P} + 12 \text{ NADP}^+ = 6 \text{ Ru-5-P} + 6 \text{ CO}_2 + 12 \text{ NADPH} + 12 \text{ H}^+$$

The oxidation proceeds through direct transfer of hydrogen from C-3 of gluconate-6-P to the *β-para* position of the nicotinamide ring of NADP$^+$ (literature in Vol. 17 of this Treatise, p. 180).

2. Enzymes of the oxidative pathway

(a) Glucose-6-phosphate dehydrogenase

This enzyme catalyses the reaction leading to the δ-lactone of 6-phosphogluconic acid (Cori and Lipmann [15]

D-Glucose-6-phosphate + NADP$^+$

= D-glucono-δ-lactone-6-phosphate + NADPH

It has received the systematic name [D-glucose-6-phosphate: NADP$^+$ 1-oxidoreductase; 1.1.1.49]. The nature of the product was indicated by the observation that it reacted with hydroxylamine to form a hydroxamic acid showing a behaviour analogous to that of δ-gluconolactone. G-6-P dehydrogenase has been the subject of many studies concerning its physical properties, the nature of its molecular forms, the regulation of its activity, its variants, the results of its deficiency, etc. This history has been retraced by Pontremoli and Grazi in Chapter IV of Vol. 17 of the present Treatise.

(b) Lactonase

As stated above, the product of the dehydrogenation of β-D-gluco-pyranose-6-P is 6-phosphoglucono-δ-lactone, while as demonstrated later (by the discovery of gluconate-6-P dehydrogenase), the participant in the pathway is the open chain of gluconate-6-P. As the hydrolysis of the lactone at neutral pH is a slow process, the existence of an enzyme catalysing this step was postulated [15]. Later Brodie and Lipmann [16] identified such enzyme, which was called lactonase, in rat liver, yeast and *Azotobacter vinelandii*.

Kawada, Kagawa, Takiguchi and Shimazono [17] have demonstrated that lactonase is formed by two protein chains, one catalysing the hydrolysis of the δ-lactone of D-gluconate and another, the one involved in the pathway concerned, hydrolysing the δ-lactone of D-gluconate-6-P. The enzyme was later more commonly called gluconolactonase. It has received the systematic name [D-glucono-δ-lactone hydrolase; 3.1.1.17].

(c) Phosphogluconate dehydrogenase

The background of the discovery of this enzyme is found in the observation made by Warburg and Christian and by Dickens independently, of the oxidative decarboxylation of gluconate-6-P by yeast extracts to CO_2 and a phosphorylated pentose in the presence of NADP. The specificity of the enzyme involved was established by the discovery of the product of the reaction, Ru-5-P by Horecker, Smyrniotis and Seegmiller [12]. The enzyme itself was isolated and crystallized (free of the isomerase and epimerase mentioned in section 4) by Pontremoli, De Flora, Grazi, Mangiarotti, Bonsignore and Horecker [18].

The enzyme now commonly called phosphogluconate dehydro-

Plate 187. Bernard Leonard Horecker.

genase has also been called 6-phosphogluconic carboxylase. Its systematic name is [6-phospho-D-gluconate : NADP⁺ 2-oxidoreductase (decarboxylating); 1.1.1.44].

(B) NON-OXIDATIVE, REVERSIBLE TRANSFER FROM PENTOSE PHOSPHATE TO HEXOSE PHOSPHATE

3. Introduction

Dische, in a paper published in 1938 [19], demonstrated that hexose phosphate is formed from pentose derivatives in red-cell haemolysates.

To quote Horecker (in Vol. 15 of this Treatise):

"Addition of adenosine to such hemolysates resulted in the appearance of both triose and hexose phosphates. Since inorganic phosphate was esterified in the course of the process, Dische [20] postulated that the first step was the phosphorolysis of adenosine to yield ribose phosphate. This observation was later confirmed and extended by Schlenk and Waldvogel [21] who used D-ribose-5-phosphate as the substrate with liver extracts and showed that D-glucose-6-P was formed" (p. 48).

That a triosephosphate acted as an intermediary in this path had been recognized, as stated above, by Dische. Experimenting with extracts of *Escherichia coli* and of yeast, Racker [22] and Sable [23] confirmed the formation of a triose phosphate in the pathway from D-ribose-5-P. These data led to suggest that the pentose phosphate was cleaved in two fragments (C_2 and C_3) which were the sources of the hexose monophosphate. As stated above, these experiments were performed with D-ribose-5-P but it was shown later that this pentose phosphate was not the ultimate pentose phosphate substrate, which can be formed from D-ribose-5-P by the participation of the two enzymes hereafter mentioned.

4. Phosphoribose isomerase and phosphoribose epimerase

Two enzymes catalyse the interconversion of D-ribose-5-P, D-ribulose-5-P and D-xylulose-5-P. Glock [24,25] showed that in the breakdown of R-5-P, a ketopentose phosphate is formed. This

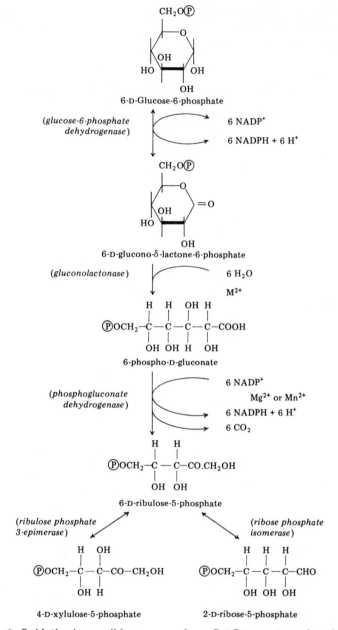

Fig. 1. Oxidative irreversible sequence from G-6-P to pentose phosphates.

ketopentose was first identified as D-xylulose-5-P by Ashwell and Hickman [26]. From the results of the action of a spleen extract on R-5-P they obtained a mixture of pentose phosphate: 80% R-5-P, 14% Ru-5-P and 6% Xu-5-P. The formation of D-xylulose-5-P from D-ribose-5-P was elucidated by the purification of two · enzymes: phosphoribose isomerase (Horecker, Smyrniotis and Seegmiller [12], Axelrod and Jang [27]), catalysing the conversion of D-ribose-5-P to D-ribulose-5-P, and D-xylulose-5-P 3-epimerase (Ashwell and Hickman [26]; Dickens and Williamson [28]; Stumpf and Horecker [29]; Hurwitz and Horecker [30]; Ashwell and Hickman [31]).

$$\text{D-Ribose-5-P} \xrightleftharpoons{\text{isomerase}} \text{D-ribulose-5-P} \xrightleftharpoons{\text{epimerase}} \text{D-xylulose-5-P}$$

The existence of ribose phosphate isomerase was pointed to by the fact that until phosphogluconate dehydrogenase was obtained in the crystalline state the reaction product always contained a mixture of D-ribulose-5-P and D-ribose-5-P (Horecker and Smyrniotis [11]). The enzyme has been purified from alfalfa by Axelrod and Jang [27].

Ribosephosphate isomerase has also been called phosphopentoisomerase or phosphoriboisomerase. It has received the systematic name [D-ribose-5-phosphate ketolisomerase; 5.3.1.6]. It catalyses the reaction:

D-Ribose-5-P = D-ribulose-5-P

The D-ribulose-5-P formed may be converted to its epimer, D-xylulose-5-P in the presence of a widely distributed enzyme (literature in Axelrod [32]) ribulosephosphate-3-epimerase (phosphoribulose isomerase). This enzyme has received the systematic name [D-ribulose-5-phosphate 3-epimerase; 5.1.3.1]. It catalyzes the reaction

D-Ribulose-5-P = D-xylulose-5-P

5. Transketolase

This enzyme was originally assayed (Horecker, Smyrniotis and Klenow [33]) with D-ribulose-5-P and D-ribose-5-P at a time when phosphoribose epimerase and phosphoribose isomerase were still unknown. Purified enzyme preparations allowed them and others to recognize that the pentose phosphate substrate of the enzyme was D-xylulose-5-P (Srere, Cooper, Klybas and Racker [34]; Horecker, Hurwitz and Smyrniotis [35]). Other phosphate esters also act as substrates, such as D-fructose-6-P (De La Haba, Leder and Racker [36]) or sedoheptulose-7-P (Horecker, Smyrniotis and Klenow [33]).

When the substrate is D-xylulose-5-P, the reaction results in a triose-P and a C_2 fragment, "active glycoaldehyde", which remains attached to the enzyme if no acceptor is added. If D-ribose-5-P is added, sedoheptulose-7-P is formed, as was shown by Horecker, Smyrniotis and Klenow [33] (for Review see Racker [37]).

The enzyme was first purified from rat liver by Horecker and Smyrniotis [38] and crystallized from yeast by Racker, De la Haba and Leder [39] who gave it the name of transketolase since it splits a ketol group from specific substrates (donors) and transfers the liberated "active glycoaldehyde" to an aldehyde (acceptor). Only when pure samples were obtained (deprived of isomerase and epimerase) was it recognized that in the formation of heptose phosphate the donor was D-xylulose-5-P and the acceptor R-5-P.

D-Xylulose-5-P + D-ribose-5-P

⇌ sedoheptulose-7-P + D-glyceraldehyde-3-P

(Srere, Cooper, Klybas and Racker [34]; Stumpf and Horecker [29]; Horecker, Smyrniotis and Hurwitz [40]; Horecker, Hurwitz and Smyrniotis [35]).

The discovery of sedoheptulose-7-P as an intermediate was made at a time when D-xylulose-5-P, the donor of the 2 carbon moiety in the presence of a transketolase, was still unknown, but the balance of 2 pentose phosphates with one D-sedoheptulose-7-P and one D-glyceraldehyde-3-P was satisfactory.

6. Mechanism of transketolase action

The prosthetic group of transketolase was identified by Racker, De La Haba and Leder [39] and by Horecker and Smyrniotis [39a] as thiamine diphosphate. Many theories explaining the activity of transketolase (Reviews: Krampitz, Suzuki and Greull [41],

Horecker in Vol. 15 of this Treatise) have been proposed. That the active group of the thiamine diphosphate is the C-2 carbon of the thiazole ring has been more recently (1958) established by Breslow [42].

The history of the establishment of the mechanism of transketolase action has been retraced by Horecker (in Vol. 15 of this Treatise).

7. Transaldolase

Finally, the long-recognized formation of hexose phosphate from pentose-5-P could be explained, together with the action of transketolase, by the action of transaldolase, transferring the upper half of the fructose-6-P molecule to a suitable aldose acceptor. This second enzyme fraction was shown by Horecker and Smyrniotis [43] and by Horecker, Smyrniotis, Hiatt and Marks [44] to contain the enzyme transaldolase, catalysing the reversible reaction

Sedoheptulose-7-P + D-glyceraldehyde-3-P

⇌ D-fructose-6-P + D-erythrose-4-P

An intermediate is formed which has been named "active dihydroxyacetone". In the presence of an acceptor such as D-glyceraldehyde-3-P, a formation of D-fructose-6-P takes place. Other components may act as acceptors. Pontremoli, Prandini, Bonsignore and Horecker [45] have detected no prosthetic group in purified preparation, the active site being a lysine residue in the protein (see Chapter II, by Horecker, of Vol. 15 of this Treatise).

Transketolase has also been shown to mediate the reversible reaction

Fructose-6-P + D-glyceraldehyde-3-P

⇌ D-xylulose-5-P + erythrose-4-P

(Smyrniotis and Horecker [46]; Racker, De la Haba and Leder [47]; Kornberg and Racker [48]).

It is also assumed that the G-6-P is resynthesised from glyceraldehyde phosphate, a process requiring a specific fructose-1,6-diphosphatase.

The isomerization of fructose-6-P to G-6-P in the presence of phosphoglucose isomerase completes the formation of G-6-P from pentose phosphate.

(C) THE PENTOSE PHOSPHATE CYCLE

The reactions catalysed by transketolase and transaldolase are essential components of the carbon cycle in photosynthesis, as will be shown in Chapter 56.

The non-oxidative pathway from pentose phosphate to hexose phosphate is also useful in the reconversion of excess pentose to hexosemonophosphates. This, in many micro-organisms, is the first step of the process of pentose fermentation and oxidation. That the oxidative sequence (from G-6-P to Ru-5-P) is not only a synthetic pathway has been recognized since Dickens [49] showed that R-5-P is utilized by yeast Lebedev's juice with the formation of CO_2 and other products. A number of data left no doubt regarding a relation between hexosephosphates and pentose phosphate formation (literature in Hollmann [50], p. 48).

As stated above, one mole of G-6-P, in animal tissues, can be oxidized into one mole of Ru-6-P and CO_2, and two moles of NADPH. The Ru-6-P after isomerization to R-5-P can be utilized in the biosynthesis of nucleic acid. But it can also be recycled to synthesize fructose-6-P. The operation of the complete cycle can be expressed as follows

(a) Oxidative sequence:

$$6 \text{ G-6-P} + 12 \text{ NADP}^+ = 6 \text{ Ru-5-P} + 6 \text{ CO}_2 + 12 \text{ NADPH} + 12 \text{ H}^+$$

(b) Non-oxidative sequence:

$$6 \text{ Ru-5-P} = 5 \text{ G-6-P} + P_i$$

The coupling of the two sequences would allow the full oxidation of 1 mole of G-6-P to 6 moles of CO_2, in six turns of a cycle, with a production of 12 moles of NADPH which can be utilized in biosynthesis, but without any direct production of ATP.

This cycle which was completely reproduced in vitro (Racker) was considered as an unconventional and accessory degradation pathway. As stated by Axelrod [32]:

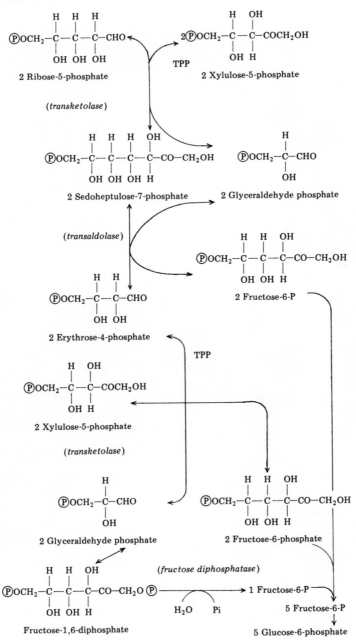

Fig. 2. Non-oxidative reversible transfer.

References p. 78

Plate 188. Efraim Racker.

"However, it should be appreciated that "other pathways" are not metabolic aberrations. They are "unconventional" only as accidents of history and of choice of experimental material" (p. 272).

The opinion remained prevalent that the pentose cycle was of very limited importance in the whole of the metabolism of carbohydrates, until, in 1953, Bloom and Stetten [51], and Bloom, Stetten and Stetten [52] introduced the use of specifically ^{14}C-labelled glucose to measure the relative amounts of glucose metabolized by glycolysis and by the pentose cycle. These authors measured the yield of ^{14}C in CO_2 produced by [1-^{14}C]glucose and [6-^{14}C]glucose using slices of rat organs. Comparing these results with those obtained with liver slices of rats, they observed that in the liver there was a much greater yield of $^{14}CO_2$ from [1-^{14}C]-glucose than from [6-^{14}C]glucose and their conclusion was that in rat liver the pentose pathway accounted for 60% and more of the CO_2 produced from glucose.

This finding changed the whole perspective and raised a great deal of interest and a number of papers were published with similar results (for literature, see Cheldelin, Wang and King [53]).

The basis of the measurement of the relative part of the pentose cycle rests on the fact that three molecules of G-6-P are oxidized each to three molecules of CO_2 and Ru-5-P. The origin of the CO_2 is in the C-1 of the glucose. In the glycolysis pathway, carbons 1 and 6 are incorporated to an equal extent into triose-P and therefore the ^{14}C yields from [1-^{14}C]glucose are identical in the products of this pathway. In the pentose cycle, only carbon 6 is incorporated into triose phosphates, since carbon 1 is converted solely to CO_2.

Another method (Abraham, Hirsch and Chaikoff [54]) relies on the yield of ^{14}C in a triose phosphate derivative such as fatty acids. The method relies on the fact that via the pentose cycle, C-1 is converted exclusively into CO_2 and [1-^{14}C]glucose can yield labelled fatty acids only via the glucose pathway.

In spite of the increased knowledge of the pentose phosphate pathway, the common biochemical thought did not reconsider the current concept of the major importance in carbohydrate metabolism conferred to glycolysis. This evaluation was in the first instance due to the dramatic development of the elucidation of the glycolytic pathway, the first occasion of isolating and purify-

ing enzymes and coenzymes and of reconstructing in vitro a complicated metabolic scheme. It was believed that glycolysis was universally the major pathway in cells. As Wood [55] points out,
three investigators, by chance, showed the dominant part played
by glycolysis in glucose catabolism. After Embden proposed his
scheme of glycolysis, phosphoglyceric acid was considered to be
the typical intermediate step of glycolysis. As Stone and Werkman
[56] surveying a group of bacteria using fluoride as inhibitor were
able to isolate phosphoglyceric acid from all, except for *Clostridium*, it was taken as evidence of the universality of glycolysis.
But we know now that phosphoglyceric acid is formed in other
pathways. A second argument was taken from the study of *E. coli*
for extensive metabolic studies, as this organism contains all the
enzymes of glycolysis [57,58]. If other bacteria had been chosen,
an enzyme of the pathway may have been found to be missing, as
shown by the discovery that *Leuconostoc mesenteroides* lacks
aldolase and triose phosphate isomerase and catabolizes glucose
through a pathway different from the classical scheme of glycolysis [59,60].

   A third factor in the emphasis on glycolysis was the study of
the degradation of glucose by *Lactobacillus casei*, performed by
Wood, Lifson and Lorber [61] in which they found that glucose
was split according to the pathway of glycolysis. If they had taken
another organism they might have found entirely different results.
Wood [55] commenting on the overemphasis on glycolysis in the
consideration of non-mitochondrial pathways of carbohydrate
metabolism concludes as follows:

"Added to this series of coincidences was the fact that the inhibitor experiments with iodoacetate, NaF, and other compounds were never fully convincing evidence of alternate pathways (cf. Racker [37] for a good discussion of
the problem). Meyerhof continually demonstrated fallacies in the evidence
for non-phosphorylating glycolysis, and the investigations with ^{32}P were not
entirely acceptable evidence of alternate pathways because the problem of
permeability and the inhomogeneity of pools could not be completely solved.
In addition opinion was swayed by the very beauty of the studies on glycolysis, which involved the isolation of pure enzymes and coenzymes and the
reconstruction in vitro for the first time of a complicated enzyme sequence."

As stated by Axelrod [32] (p. 219)

"In general, the isotope experiments strongly support the premise that an
operative pentose pathway is widely distributed. However even the most

refined experiments with isotopically labelled metabolites frequently yielded results which are offered with reservation and are acceptable only as qualitative guides. In general, results which indicate exclusive participation by the fermentative pathway are regarded as unequivocal. When, however, as is often the case, the pentose phosphate pathway, either oxidative or nonoxidative, shares the hexose phosphate with the glycolytic pathway, it is difficult to make a quantitative interpretation of such experiment. Difficulties arise because of exchange reactions, compartmentalization, redistribution of labelled carbon by the nondehydrogenase portion of the pentose phosphate pathway and various other factors".

That the phosphate cycle operates in tissues engaged in active biosynthesis, such as the mammary gland has been well documented (Wood, Schambye and Peters [62]) and a number of studies have been devoted to the presence of an activity of the cycle in different kinds of tissues, for instance in the lens, which lacks respiratory enzymes (Kinoshita, Masurat and Helfant [63]).

As stated by Horecker (p. 148):

"In those tissues where the pentose phosphate pathway appears to function as a cyclic mechanism, this can be related to the requirement of the tissue for the NADPH for reduction reactions. If the quantity of pentose phosphate formed from glucose-6-phosphate by the oxidation reaction exceeds the requirements of the tissue for nucleic acid and nucleotide synthesis, this excess pentose must be converted back to hexosemonophosphate by the nonoxidative pathway".

It may be concluded, as Pontremoli and Grazi state at the beginning of Chapter IV of Vol. 17 of this Treatise, that

"The pentose phosphate cycle has essentially two functions: (a) the production of NADPH which is utilized in biosynthetic processes; (b) the production of pentose and erythrose as building blocks for biosynthetic reactions." (p. 184).

References p. 78

REFERENCES

1   O. Warburg, F. Kubowitz and W. Christian, Biochem. Z., 227 (1930)
     245.
2   O. Warburg and W. Christian, Biochem. Z., 238 (1930) 131.
3   O. Warburg and W. Christian, Biochem. Z., 242 (1931) 206.
4   G.R. Julian, R.G. Wolfe and E.J. Reithel, J. Biol. Chem., 236 (1961)
     754.
5   A. Yoshida, J. Biol. Chem., 241 (1966) 4966.
6   O. Warburg and W. Christian, Biochem. Z., 287 (1936) 440.
7   O. Warburg and W. Christian, Biochem. Z., 292 (1937) 287.
8   F. Lipmann, Nature, 138 (1936) 588.
9   F. Dickens, Biochem. J., 32 (1938) 1626.
10   S.S. Cohen and D.B.M. Scott, Science, 111 (1950) 543.
11   B.L. Horecker and P.Z. Smyrniotis, Arch. Biochem., 29 (1950) 232.
12   B.L. Horecker, P.Z. Smyrniotis and J.E. Seegmiller, J. Biol. Chem., 193
     (1951) 383.
13   J.E. Seegmiller and B.L. Horecker, J. Biol. Chem., 194 (1952) 261.
14   D.B. Scott and S.S. Cohen, Biochem. J., 65 (1957) 686.
15   O. Cori and F. Lipmann, J. Biol. Chem., 194 (1952) 417.
16   A.F. Brodie and F. Lipmann, J. Biol. Chem., 212 (1955) 677.
17   M. Kawada, Y. Kagawa, H. Takiguchi and N. Shimazono, Biochim. Bio-
     phys. Acta, 57 (1962) 404.
18   S. Pontremoli, A. De Flora, E. Grazi, G. Mangiarotti, A. Bonsignore and
     B.L. Horecker, J. Biol. Chem., 236 (1961) 2975.
19   Z. Dische, Naturwissenschaften, 26 (1938) 252.
20   Z. Dische, in W.D. McElroy and B. Glass (Eds.), Phosphorous Metabo-
     lism, Vol. 1, Baltimore, 1951, p. 171.
21   F. Schlenk and M.J. Waldvogel, Arch. Biochem. Biophys., 12 (1947)
     181.
22   E. Racker, in W.D. McElroy and B. Glass (Eds.), Phosphorous Metabo-
     lism, Vol. 1, Baltimore, 1951, p. 147.
23   H.Z. Sable, Biochim. Biophys. Acta, 8 (1952) 687.
24   G.E. Glock, Biochem. J., 52 (1952) 575.
25   G.E. Glock, Nature (London), 170 (1952) 162.
26   G. Ashwell and J. Hickman, J. Am. Chem. Soc., 76 (1954) 5889.
27   B. Axelrod and R. Jang, J. Biol. Chem., 209 (1954) 847.
28   F. Dickens and D.H. Williamson, Nature (London), 176 (1955) 400.
29   P.K. Stumpf and B.L. Horecker, J. Biol. Chem., 218 (1956) 753.
30   J. Hurwitz and B.L. Horecker, J. Biol. Chem., 223 (1956) 993.
31   G. Ashwell and J. Hickman, J. Biol. Chem., 226 (1957) 65.
32   B. Axelrod in D.M. Greenberg (Ed.), Metabolic Pathways, Vol. 1, 3rd
     ed., New York, 1967.
33   B.L. Horecker, P.Z. Smyrniotis and H. Klenow, J. Biol. Chem., 205
     (1953) 661.
34   P.A. Srere, J.R. Cooper, V. Klybas and E. Racker, Arch. Biochem. Bio-
     phys., 59 (1955) 535.

35 B.L. Horecker, J. Hurwitz and P.Z. Smyrniotis, J. Am. Chem. Soc., 78 (1956) 692.

36 G. De La Haba, I.G. Leder and E. Racker, J. Biol. Chem., 214 (1955) 409.

37 E. Racker, Adv. Enzymol., 15 (1954) 141.

38 B.L. Horecker and P.Z. Smyrniotis, J. Am. Chem. Soc., 74 (1952) 2123.

39 E. Racker, G. de la Haba and I.G. Leder, J. Am. Chem. Soc., 75 (1953) 1010.

39a B.L. Horecker and P.Z. Smyrniotis, J. Am. Chem. Soc., 75 (1953) 1009.

40 B.L. Horecker, P.Z. Smyrniotis and J. Hurwitz, J. Biol. Chem., 223 (1956) 1009.

41 L.O. Krampitz, I. Suzuki and G. Greull, Fed. Proc., 20 (1961) 971.

42 R. Breslow, J. Am. Chem. Soc., 80 (1958) 3719.

43 B.L. Horecker and P.Z. Smyrniotis, J. Biol. Chem., 212 (1955) 811.

44 B.L. Horecker, P.Z. Smyrniotis, H.H. Hiatt and P.A. Marks, J. Biol. Chem., 212 (1955) 827.

45 S. Pontremoli, B.D. Prandini, A. Bonsignore and B.L. Horecker, Proc. Natl. Acad. Sci. USA, 47 (1961) 1942.

46 P.Z. Smyrniotis and B.L. Horecker, J. Biol. Chem., 218 (1956) 745.

47 E. Racker, G. de la Haba and I.G. Leder, Arch. Biochem. Biophys., 48 (1954) 238.

48 H.L. Kornberg and E. Racker, Biochem. J., 61 (1955) iii.

49 F. Dickens, Biochem. J., 32 (1938) 1645.

50 S. Hollmann, Non Glycolytic Pathways of Metabolism of Glucose, New York and London, 1964 (translated and revised by O. Touster).

51 B. Bloom and D. Stetten Jr., J. Am. Chem. Soc., 75 (1953) 5446.

52 B. Bloom, M.R. Stetten and D. Stetten Jr., J. Biol. Chem., 204 (1953) 681.

53 V.H. Cheldelin, C.H. Wang and T.E. King, in M. Florkin and H.S. Mason (Eds.), Comparative Biochemistry, Vol. 3, New York, 1962.

54 S. Abraham, P.F. Hirsch and I.L. Chaikoff, J. Biol. Chem., 211 (1954) 31.

55 R.G. Wood, Physiol. Rev., 35 (1955) 841.

56 R.W. Stone and C.H. Werkman, Biochem. J., 31 (1937) 1516.

57 M.F. Utter and C.H. Werkman, J. Bacteriol., 41 (1941) 5.

58 J.L. Still, Biochem. J., 34 (1940) 1374.

59 R.D. De Moss, R.C. Bard and I.C. Gunsalus, J. Bacteriol., 62 (1951) 499.

60 I.C. Gunsalus and M. Gibbs, J. Biol. Chem., 194 (1952) 871.

61 H.G. Wood, N. Lifson and V. Lorber, J. Biol. Chem., 159 (1945) 475.

62 H.G. Wood, P. Schambye and G.H. Peeters, J. Biol. Chem., 226 (1957) 1023.

63 J.H. Kinoshita, T. Masurat and M. Helfant, Science, 122 (1955) 72.

Chapter 56

The photosynthetic cycle of carbon reduction

1. First experiments with ^{11}C. Demise of the formaldehyde theory. Discovery of CO_2 fixation in the dark

Many examples of fixation of carbon dioxide in vitro are well known to the chemist. We may for instance mention the fixation of CO_2 in water in the form of carbonic or bicarbonic acid, the absorption of CO_2 by alcohols or amines, the carboxylation of phenols (one example being the preparation of salicylic acid); etc. But, until the summer of 1937 it was a common belief that, among living creatures, only photosynthetic plants and chemosynthetic autotrophs were able to utilize CO_2 as a material for biochemical synthesis. It was, as stated above, the current view that under the influence of light, carbon dioxide was the precursor of formaldehyde which was itself considered as the precursor of the carbohydrates biosynthesized in the photosynthetic process in plants.

Isotopes were first introduced in the field of photosynthesis when, in the summer of 1937, at Berkeley, Ernest Lawrence asked Kamen to cooperate with I.L. Chaikoff, in utilizing the radioactive isotope ^{11}C (half-life: 21 min) in a study of the labelled carbohydrates produced in the course of photosynthesis. This made it compulsory to work in the vicinity of a cyclotron, either near the cyclotron of Lawrence at Berkeley or at Boston (near the Harvard cyclotron). Ruben and Hassid were also introduced into the team *.

_____

* As it was stated in the Introduction to Vol. 31, the hydrogen isotope deuterium had already been used in biochemical studies, by Schoenheimer and Rittenberg.

Plate 189. Martin David Kamen.

To quote from Kamen [1]:

"To make a long story short, we soon discovered that the expected production of the labelled glucose was not proceeding according to the textbook suggestions but instead that there was a massive short term fixation of labeled carbon from CO_2 into unknown products and that, most surprisingly, the primary chemistry involved a *dark* (reversible) thermal fixation of labeled carbon in a carboxyl function with subsequent light-activated reduction of the labeled carboxyl to more reduced states of carbon" (p. 99).

A paper by Ruben, Hassid and Kamen [2] reported in 1939 that the ^{11}C-labelled CO_2 which barley plants were allowed to assimilate in the light or in the dark was incorporated in carbohydrates in the dark as well as in the light. However, if the leaves had been kept for three hours in the dark and then allowed to take up $^{11}CO_2$, no radioactive carbon was incorporated. This pointed to the accumulation of a CO_2 acceptor in preilluminated material. Ruben, Kamen and Hassid [3] showed that during the photosynthesis of barley leaves, in the presence of $^{11}CO_2$, formaldehyde does not become labelled. Ruben and Kamen [4] also experimented on the alga *Chlorella* and found that in the first (not yet identified) assimilation products, ^{11}C was mostly incorporated in a carboxyl.

These experiments with $^{11}CO_2$ changed the face of the theory of photosynthesis, particularly in the demise of the formaldehyde theory and the surprising discovery of CO_2 fixation in the absence of light.

But two factors hampered these early studies. One of them was the very short half-life (21 min) of the radioisotope ^{11}C. This, as stated above, did limit the utilization of the isotope to the vicinity of its producer cyclotron.

On the other hand, suitable methods were lacking for the separation of the components of the complex mixture of labelled metabolites resulting from an exposure to radiocarbon after only a few seconds.

In 1943, Ruben [5] proposed that the formation of sugar in photosynthesis was a completely dark, chemosynthetic process depending on only two products formed in the light reaction: NADPH and ATP. In Ruben's scheme, two phases are considered in dark sugar biosynthesis. In the first, a carboxylative phase, CO_2 carboxylates an (unknown) acceptor. This phase depends on the

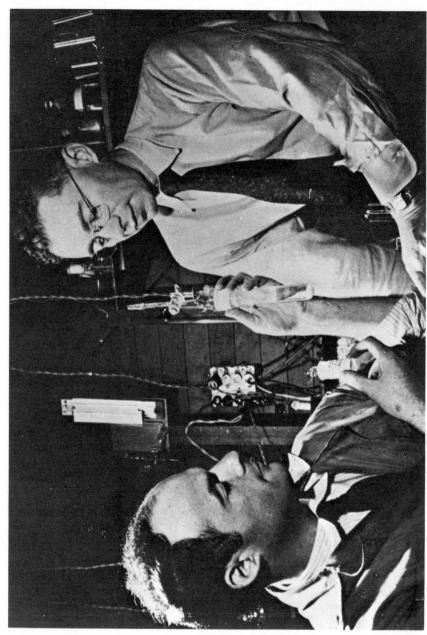

Plate 190. Sam Ruben (left) and William Zev Hassid (right).

presence of ATP. The second phase is a reductive process in which a carboxyl group is reduced by NADPH in the presence of ATP. Ruben was elaborating on suggestions presented by Thimann [6] and by Lipmann [7]. Thimann, in 1938, considered it probable that CO_2 combined with pyruvic acid to produce oxaloacetic acid, or perhaps with lactic acid to produce malic acid. The so-called reduction of CO_2 in photosynthesis would then be the reduction of a carboxyl group. That the suggestions made by Ruben in 1943, on the basis of the data obtained with ^{11}C, received experimental support during the following decades will appear in the following sections.

It may be underlined that, in this development, two main lines of evidence were followed. In one of these lines the experimental evidence depended on the kinetics of the labelling with ^{14}C, of compounds in the living cell, following the introduction of labelled substrates. Such a method must be carefully evaluated with respect to artifacts, and, as stated above, accomplished with living material. That isolated or broken chloroplasts, useful in revealing biochemical possibilities, cannot bring in any information concerning the kinetics of carbon reduction, is the consequence of the depressed efficiency of such systems with respect to CO_2 reduction.

On the other hand, the other line of evidence, the enzymatic one, has depended on the recognition of enzymatic activities in soluble or particulate fractions obtained from broken photosynthesizing cells and on the reconstitution in vitro of the systems involved.

2. Identification of an ATP-dependent carboxylative phase

(a) Discovery of ^{14}C and recognition of phosphoglyceric acid as an early intermediate

The radioactive carbon isotope ^{14}C was discovered in 1940 by Ruben and Kamen [8,9]. They showed that it could be obtained by an (n, p) reaction from ^{13}N. It became possible, due to the development of high neutron fluxes in nuclear reactors, to obtain large amounts of this radioactive isotope, with a half-life of 5700 years. While approaching the upper limit of the level at which the number of disintegrations per minute become too small to be detected, ^{14}C is still within the biochemically useful range.

As was stated in the Introduction to Part III (p. 13), ^{14}C was first used by Ruben and Kamen in 1945, in a study of fatty acid catabolism. It was introduced in photosynthetic studies by Benson and Calvin [10] in 1947, in experiments on the dark reduction (with and without preillumination). These authors found that, in suspensions of the alga *Chlorella*, the fixation was greatly stimulated by preillumination, immediately before dark exposure to $^{14}CO_2$. This confirmed the concept, arrived at by Ruben on the basis of experiments with ^{11}C, of an accumulation of a CO_2 acceptor during preillumination.

Benson and Calvin, in their experiments on $^{14}CO_2$ fixation in the dark (with or without pre-illumination) compared two *Chlorella* suspensions: one (I) kept in the dark for 8 h in 4% ordinary carbon dioxide and the other (II) strongly illuminated for 1 h in CO_2-free nitrogen. Both were then exposed for 5 min to $^{14}CO_2$ in the dark, and the cells were killed. The uptake of tagged carbon was five times greater in the pre-illuminated suspension and the distribution of ^{14}C was different. In suspension I, 70% of the activity was found in succinic acid, 15% in amino acids and 9% in substances absorbed on anion-exchange resins. In (pre-illuminated) suspension II, the largest portion of the activity (30%) was in the amino acids (mostly alanine), 25% in a fraction extractable from water by ether at pH 1 and 10% in anionic substances. Only 6% was found in succinic acid, 1% in fumaric acid and 1.5% in sugars.

The diagrams for the ^{14}C uptake were interpreted by the authors as resulting from the slow uptake due to a reversal of the respiratory pathway, while they considered the fast uptake, which occurs only after pre-illumination, as part of the photosynthetic pathway proper and involving the production not only of a CO_2 acceptor, but also of a "surviving reductant".

In the experiments of Calvin and Benson [11] on $^{14}CO_2$ fixation in the light, algae which had been photosynthesizing in ordinary CO_2 were briefly exposed to $^{14}CO_2$ and killed. After $^{14}CO_2$ exposure for 30 sec in light, the total ^{14}C fixation was 3—6 times higher than after 1 min darkness after "saturating" pre-illumination. The tracer was distributed in practically all fractions, including fatty acids and amino acids, but there was a marked increase in the sugar fraction and strong accumulation in the ether-insoluble anionic fractions (specifically in the subfraction elutable by ammonia from A-3 duolite resin, which constituted the main part of target material formed in 30 sec exposure in light). How he was led to suppose that this fraction could be phosphoglyceric acid has been told by Calvin in a private communication to A.B. Garrett [12], showing a good example of the interplay of biological and chemical knowledge in the mind of a gifted experimenter who, at the same time, is a competent chemist. Calvin tells how, one day while in his car, waiting in front of a supermarket for Mrs. Calvin, he formed the hypothesis that the highly tagged material could be phosphoglyceric acid.

To quote Calvin:

"We knew that the major compound formed in the early phases was very sticky on an ion exchange resin. By sticky I mean it was very hard to wash out of an ion exchanger, whereas simple carboxylic acids, sulfates and phosphates would wash out relatively easily. This sticky material required strong acid or strong base to displace it from such an ion exchange column. The idea that it was a carboxylic acid of some sort had already been evolved, but no simple carboxylic acid was as tightly held as this material was.

"This gave rise to the idea that perhaps it had more than one holding point on it, that is, more than one anionic center. Now if it were a carboxylic acid with more than one anionic center, it would presumably be another carboxylic acid or some other anionic center. It seemed unlikely that it might be a di- or poly-carboxylic acid since sugars and citric acid would wash off the column even ahead of this material. Besides, if the CO_2 was entering this molecule first, it was reasonable to suppose it was entering the carboxyl groups and, if we put two carboxyl groups, it complicated the matter. Therefore, the other anionic center which would make it sticky should not be a carboxyl group, and the obvious choice for it was a phosphate.

"I think by that time we had already the information that the material contained phosphorus by virtue of the tracer phosphorus combination with tracer carbon experiments. The simplest point at which carbon could enter into a phosphorylated compound that was known to us, at that time at least, would be a reversal of the oxidation of a phosphorylated pyruvic acid to give a three-carbon compound in which the carboxyl group arose from the CO_2 and the phosphorylated two-carbon piece arose from something resembling acetyl phosphate. The only compound of this general type that had the enormous stability which our unknown material had was phosphoglyceric acid, and a few tests after this idea arose confirmed that this was indeed what we had" (pp. 234—235).

Benson and Calvin, in fact, showed the bulk of the material eluted by ammonia from A-3 duolite resin to be phosphoglyceric acid.

References p. 106

Plate 191. Melvin Calvin.

The compound was chemically identified by the preparation of crystallized p-bromophenacetylglycerate after hydrolysis by 1 N HCl. It was logical, after the important discovery of 3-phosphoglyceric acid as the first stable product of CO_2 fixation, to search for a C_2 precursor of it. While the acceptor finally turned out to be a pentose phosphate, its real nature was discovered in the course of the search for a C_2 acceptor.

On the other hand, the recognition "that the sticky stuff had to be something with two handles" and had to be regenerable, led Calvin to the idea of a simple cycle.

At the time when they had not yet made a distinction between "photosynthetic" and "respiratory" uptake, and after they had found tagged succinic, malic and fumaric acid, Calvin and Benson [11] figured out the pathway as a reversal of the C_3-C_4 decarboxylation cycle of Szent-Györgyi (see Chapter 30).

In Chicago, Gaffron, Fager and Brown [13] reported a continuous dark uptake of $^{14}CO_2$ by algae, for periods of 12 h and more. In dark fixation after preillumination, they found no tagged hexosephosphate or sucrose. They considered therefore that preillumination merely led to an increase in CO_2 acceptor but discarded the formation of "reducing power". The concept of "surviving reductant" became at the time a subject of discussions which the reader will find retraced by Rabinowitch [14].

As Gaffron et al. [13], for reasons unknown, had only found insignificant amounts of phosphoglyceric acid among the products of photosynthesis, its consideration by Calvin as the first stable product of CO_2 fixation was opposed by the Chicago school. However, in 1950, this position was abandoned and Fager, Rosenberg and Gaffron [15] confirmed the concept.

Even after this conversion, the identification of 3-phosphoglyceric acid as the first photosynthesis product was criticized by Utter and Wood [16] who suggested that the tagging resulted from an exchange rather than of a net carboxylation. But the methods available before 1948 were still difficult, time-consuming and often led to contradictory results. Therefore no definite conclusion could be reached before new methods were available.

Plate 192. A.A. Benson.

(b) Introduction of paper chromatography

This method was developed by Consden, Gordon and Martin [17] in 1944 and was first used for separation of amino acids. It was derived from the methods of partition chromatography on inert support for the separation of amino acids, developed in England since 1941 by Martin and Synge [18]. Fink and Fink [19] suggested using the method of paper chromatography in combination with the radioactive tracer technique in photosynthesis studies. But the first application was accomplished by Stepka, Benson and Calvin [20] in a study of ^{14}C-tagged amino acids from the algae *Chlorella* and *Scenedesmus* and the method has been extended to other tagged products of photosynthesis by Calvin and Benson [21] and by Benson, Bassham, Calvin, Goodale, Haas and Stepka [22].

Paper chromatography, associated with autoradiography in the form of radiograms, made it possible to experiment with small amounts of material and to unfold the variety of compounds extracted from suspensions of algae or from plant leaves.

In the experiments of Benson et al. [22], *Scenedesmus* suspensions, after a period of steady photosynthesis in 1—4% ordinary carbonate for 0.5 to 1 h, were exposed to $^{14}CO_2$ between 5 and 90 sec in light and dropped in hot 80% ethanol. The alcoholic extract was concentrated to 2 ml. To the right bottom corner of the paper sheet an amount corresponding to 0.01—0.20 ml of concentrated extract was applied. For the preparation of the paper chromatogram, water-saturated phenol was used as basic solvent and a mixture of butanol with propionic acid and water as acidic solvent. In such a chromatogram, cationic compounds move towards the left, anionic compounds towards the upper right corner, lipids and lipophilic compounds towards the upper left corner, while sugars, phosphate esters and other neutral, hydrophilic compounds stay in the vicinity of the right bottom corner.

In the upper right region (carboxylic acid field) spots corresponding to malic, succinic, glycolic and fumaric acids were identified. After a period of 15 sec or 1 min, the radiogram obtained showed three spots which, from the bottom upward were identified as "hexose diphosphate", "hexose monophosphate" and "phosphoglyceric acid". If the time of assimilation was reduced to 5 sec or less, only the uppermost, "phosphoglyceric acid" appeared. The corresponding spot was isolated and shown to consist of phosphoglyceric acid. The role of phosphoglyceric acid as the first product of the carboxylation process was firmly established. The early appearance of phosphopyruvic acid and of biose phosphates and hexose phosphate was clarified.

The authors also performed degradation experiments to determine the distribution of ^{14}C within phosphoglyceric acid and other early products. They showed that the carboxyl group in phosphoglyceric acid, as well as in malic acid and alanine is preferentially or exclusively labelled initially and confirmed the gradual approach to equidistribution which had already been noted before. ^{14}C first appeared in the γ, or 3-position (carboxyl) of phosphoglyceric acid but it also appeared rapidly in the α- and β-position. Bassham, Benson and Calvin [23] continued the study of the intramolecular distribution of ^{14}C in phosphoglyceric acid and confirmed that the isotope first appears only in the carboxyl carbon and later equally in the α- and β-carbons. After 60 sec the distribution of ^{14}C among carboxyl, α-carbon and β-carbon was about in the ratio 2 : 1 : 1.

This suggests that only the carboxyl C of phosphoglyceric acid is coming directly from a carboxylation; while the α- and β-carbons are derived from the carboxyl atom via a cyclic process.

(c) Ribulose and sedoheptulose as intermediates

In the lower right corner where the spots of phosphoenolpyruvic, phosphoglyceric and phosphoglycolic acid, triose phosphate and several hexose phosphates were located two spots remained unexplained, though their importance increased with decreasing exposure to $^{14}CO_2$. This indicated that these spots belonged to early intermediates.

Benson, Bassham, Calvin, Hall, Hirsch, Kawaguchi, Lynch and Tolbert [24] isolated the compounds corresponding to the unidentified spots in the form of phosphate esters. When hydrolysed, they gave two sugars, ribulose and sedoheptulose. At the time sedoheptulose had already been shown to accumulate during photosynthesis in several succulent plants, e.g. *Sedum* (Bennett-Clark [25]; Proner [26]; Nordal and Klevstrand [27]). Ribulose, which was identified as a pentose by reduction and by co-chromatography with the pure synthetic compound, had not been recognized as a constituent of photosynthetic organisms. Two years earlier, in 1950, Horecker and Smyrniotis [28] had obtained from bacteria a ketose phosphate fraction which they identified as ribulose-5-phosphate (Ru-5-P). In the discussion of their results, Benson et al. [24], influenced by the current concept of the fixation of CO_2 on a two carbon acceptor, suggested that the C_5 and C_7 intermediates may be the source of such an acceptor.

"The close relationship between the structure of sedoheptulose and that of D-ribulose strongly suggests a synthetic relationship. The configuration of C-3 ˙ and C-4 of ribulose is identical with that of C-5 and C-6 of sedoheptulose. The recent results of Rappoport, Barker and Hassid [29] and Lampen, Gest, and Sowden [30] require the ultimate cleavage of an aldose by an acyloin type of reaction. These authors pointed out that such cleavage most likely involves a 2-ketose intermediate. Similar cleavage of sedoheptulose would give glycolaldehyde (diose) and D-ribose, which could isomerize to D-ribulose. Subsequent cleavage of the pentose would give diose and a triose. It appears reasonable from the compounds available in plants and the apparently universal aldolase activity in leaves [31] that diose and dihydroxyacetone could condense to form xylulose, as suggested by Hough and Jones [32]. "None of these sugars is stereochemically related to glucose by a simple sequence of reactions. One of the functions of these compounds may be to serve as source of 2-carbon molecules capable of accepting carbon dioxide to form phosphoglycerate during photosynthesis. The concurrent synthesis of fructose and sedoheptulose may represent steps in carbohydrate synthesis and in regeneration of the required carbon dioxide acceptors respectively" (pp. 713—714).

In the summary, referring to the biosynthetic relationships of sedoheptulose and ribulose, the authors conclude:

"It is suggested that they are not directly involved in hexose synthesis but may serve as sources of C_2 carbon dioxide acceptors required during photosynthesis" (p. 715).

The situation of their respective spots on the radiogram suggested that ribulose is in the form of a diphosphate and sedoheptulose as a monophosphate.

Calvin and Massini [33] performed kinetic studies of the steady-state levels of phosphoglyceric acid and the sugar diphosphates in *Scenedesmus*, determining absolute and relative amounts of labellings of different compounds as function of time. When illumina-

tion was interrupted, they observed a rise in the level of phosphoglyceric acid and a fall of sugar diphosphate. This reinforced the
notion of the origin of the C$_2$ acceptor from ribulose diphosphate.
On the other hand, when illumination was continued but [14]CO$_2$
removed, Bassham, Benson, Kay, Harris, Wilson and Calvin [34]
obtained a fall in phosphoglyceric acid and a rise in ribulose
diphosphate.

*(d) Phosphoglyceric acid formation with cell-free enzyme
preparations*

Using a spinach "chloroplast" preparation, Fager [35] obtained a
labelling of phosphoglyceric acid from [14]CO$_2$ in the presence of a
heat-stable extract of spinach leaves or of algae. This heated
extract could not be replaced by a mixture of ribulose and sedoheptulose phosphates.

To quote Fager [35] (p. 271):

"If, as suggested in the preceding paragraph, no reduction is involved in the
carboxylation reaction, the acceptor formed by pre-illumination being fully
reduced, then there will be several possibilities for the acceptor of which
perhaps the most likely may be vinyl phosphate, or a ketose (perhaps ribulose)
phosphate. The evidence available at present is against the participation of
either of these compounds as acceptors in the system being studied".

The mention of ribulose phosphate as CO$_2$ acceptor is referred by
Fager [35] to a paper given by Bassham at an informal session on
photosynthesis during a meeting of the American Society of Plant
Physiologists (6—10 September 1953).

Wilson and Calvin [36] obtained results which strongly suggested that ribulose diphosphate is the carboxylation substrate in
the formation of phosphoglyceric acid. For instance, when, following a period of photosynthesis with [14]CO$_2$, the CO$_2$ pressure
was lowered to a few thousands of a percent, a rapid rise in [^{14}C]-
ribulose diphosphate and a rapid drop in [^{14}C]phosphoglyceric
acid were observed.

The evidence for an enzymatic system that formed phosphoglyceric acid from CO$_2$ and pentose phosphate was first reported
by Horecker [36a] in 1954.

Weissbach, Smyrniotis and Horecker [37] experimenting on

extracts of spinach leaves with ribose-5-phosphate (R-5-P) as substrate, showed that $^{14}CO_2$ was fixed in the carboxyl of phosphoglyceric acid. When they used as substrate $[1-^{14}C]$R-5-P the label was mostly found in the β-carbon of phosphoglyceric acid. If R-5-P and ATP were added to the soluble extract, in the absence of CO_2, ribulose diphosphate was obtained. In the same issue of the *Journal of the American Chemical Society* appeared a paper by Quayle, Fuller, Benson and Calvin [38] showing that some extracts of *Chlorella* (after removal by centrifugation of cell wall material and whole cells) would fix CO_2 in the absence of light when ribulose diphosphate was added.

Fractionating the spinach extract, Weissbach, Smyrniotis and Horecker [39] obtained an enzyme system forming ribulose diphosphate from R-5-P and ATP. Hurwitz, Weissbach, Horecker and Smyrniotis [40] demonstrated that the formation of ribulose diphosphate is due to two enzymes catalysing the reactions

R-5-P $\xrightleftharpoons{\text{phosphoriboisomerase}}$ Ru-5-P

Ru-5-P + ATP $\xrightarrow{\text{phosphoribulokinase}}$ RuDP + ADP

Phosphoriboisomerase had been isolated from yeast by Horecker, Smyrniotis and Seegmiller [41], purified from alfalfa by Axelrod and Jang [42] and from spinach leaf extracts by Hurwitz et al. [40]. The denomination of phosphoribulokinase, given to the enzyme by Horecker et al. [41] is recommended as common name in EN, which mentions as other names: phosphopentokinase and ATP → Ru-5-P transphosphatase. Phosphoribulokinase was a new enzyme. Its systematic name is [ATP : D-ribulose-5-phosphate 1 phosphotransferase; 2.7.1.19]. It has been purified from spinach leaves by Hurwitz et al. [40] and by Jakoby, Brummond and Ochoa [43]. As stated in the summary of the paper by Hurwitz et al. [40],

"The new enzyme, phosphoribulokinase, is specific for ribulose-5-phosphate and ATP. It is activated by divalent metal ions, particularly Mg^+, and is inhibited by sulfhydryl-binding agents. In the presence of excess ATP, ribulose-5-phosphate is completely utilized, and equivalent quantities of ADP and ribulose diphosphate are formed".

Purified preparations of phosphoribulokinase have been shown by Weissbach et al. [39] to have no activity with R-5-P unless isomerase is added. Horecker, Hurwitz and Weissbach [44] have isolated RuDP from reaction mixtures obtained by incubating R-5-P and ATP with phosphoisomerase and phosphoribulokinase. These authors have established the identity of RuDP as D-ribulose-1,5-diphosphate through the preparation of this compound by enzymatic methods. Little doubt remains as to its function as the primary acceptor in photosynthesis.

(e) Enzymatic formation of phosphoglyceric acid from ribulose diphosphate and carbon dioxide

Highly purified enzyme preparations catalysing the reaction

$$RuDP + CO_2 \rightarrow 2 PGA$$

have been obtained from spinach leaves by Racker [45]; Weissbach, Horecker and Hurwitz [46] and Jakoby et al. [43]. The enzyme has been called carboxylation enzyme by Horecker et al. [44], and carboxydismutase by Calvin, Quayle, Fuller, Mayaudon, Benson and Bassham [47]. The common name ribulosediphosphate carboxylase (similar to ribulose diphosphate carboxylase used by Utter [48] is recommended in EN which gives to the enzyme the systematic name [3-phospho-D-glycerate carboxy-lyase (dimerizing); 4.1.1.39]. Several mechanisms have been proposed for the action of the enzyme (see Utter [48]; Vishniac, Horecker and Ochoa [49]).

The sequence of discoveries recorded in the present section has led to the identification of the ATP-dependent carboxylative phase of CO_2 assimilation in photosynthesis. This was the result of the emulation which reigned between several schools of American biochemists and their leaders: Calvin and Benson in Berkeley, Ochoa in New York, Horecker in Bethesda and Racker in New Haven. The success of this enterprise was greatly due to the introduction of the use of ^{14}C and of paper chromatography, but enzymatic methodology has contributed to the demonstration that the entry of CO_2 into the metabolism of photosynthetic cells depends on the phosphorylation of ribulose monophosphate by ATP to ribulose diphosphate which is carboxylated by CO_2 and cleaved into two molecules of 3-phosphoglyceric acid.

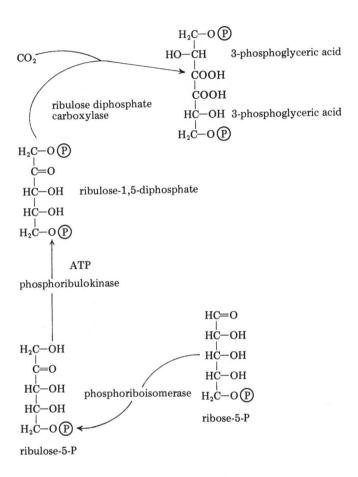

Fig. 3. From ribose-5-phosphate to 3-phosphoglyceric acid.

3. Identification of an ATP- and NADPH-dependent reductive phase

(a) Kinetic studies

That this phase consists in a reduction of 3-phosphoglyceric acid to triose phosphate by a reversal of glycolysis had already been suggested by Calvin and Benson [11] in 1948. The schema of Fig. 4 was established through kinetic experiments in which the absolute and relative amounts of labelling on different compounds was systematically studied as a function of time. The methodology for kinetic studies developed in Calvin's laboratory has been the product of a long inquiry pursued by a team which varied in number of members in the course of a decade. The method consists in setting up a steady state in a suspension of green algae in which radiocarbon is fed at a given time over a period of some seconds to several minutes; after which the algae are killed by dropping them into hot alcohol. The alcohol extract is examined by paper chromatography and autoradiography and the intermediary compounds are degraded. If that kind of experiment is repeated for various exposure times, curves can be drawn showing the rate of entry of radiocarbon into the various positions. A kinetic analysis of the curves obtained permits the definition of the path followed by the carbon through the metabolic network. But of course the introduction of $^{14}CO_2$ or $H^{14}CO_3^-$ into an actively photosynthesizing suspension of algae (or leaf) in the presence of ordinary CO_2 should not disrupt the conditions of steady state. For instance the change should not be accompanied by a change of pressure. If such a change occurs, changes in the concentrations of intermediates of the metabolic network may lead to non steady-state flow of carbon from one compound to another.

The conditions of a correct realization of the kinetic experiments which allowed Calvin and his collaborators to unravel the pathway of carbohydrate biosynthesis in chloroplasts are discussed by Bassham and Calvin [50].

Triose phosphates and hexose phosphates are among the early products and the degradation of the hexoses showed a distribution of ^{14}C similar to that found in phosphoglyceric acid, the first product of carboxylation in photosynthesis.

Benson, Calvin, Haas, Aronoff, Hall, Bassham and Weigl [51] degraded, after 15 sec of photosynthesis in the presence of $^{14}CO_2$ in barley shoots, the hexose derived from the hexose phosphate formed and observed that the highest activity (52%) was found in C-3 and C-4. Considerably less was present in C-2 and C-5 (25%) and in C-1 and C-6 (24%). Thus, half of the activity was in the carboxyl group and the other half was equally split between the other carbon atoms. From the same experiment, much the same

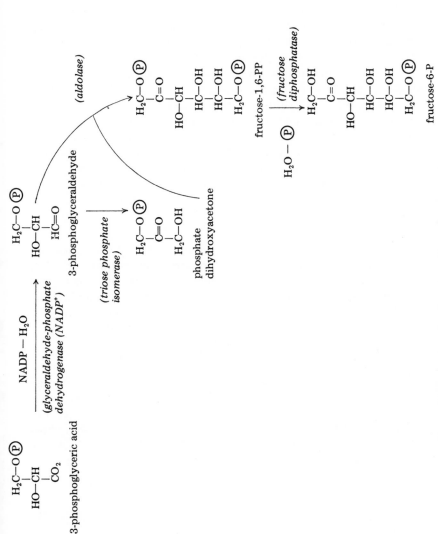

Fig. 4. From 3-phosphoglyceric acid to fructose-6-P.

distribution was obtained in the glyceric acid prepared from the phosphoglyceric acid biosynthesized. The conclusion was that the six-carbon piece is made by the two threes by putting the two carboxyl groups together, the two carbons which were carboxyl originally appearing in the middle of the hexose molecule. A similar correspondence between the C atoms of glyceric acid and hexose was maintained when compared during periods of 2—60 sec. The reduction process is a reversal of the aldolase split of fructose diphosphate in glycolysis. Here, as enzymatic studies showed, the phosphoglyceric acid is reduced, in the presence of glyceraldehydephosphate dehydrogenase and ATP, into glyceraldehyde with a product of the photochemical reaction, NADPH (see Chapter 38). Glyceraldehyde is then, in the presence of triose phosphate isomerase, isomerized in P-dihydroxyacetone.

(b) Enzymes of the reductive phase

Glyceraldehyde-phosphate dehydrogenase (NADP⁺). The presence of this enzyme in plants was first reported by Stumpf [52,53] who called it triose phosphate dehydrogenase. Its systematic name is [D-glyceraldehyde-3-phosphate : NADP⁺ oxidoreductase; 1.2.1.9].

Triose phosphate isomerase. The activity corresponding to this enzyme (other common names recorded by EN, p. 307: phosphotriose isomerase, triose phosphate mutase) has first been detected in pea seed extracts by Tewfic and Stumpf [54]. The systematic name of the enzyme is [D-glyceraldehyde-3-phosphate ketol isomerase; 5.3.1.1]. It catalyses the reaction D-glyceraldehyde-3-phosphate = dihydroxyacetone phosphate. The enzyme has been obtained in the crystallized state by Meyer-Arendt, Beisenhertz and Bücher [55].

Aldolase (see Chapter 22, p. 113). It has been detected in plant tissues by Stumpf [52] and by Hough and Jones [56].

Fructose diphosphatase. This specific enzyme, known to exist in mammals, has been detected in pea leaves by Gibbs and Horecker [57]. It has been purified from spinach by Racker [45].

4. Regenerative phase

As stated above, Calvin's studies on the fixation of CO_2 into sugar phosphate identified ribulose diphosphate as the precursor of phosphoglyceric acid, but the mechanism whereby this precursor was regenerated remained obscure. Indeed, Calvin and his group did not first visualize heptulose and ribulose "as directly involved in hexose synthesis". The key was provided by the demonstration that with rat liver (Horecker, Gibbs, Klenow and Smyrniotis [57a]) and pea leaf (Gibbs and Horecker [57]) extracts, sedoheptulose-7-P was an intermediate in the conversion of pentose phosphate to hexose phosphate by the reversible reactions catalysed by transketolase and transaldolase (see Chapter 55). The other key was provided by Racker's group when they showed that fructose-6-P was a substrate for transketolase (Racker, De la Haba and J.G. Leder [57b]) which made it obvious that the conversion of pentose phosphate to hexose phosphate was indeed reversible.

Bassham et al. [34] have studied the mechanism of the regeneration of ribulose diphosphate. They found that several compounds were simultaneously formed after PGA and triose and identified them as three-, five-, six-, and seven-carbon sugars. By a study of the order of appearance of radioactive carbon in these compounds, the authors, realized that the C_5 compounds could have several origins. To quote Calvin [58]:

"By taking two carbons off the tops of the C_7 and adding them on to a three-carbon piece labeled as is phosphoglyceraldehyde, we would get two five-carbon pieces, one ribulose and one ribose and their label would be thus distributed. The average of them would be the actual one found. This, then, gave us a clue as to the origin of the ribose and ribulose phosphates which we were finding, namely by a transketolase reaction of the sedoheptulose phosphate with the triose phosphate to give the two pentose phosphates. These can be interconverted by suitable isomerization. Thus we have the pentose formed from heptose and triose.

We know how the hexose is formed, from two trioses. The question then remains, where does the heptose come from? And here, again, a similar detailed and careful analysis of the carbon distribution as a function of time within the heptose molecule was made, and this was a long term job. It finally led to the realization that the heptose must have been made from the combination of a four-carbon with a three-carbon piece. And the question arose them; where does the properly labeled four-carbon piece come from? It could only come by splitting the C_6 (hexose) into a C_4 and a C_2. Figure 10 shows how this scheme was arrived at. Here are shown the two trioses, which as you know can make a hexose; reacting one triose with a hexose to form a pentose and under the influence of transketolase a tetrose; the tetrose then reacting with another triose, under the influence of the enzyme aldolase, to form the heptose, and giving the proper distribution of carbon" (pp. 213—214).

$$
\begin{array}{ccc}
\text{CHO*} & \text{CH}_2\text{OH} & \xrightarrow{\text{transketolase}} \\
| & | & \\
\text{HCOH} + & \text{C=O} & \\
| & | & \\
\text{H}_2\text{CO}\textcircled{P} & \text{HOCH*} & \\
\Updownarrow & | & \\
& \text{HCOH*} & \\
& | & \\
& \text{HCOH} & \\
& | & \\
& \text{H}_2\text{CO}\textcircled{P} &
\end{array}
\qquad
\begin{array}{cc}
\text{CH}_2\text{OH} & \text{CHO*} \\
| & | \\
\text{C=O} & \text{HCOH*} \\
| & | \\
\text{HCOH*} + & \text{HCOH} \\
| & | \\
\text{HCOH} & \text{H}_2\text{CO}\textcircled{P} \\
| & \\
\text{H}_2\text{CO}\textcircled{P} &
\end{array}
$$

$$
\begin{array}{cc}
\text{H}_2\text{CO}\textcircled{P} & \text{CHO*} \\
| & | \\
\text{C=O} + & \text{HCOH*} \\
| & | \\
\text{H}_2\text{COH*} & \text{HCOH} \\
& | \\
& \text{H}_2\text{CO}\textcircled{P}
\end{array}
\xrightarrow{\text{aldolase}}
\begin{array}{c}
\text{H}_2\text{CO}\textcircled{P} \\
| \\
\text{C=O} \\
| \\
\text{HOCH*} \\
| \\
\text{HCOH*} \\
| \\
\text{HCOH*} \\
| \\
\text{HCOH} \\
| \\
\text{H}_2\text{CO}\textcircled{P}
\end{array}
$$

The chemical background of this study situating the regenerative phase at the level of an adaptation of the pentose phosphate cycle (Chapter 55) was based on the new investigations of Horecker [59] in the chemistry of pentoses and heptoses.

From the point of view of biochemical evolution it may be stated that the adaptation of the pentose phosphate pathway to the biosynthesis of carbohydrates from CO_2 in green plants is realized by the acquisition by chloroplasts of ribulose diphosphate carboxylase, Ru-5-P kinase and the NADP-linked triose phosphate dehydrogenase.

5. Synthesis of hexose phosphate from CO_2 in the dark in a model multi-enzyme system

This was realized by Racker [60] who obtained in the dark a synthesis, driven by NADP and ATP, of hexose phosphate from CO_2. His model system consists of enzymes of glycolysis obtained from rabbit muscle and yeast and enzymes of the pentose cycle obtained from spinach leaves.

6. The reductive carbohydrate cycle

As stated above, it was in 1943 that Ruben formulated the concept according to which sugar formation in photosynthesis is a dark chemosynthetic process depending on two products formed by light reactions: NADP and ATP (see Chapter 38, on photosynthetic phosphorylation).

Ruben also proposed to distinguish two phases in the dark synthesis of sugar. The first phase is a carboxylative phase, depending on ATP generated by photosynthetic phosphorylation, and CO_2, carboxylating an acceptor. The second phase is, in the theory proposed by Ruben, a reductive phase in which, in the presence of ATP, a carboxyl group is reduced by a reduced pyridine nucleotide.

Ruben's scheme has during the 15 following years been supported as stated above. The ATP-dependent carboxylative phase was identified. It was shown that the entry of CO_2 depends on the phosphorylation of ribulose monophosphate by ATP to ribulose diphosphate, which is carboxylated by CO_2 and cleaved in two molecules of 3-phosphoglyceric acid, reduced to triose phosphate.

As Ruben had foreseen, a carboxylative and a reductive phase were identified. In addition, analysis revealed the presence of components of the pentose phosphate cycle (Chapter 55) in photosynthetic tissues. This finding led to the concept of a third, regenerative phase in CO_2 assimilation.

The coexistence of carboxylative, reductive and regenerative phases led Calvin to put together the operation of a cyclic feedback system which he presented (Fig. 5) to the 3rd Congress of Biochemistry (Brussels, 1955) in a lecture entitled "The photosynthetic carbon cycle" [58]. Most of the work he presented had been documented in a paper by Bassham, Benson, King, Harris, Wilson and Calvin [61].

In 1955, in an article in *Annual Reviews of Biochemistry*, Horecker and Mehler [61a] laid out the entire sequence of enzymatic reactions. The reader will note that in Calvin's cycle sedoheptulose diphosphate is considered as an intermediate while in the cycle of Horecker and Mehler, sedoheptulose-7-P is considered as the key intermediate.

Rather than a photosynthetic cycle, it is more correct to call the cycle a reductive carbohydrate cycle (cf. Racker [62]). The essential

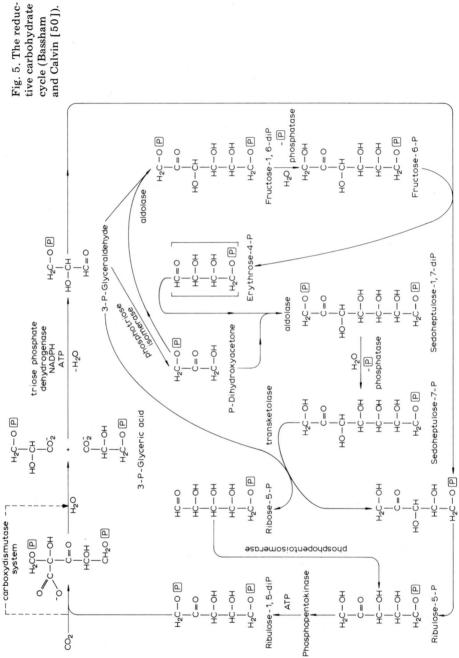

Fig. 5. The reductive carbohydrate cycle (Bassham and Calvin [50]).

enzyme of the cycle is ribulose-di-P-carboxylase which is not only present in photosynthetic systems but also, for instance, in heterotrophs like *E. coli* (Vishniac, Horecker and Ochoa [63]; Racker [60]; Fuller and Gibbs [64].

On the other hand, the whole reductive carbohydrate cycle was identified in the *non-photosynthetic* sulphur bacterium *Thiobacillus denitrificans* (Trudinger [65]; Aubert, Milhaud and Millet [66]). By these observations, CO_2 assimilation was removed from the domain of photosynthetic reactions proper, in which light is converted into chemical energy (photosynthetic phosphorylation, see Chapter 38).

The concept according to which the reductive carbohydrate cycle is the same in photosynthetic and non-photosynthetic cells made it necessary to investigate the pathway, not only in model reconstituted systems but at the site where sugar synthesis operates in green plants.

The history of the identification of this site began with its location in chloroplasts and this view was once widely held by plant physiologists (Sachs [67]; Pfeffer [68]).

Contradictory evidence led to its demise (R. Hill [69]; Brown and Franck [70]; Benson and Calvin [71]; Rabinowitch [72]; Lumry, Spikes and Eyring [73].

The notion that chloroplasts were the seat of sugar synthesis was formulated anew in 1954, on the basis of experiments with tracer carbon and improved methods of isolating functional chloroplasts from leaves, by Arnon and his colleagues (Arnon, Allen and Whatley [74]; Arnon [75]; Allen, Arnon, Capindale, Whatley and Dunham [76]; Arnon, Allen and Whatley [77].

Trebst, Tsujimoto and Arnon [78] showed that within the chloroplast, the enzymes of the reductive carbohydrate cycle were localized in the water-soluble portion and were capable, when physically separated from the chlorophyll-containing particles, to carry out the biosynthesis of sugar from CO_2 in the dark.

REFERENCES

1   M.D. Kamen, Mol. Cell. Biochem., 5 (1974) 99.
2   S. Ruben, W.Z. Hassid and M.D. Kamen, J. Am. Chem. Soc., 61 (1939) 661.
3   S. Ruben, M.D. Kamen and W.Z. Hassid, J. Am. Chem. Soc., 62 (1940) 3443.
4   S. Ruben and M.D. Kamen, J. Am. Chem. Soc., 62 (1940) 3451.
5   S. Ruben, J. Am. Chem. Soc., 65 (1943) 279.
6   K.V. Thimann, Science, 88 (1938) 506.
7   F. Lipmann, Adv. Enzymol., 1 (1941) 148.
8   S. Ruben and M.D. Kamen, Proc. Natl. Acad. Sci. USA, 26 (1940) 418.
9   S. Ruben and M.D. Kamen, J. Appl. Phys., 12 (1941) 321.
10 A. Benson and M. Calvin, Science, 105 (1947) 648.
11 M. Calvin and A.A. Benson, Science, 107 (1948) 476.
12 A.B. Garrett, The Flash of Genius, Toronto, 1963, p. 234.
13 H. Gaffron, E.W. Fager and A. Brown, in Symposium on the Uses of Isotopes in Biology and Medicine, Madison, Wisc., 1947.
14 E.I. Rabinowitch, Photosynthesis and Related Processes, Vol. II, Part 2, New York, 1956.
15 E.W. Fager, J.L. Rosenberg and H. Gaffron, Fed. Proc., 9 (1950) 535.
16 M.F. Utter and H.G. Wood, Adv. Enzymol., 12 (1951) 41.
17 R. Consden, A.H. Gordon and A.J.P. Martin, Biochem. J., 38 (1944) 224.
18 A.J.P. Martin and R.L.M. Synge, Biochem. J., 35 (1941) 91, 1358.
19 R.M. Fink and K. Fink, Science, 107 (1948) 253.
20 W. Stepka, A.A. Benson and M. Calvin, Science, 108 (1948) 304.
21 M. Calvin and A.A. Benson, Science, 109 (1949) 140.
22 A.A. Benson, J.A. Bassham, M. Calvin, T.C. Goodale, V.A. Haas and W. Stepka, J. Am. Chem. Soc., 72 (1950) 1710.
23 J.A. Bassham, A.A. Benson and M. Calvin, J. Biol. Chem., 185 (1950) 781.
24 A.A. Benson, J.A. Bassham, M. Calvin, A.G. Hall, H.E. Hirsch, S. Kawaguchi, V. Lynch and N.E. Tolbert, J. Biol. Chem., 196 (1952) 703.
25 T.A. Bennett-Clark, New Phytol., 32 (1933) 128.
26 M. Proner, Bull. Sci. Pharmacol., 43 (1936) 7.
27 A. Nordal and R. Klevstrand, Acta Chem. Scand., 5 (1961) 85.
28 B.L. Horecker and P.Z. Smyrniotis, Arch. Biochem., 29 (1950) 232.
29 D.A. Rappoport, H.A. Barker and W.Z. Hassid, Arch. Biochem. Biophys., 31 (1951) 326.
30 J.O. Lampen, H. Gest and J.C. Sowden, J. Bacteriol., 61 (1951) 97.
31 S. Tewfik and P.K. Stumpf, Am. J. Bot., 36 (1949) 567.
32 L. Hough and J.K.N. Jones, J. Chem. Soc., (1951) 1122.
33 M. Calvin and P. Massini, Experientia, 8 (1952) 445.
34 J.A. Bassham, A.A. Benson, L.D. Kay, A.Z. Harris, A.T. Wilson and M. Calvin, J. Am. Chem. Soc., 76 (1954) 1760.
35 E.W. Fager, Biochem. J., 57 (1954) 264.
36 A.T. Wilson and M. Calvin, J. Am. Chem. Soc., 77 (1953) 5984.

36a B.L. Horecker, Fed. Proc., 13 (1954) 711.
37   A. Weissbach, P.Z. Smyrniotis and B.L. Horecker, J. Am. Chem. Soc., 76 (1954) 3611.
38   J.R. Quayle, R.C. Fuller, A.A. Benson and M. Calvin, J. Am. Chem. Soc., 76 (1954) 3610.
39   A. Weissbach, P.Z. Smyrniotis and B.L. Horecker, J. Am. Chem. Soc., 76 (1954) 5572.
40   J. Hurwitz, A. Weissbach, B.L. Horecker and P.Z. Smyrniotis, J. Biol. Chem., 218 (1956) 769.
41   B.L. Horecker, P.Z. Smyrniotis and J.E. Seegmiller, J. Biol. Chem., 193 (1951) 383.
42   B. Axelrod and R. Jang, J. Biol. Chem., 209 (1954) 847.
43   W.B. Jakoby, D.O. Brummond and S. Ochoa, J. Biol. Chem., 218 (1956) 811.
44   B.L. Horecker, J. Hurwitz and A. Weissbach, J. Biol. Chem., 218 (1956) 785.
45   E. Racker, Nature (London), 175 (1955) 249.
46   A. Weissbach, B.L. Horecker and J. Hurwitz, J. Biol. Chem., 218 (1956) 795.
47   M. Calvin, R. Quayle, R.C. Fuller, J. Mayaudon, A.A. Benson and J.A. Bassham, Fed. Proc., 14 (1955) 188.
48   M.F. Utter, in P.D. Boyer, H. Lardy and K. Myrbäck (Eds.), The Enzymes, 2nd ed., New York and London, 1961.
49   W. Vishniac, B.L. Horecker and S. Ochoa, Adv. Enzymol., 19 (1957) 1.
50   J.A. Bassham and M. Calvin, The Path of Carbon in Photosynthesis, Englewood Cliffs, N.J., 1957.
51   A.A. Benson, M. Calvin, V.A. Haas, S. Aronoff, A.G. Hall, J.A. Bassham and J.W. Weigl, in J. Frank and W.E. Loomis (Eds.), Photosynthesis in Plants, Ames, Iowa, 1949, p. 381.
52   P.K. Stumpf, J. Biol. Chem., 176 (1948) 233.
53   P.K. Stumpf, J. Biol. Chem., 182 (1950) 261.
54   S. Tewfik and P.K. Stumpf, J. Biol. Chem., 192 (1951) 519.
55   E. Meyer-Arendt, G. Beisenherz and T. Bücher, Naturwissenschaften, 40 (1953) 59.
56   L. Hough and J.K.N. Jones, J. Chem. Soc. (London) (1953) 342.
57   M. Gibbs and B.L. Horecker, J. Biol. Chem., 208 (1954) 813.
57a B.L. Horecker, M. Gibbs, H. Klenow and P.Z. Smyrniotis, J. Biol. Chem., 207 (1954) 393.
57b E. Racker, G. de la Haba and J.G. Leder, Arch. Biochem. Biophys., 48 (1954) 238.
58   M. Calvin, in C. Liebecq (Ed.), Union Internationale de Biochimie; Conférences et Rapports présentés à 3e Congrès International de Biochimie, Bruxelles, 1—6 août, 1955, Liège, 1956, p. 211.
59   B.L. Horecker, J. Cell. Comp. Physiol., 41 (1952) Suppl. 1, p. 137.
60   E. Racker, Nature (London), 175 (1955) 249.
61a B.L. Horecker and A.H. Mehler, Annu. Rev. Biochem., 24 (1955) 207.

62   E. Racker, Arch. Biochem. Biophys., 69 (1957) 300.
63   N. Vishniac, B.L. Horecker and S. Ochoa, Adv. Enzymol., 19 (1957) 1.
64   R.C. Fuller and M. Gibbs, Plant Physiol., 34 (1959) 324.
65   P.A. Trudinger, Biochem. J., 64 (1956) 274.
66   J.P. Aubert, G.M. Milhaud and J. Millet, Ann. Inst. Pasteur, 92 (1957) 515.
67   J. Sachs, Lectures on the Physiology of Plants, London, 1887.
68   W. Pfeffer, Physiology of Plants, London, 1900.
69   R. Hill, Symp. Soc. Exp. Biol., 5 (1951) 223.
70   A.H. Brown and J. Franck, Arch. Biochem. Biophys., 16 (1948) 55.
71   A.A. Benson and M. Calvin, Annu. Rev. Plant Physiol., 1 (1950) 25.
72   E. Rabinowitch, Annu. Rev. Plant Physiol., 3 (1952) 229.
73   R. Lumry, J.D. Spikes and H. Eyring, Annu. Rev. Plant Physiol., 5 (1958) 18.
74   D.I. Arnon, M.B. Allen and F.R. Whatley, Nature (London), 174 (1954) 394.
75   D.I. Arnon, Science, 122 (1955) 9.
76   M.B. Allen, D.I. Arnon, J.B. Capindale, F.R. Whatley and J.L. Dunham, J. Am. Chem. Soc., 77 (1955) 4149.
77   D.I. Arnon, M.B. Allen and F.R. Whatley, Biochim. Biophys. Acta, 20 (1956) 449.
78   A.V. Trebst, H.Y. Tsujimoto and D.I. Arnon, Nature (London), 182 (1958) 351.

Chapter 57

CO_2 fixation in heterotrophs. Gluconeogenesis

1. Introduction

Not only do such tissues as liver and kidney cortex possess the enzymatic arsenal operating in the process of glycolysis, one of the devices for free energy supply in organisms, but also the enzymatic potential for glucose biosynthesis from non-carbohydrate precursors, designated as gluconeogenesis, an important pathway for providing the organisms with glucose.

Gluconeogenesis is an important biosynthetic process (see Krebs [1]). The high gluconeogenetic capacity, mentioned above, of liver and kidney is in contrast to their low glycolytic power.

Nutrition studies were at the origin of the concept. They showed conclusively that in diabetes the glucose may find its origin in proteins, i.e. in amino acids. For instance, Stiles and Lusk [2] gave a diabetic dog a mixture of amino acids resulting from a digestion of meat by pancreatic juice and found an increase in glucose elimination corresponding to 40% of the amount of amino acids administered. The concept was reinforced by observations on phlorizinized dogs. Knopf [3], experimenting on such a dog fed on meat, observed, after adding to the diet 50 g of asparagine, an increase of glucose elimination. Also working on phlorizinized dogs, Ringer and Lusk [4] showed that glycine and alanine may be completely converted into glucose. Moreover, lactic acid was observed to be converted into glucose (Mandel and Lusk [5]).

The participation of pyruvate in gluconeogenesis has been the subject of considerable discussion. The matter was settled when Ringer [6] on the one hand and Dakin and Janney [7] on the other showed that the sodium salt of pyruvic acid given by mouth to diabetic animals under suitable conditions gives rise to an excre-

Plate 193. Chester Hamlin Werkman.

tion of glucose. Obstacles to a glucose biosynthesis through a reversal of the glycolysis pathway were recognized in the existence of irreversible steps for instance the impossibility of a reversal of the pyruvate kinase reaction (XI in Fig. 7, p. 149 of Vol. 31) because of its large positive standard free energy change.

On the pathway of glucose biosynthesis, phosphoenolpyruvate is obtained by another path starting with a fixation of CO_2 by combination with pyruvate.

This possibility was not made evident until 1936, since it was considered until then that incorporation of carbon dioxide into cell constituents was a monopoly of autotrophic organisms — plants and a small group of photosynthetic and chemosynthetic bacteria. Admittedly carbon dioxide was known as a source of carbon in urea synthesis. It was also known that a number of bacteria reduce carbon dioxide, for example to methane and formate.

"But", Krebs [8] writes, "none of these reactions represented a fixation or "assimilation" of carbon dioxide, because they do not lead to the incorporation of carbon dioxide into the cell structure: urea, methane and formate are metabolic end-products" (p. 79).

Only one exception to this epistemological obstacle was known in the case of the views of A.F. Lebedev stated in 1921. Lebedev, in a discussion of the relationships between photosynthesis, chemosynthesis and heterotrophic synthesis, stated

"The aggregate of these facts, together with methodological considerations, leads us to conclude that: (1) the accepted divisions of organisms into two groups, based on their ability to assimilate carbon from carbon dioxide, cannot be considered demonstrated; and (2) already established facts permit us to suppose that even under so called heterotrophic conditions the assimilation of carbon is accomplished as under autotrophic conditions.

"And though the new hypothesis cannot be considered proved by the analysis of already existing knowledge, in view of the local lack of foundation of the accepted view, it deserves deep attention and study". (A.F. Lebedev, translation in Am. Rev. Sov. Med., 5 (1947—48) 15, quoted from H.G. Wood [9].)

2. Fixation of CO₂ in heterotrophic bacteria

The fixation of CO_2 in heterotrophic bacteria was discovered in 1936 by Wood and Werkman [10] studying the fermentation of glycerol in propionic acid bacteria. The main end products of this fermentation are propionic and succinic acids. The authors found that succinate, which is an oxidation product with respect to glycerol, is produced but no product more reduced than glycerol occurred. Propionate ($C_3H_6O_2$) is at the same oxidation level as glycerol ($C_3H_8O_3$), i.e., $C_3H_8O_3$-H_2O equals $C_3H_6O_2$ (see Johnson et al. [10a]). The authors found that they recovered less carbonate than the calcium carbonate they had added in order to neutralize the acids formed during the fermentation. Wood and Werkman [10] concluded that the propionic acid bacteria utilize carbon dioxide, and that its reduction and conversion into inorganic compounds permitted the formation of the oxidized product, succinate. The fact that the oxidation reduction balances were satisfactory when the reduced CO_2 was taken into account was most convincing. In addition, in some of the fermentations, they found somewhat more carbon in the products than could be accounted for by the fermented glycerol. Wood and Werkman [11] published data from which they concluded that the succinic acid formed and the carbon dioxide used were approximately equimolar. They suggested that pyruvic acid "may be the point of entry of CO_2" with a formation of succinic acid. While these experiments were accomplished in the Bacteriology section of Iowa Agricultural Experiment Station in Ames, Iowa, they were published in the British *Biochemical Journal*. On this point, Wood had commented as follows in a letter to Edsall and Krebs [9]:

". . . it was C.H. Werkman's view and perhaps with some reason, that papers on bacterial metabolism were not entirely welcome by the editors of the *Journal of Biological Chemistry*. At any rate most of our papers at the time were sent to the *Biochemical Journal*, in part I'm sure because the center of study of intermediary metabolism prior to World War II was in Europe and not in the USA" (p. 91).

In their paper of 1938, Wood and Werkman [11] were already considering that the fixation of CO_2 was a general phenomenon. This is shown by the following quotation from their paper:

"Krebs and Johnson (1937) have recently shown that citric acid is synthesized by avian tissues from oxaloacetic acid and some unknown compound. It is possible that this synthesis involves utilization of CO_2" [11].

In his letter to Edsall and Krebs [9], Wood adds:

"We made this statement in part because of our views on the synthesis of citrate by molds and in part because we considered that CO_2 was fixed by C_3 plus C_1 addition with pyruvate being the point of entry. We were of the opinion that CO_2 fixation probably accounted for the very high yields of citrate formed by molds; a yield which was greater than the theoretically possible if it was synthesized from C_2 compounds formed from C_3 compounds, i.e., unless the resulting C_1 compounds were utilized in turn" (p. 92).

The concept of a fixation of CO_2 by propionic bacteria received a "mixed reception", as expressed in a review by Van Niel [12] as follows:

"Wood and Werkman claim that carbon dioxide is reduced during the fermentation of glycerol by propionic acid bacteria. The published results cannot, however, be considered conclusive, although the data do seem to favor their claim".

In the same review, Van Niel accepted the evidence that *Escherichia coli* reduces CO_2 to formate, as presented by Woods [13] and that methane-producing bacteria reduce CO_2 to methane (Barker [14]). In his letter to Edsall and Krebs [9], Wood comments on this aspect as follows:

"Probably the reluctance to accept the fixation of CO_2 by the propionic acid bacteria arose from the fact that it went one step further than the utilization described by Woods and by Barker in that creation of a carbon to carbon linkage with CO_2 was involved, whereas the conversion to formate and methane involved CO_2 only as a hydrogen acceptor. The creation of a carbon to carbon linkage with CO_2 as found in the propionic acid bacteria was truly considered an exclusive property of the photo- and chemosynthetic autotrophs" (p. 92).

This epistemological obstacle persisted until the confirmation of the net fixation of CO_2 in propionic acid bacteria by Phelps, Johnson and Peterson [15] in 1939, and by Carson and Ruben [16] in 1940. The latter authors also concluded that the $^{11}CO_2$ fixed is converted not only to succinate but to propionate, which is an important result.

During the same year, Carson, Foster, Ruben and Kamen [17] conducted further experiments on the uptake of $^{11}CO_2$ by propionic acid bacteria during dissimilation of glycerol, using alkaline

Plate 194. Harland Goft Wood.

permanganate degradation of the labelled propionate formed to ascertain the distribution of labelled carbon. It appeared that roughly three times as much label was recovered in oxalate, as compared with carbonate. The presence of C* in both the oxalate and carbonate fractions suggested that the CO_2 was fixed not only in the carboxyl group but also in the α and β carbons. The authors recognized that the degradation might not be reliable, stating.

"It is likely, however, that in the $KMnO_4$ oxidation of the propionate, the carbonate liberated is not always the original carboxyl. Thus, radioactive oxalate may be obtained from propionate containing C* only in the carboxyl group" [17].

They, therefore, degraded the propionate by dry distillation of the barium salt (350°C) to obtain barium carbonate and diethyl ketone. If there was a uniform distribution of C* in the propionate the yields should have been 83.4% and 16.6%, respectively. They found 88% of the C* in the ketone and 12% in the barium carbonate. They stated,

"In view of these results it seems reasonable to conclude that all the carbons in the propionic acid are labelled, although by the latter method it is not possible to distinguish between the α and β carbons."

"It seems that an appreciable fraction of the propionic acid has been synthesized from C^*O_2 rather than from a simple transformation of glycerol without degradation of the carbon skeleton. In other words, CO_2 is reduced to propionic acid with an organic compound (in this case glycerol) acting as the ultimate reducing agent. The reduction of CO_2 by all living cells (photosynthetic and non-photosynthetic) for synthesis of cellular constituents or excretory products is strongly supported by the present findings."

However, these results were not confirmed by Wood, Werkman, Hemingway and Nier [18]. They converted the propionate formed from glycerol in the presence of $^{13}CO_2$ to lactate and degraded the lactate to obtain the carboxyl-, α and β carbons as separate fractions. The excess ^{13}C was found exclusively in the carboxyl group. Thus, the $^{13}CO_2$ is fixed by the propionic acid bacteria only in the carboxyl group. Wood et al. [18a] then chemically synthesized propionate containing excess ^{13}C in the carboxyl group and degraded it by alkaline permanganate oxidation and by pyrolysis of the barium salt. They found that the alkaline permanganate oxidation was not a reliable procedure for obtaining the carboxyl carbon. The excess ^{13}C was about equally distributed in the carbonate and oxalate. However, the pyrolysis of barium propionate (450°C) was found to be a reliable method of degradation. In the same year, Nahinsky and Ruben [18b] and later Nahinsky et al. [18c], likewise, showed that the permanganate oxidation is not reliable and Carson et al. [18d] proved by dichromate oxidation of propionate containing fixed CO$_2$ that the C* is exclusively in the carboxyl group.

In 1940 Wood and Werkman [19] reported on fixation of CO$_2$ by propionic bacteria in the presence of a variety of substrates. The dissimilation of mannitol, adonitol, erythritol and rhamnose under an atmosphere of CO$_2$ was reported to be accompanied as well as that of glycerol by a definite uptake of CO$_2$.

These authors stated that

"a possible mechanism accounting for CO$_2$ utilization as well as succinic acid formation involves the addition of CO$_2$ to pyruvic acid to form oxaloacetic acid, followed by reduction to malic acid, dehydration to fumaric acid and finally reduction to succinic acid".

Wood, Werkman, Hemingway and Nier [20] showed with help of isotopic carbon dioxide that the utilized carbon is found in the carboxyl of succinic acid.

The scheme of CO$_2$ fixation in propionic acid bacteria, with glycerol as substrate could be formulated as follows

$$
\begin{array}{ccccccccccc}
 & & & & \text{COOH} & & \text{COOH} & & \text{COOH} & & \text{COOH} \\
 & & & & | & & | & & | & & | \\
\text{CH}_2\text{OH} & & \text{CH}_3 & & \text{CH}_2 & & \text{CH}_2 & & \text{CH} & & \text{CH}_2 \\
| & \xrightarrow{-4\,H} & | & \xrightarrow{+\,CO_2} & | & \xrightarrow{+2\,H} & | & \xrightarrow{-H_2O} & \| & \xrightarrow{+2\,H} & | \\
\text{CHOH} & & \text{CO} & & \text{CO} & & \text{CHOH} & & \text{CH} & & \text{CH}_2 \\
| & & | & & | & & | & & | & & | \\
\text{CH}_2\text{OH} & & \text{COOH} & & \text{COOH} & & \text{COOH} & & \text{COOH} & & \text{COOH} \\
\\
\text{Glycerol} & & \text{Pyruvic} & & \text{Oxalo-} & & \text{Malic} & & \text{Fumaric} & & \text{Succinic} \\
 & & \text{acid} & & \text{acetic} & & \text{acid} & & \text{acid} & & \text{acid} \\
 & & & & \text{acid} & & & & & &
\end{array}
$$

The studies reported in this section had been accomplished with the stable isotope ^{13}C in the case of Wood, Werkman and their collaborators and with ^{11}C (half-life 21 min) in the case of Ruben and Kamen.

3. Carbon dioxide fixation in pigeon liver preparations

At the time the fixation by propionic bacteria had been described, as stated above, but the mechanism had not been clarified. Krebs [8] reports as follows on his first approach to CO$_2$ fixation in animal tissues.

"The discovery of carbon dioxide fixation arose from studies of the fate of pyruvic acid in pigeon liver. In 1939, slices of this tissue were found to be capable of synthesizing glutamine in the presence of pyruvate provided that ammonium chloride had also been added to the pigeon liver preparation. As glutamine is also formed by pigeon liver from ammonium oxoglutarate and from glutamate it was thought that these two substances might be the intermediate stages in the synthesis of glutamine from ammonium pyruvate" (p. 80).

At the time, E.A. Evans Jr. had come from Chicago to work in the laboratory of Krebs. There, Evans showed that when sodium pyruvate is added to respiring pigeon liver, relatively large quantities of α-oxoglutarate were formed [21].
To quote Krebs [8]

"At this time it was already known, from the work on the tricarboxylic acid cycle, that animal tissues form oxoglutarate when pyruvate and oxaloacetate, or a precursor of oxaloacetate, are available. The reactions were formulated thus:

oxaloacetate + pyruvate → citrate + O$_2$

citrate → α-oxoglutarate + CO$_2$

In contrast to other tissues, for example muscle, pigeon liver formed oxoglutarate without the addition of oxaloacetate. Evans therefore came to the conclusion that α-oxoglutarate can arise in pigeon liver in two ways, first according to the tricarboxylic acid cycle and second by way of an unknown mechanism not requiring a 4-carbon dicarboxylic acid.
    After lengthy deliberations, Evans and I decided not to suggest in the discussion of the work a pathway leading from pyruvate to oxoglutarate involving CO$_2$ fixation as Evans was rather sceptical about the possibility of CO$_2$ fixation and because there was no direct evidence in support" (p. 80).

References p. 138

Plate 195. Earl A. Evans Jr.

Commenting on his stay in Sheffield, Evans [22] writes in a letter to Krebs:

"You will remember that I spent the bulk of my time to demonstrate an insulin effect on pyruvate (carbohydrate) metabolism in pigeon liver. While effects were observed, they could not be repeated at will. The CO$_2$ work came about from my observation that there was increased pyruvate utilization in pigeon liver when a bicarbonate buffer was used. It was during the discussion of these results that the possibility of a Wood-Werkman reaction was suggested" (p. 90).

After Evans had, in March 1940, returned to Chicago, Krebs continued the search for evidence in favour of CO$_2$ fixation in animals and in the same year (1940) a paper by Krebs and Eggleston [23] appeared concluding that the only satisfactory explanation for the fate of pyruvate was the formation of oxaloacetate from pyruvate and CO$_2$. Additional evidence came from the fact that in pigeon liver preparations the rate of pyruvate consumption depends on the concentration of bicarbonate and CO$_2$. It was clear that the evidence for the fixation of CO$_2$ could be afforded by using the isotopic tracer techniques. There were, at this period when radioisotopes could only be obtained by means of a cyclotron, two such apparatuses in Britain (Cambridge, Liverpool). They were fully occupied with war work. Krebs knew that the laboratory of Baird Hastings at Harvard was endowed with the facilities necessary to accomplish metabolic research with the short-lived isotope ^{11}C.

In his letter of November 5, 1940 to Hastings, asking him hospitality in his laboratory for the period between Easter and October 1941 Krebs [24] states his problem as follows:

"The problem, which I believe is of considerable interest, arose in the course of experiments on the metabolism of pyruvic acid. We have data suggesting the occurrence of a synthesis of oxaloacetic acid from pyruvic acid and CO$_2$

$$CO_2 + CH_3.CO.COOH \rightarrow COOH.CH_2CO.COOH$$

The data are chiefly derived from experiments on pigeon liver, but it seems that the reaction occurs in many other tissues. There is also evidence in favour of its occurrence in *Staphylococcus*, *B. coli* and propionic acid bacteria. For the latter organism its occurrence has also been suggested by Werkman, as you will be aware ... Although the evidence is, in my view, very strong, it is not conclusive as far as animal tissues are concerned. The missing link is direct demonstration of CO$_2$ utilization in animal tissues and I think

Plate 196. Albert Baird Hastings.

this can only be done with the help of carbon isotopes. Since oxaloacetic acid reacts very rapidly it will probably be impossible to isolate it, but the substances derived from it (malic, fumaric, citric, α-ketoglutaric acids, see *Biochem. J.*, 34, 1234) which are expected to contain the active carbon can be isolated with comparative ease within the brief life period of ^{11}C" (p. 83).

Hastings [25] replied to Krebs on 4 January 1941, offering him hospitality and remarking that, late in 1940, in the *Journal of Biological Chemistry* a note had appeared in the meantime by Evans and Slotin on the pyruvic acid—CO_2 reaction

"Although, he says, this is the problem of which you spoke and, indeed is one that we had considered trying out ourselves, I am sure that there are many other aspects of the CO_2 cycle on which we could work together profitably" (p. 84).

4. Evans and Slotin's experiments demonstrating the participation of CO_2 in the biosynthesis of α-oxoglutarate

In his letter to Krebs, December 28, 1973, Evans [22] writes:

"When I returned to Chicago from the laboratory of Krebs in Sheffield I had no intention of following up the work until the physicists (notably Louis Slotin) inquired if any of the biologists could use the isotopic material that was produced in the cyclotron. The fixation experiments were carried out in very short order. I had, of course, no idea that you were planning to continue the work nor that you had written to Hastings about working in his laboratory. If I remember correctly the protocols from the bicarbonate experiments were turned over to you since custom officials in Whitehall objected to my carrying large quantities of data paper filled with mysterious figures when I left England to get back to the United States by way of Italy. However, the protocols may be here in the mass of paper that has accumulated in the last 30 years and I can look for them if you think it necessary.

"Incidentally, I wrote what I believe to be a correct account of the matter in the biographical sketch of you in *Science* (118 (1953) 711) at the time you won the Nobel Prize. The priority of the Kamen work, which was unknown to Slotin and me, was acknowledged in several of the earlier publications" (p. 90).

It is interesting to note that Wood missed the discovery of CO_2 fixation in pigeon liver as he related in a lecture published in 1972 [26] and it is of interest to know why he missed this important discovery. As he briefly states in a letter to J.T. Edsall and H.A.

Krebs [9]:

"Our hats went off to E.A. Evans Jr. and L. Slotin when they showed that $^{11}CO_2$ is converted to α-ketoglutarate by pigeon liver during the metabolism of pyruvate. We had foreseen this discovery but had missed our opportunity by not collecting and analyzing the CO_2 formed during the permanganate oxidation of the labeled α-ketoglutarate" (p. 932).

5. Ruben and Kamen's experiments

But even if unknown to Evans and Slotin when they published their note in the *Journal of Biological Chemistry*, Ruben, Kamen and their coworkers had already shown that CO_2 can be incorporated into organic molecules by a variety of living systems, including rat liver and pigeon liver preparations.

Kamen [27,28] has said how he and Ruben came to accomplish these highly significant experiments and why they did not give them the attention they deserved. Kamen states the work began because Ernest Lawrence was anxious to get support for further development of the Berkeley cyclotron and the only place from which he could expect support was the medical foundations. At the time physics had not yet won the prestige it got later as a primordial science.

Kamen, an accomplished radiochemist was, at the start of 1937, in charge of production of radioisotopes by the cyclotron for the use by a number of investigators. S. Ruben, a graduate student, was collaborating with Kamen on a research problem of induced radioactivity (neutron-induced transmutation of rare earth elements). Ruben presented to a young assistant professor of physiology at Berkeley, I.L. Chaikoff, the idea that the short life of ^{11}C would not handicap tracer studies on carbohydrate metabolism. Lawrence requested Kamen to work with Chaikoff and Ruben. An instructor of the Berkeley department of Soil Science, W.Z. Hassid joined the team as an expert carbohydrate chemist.

To quote Kamen [28]:

"Ruben's idea was to take some CO_2 and give it to some green plants, shine light on it, and then extract the glucose and give it to the rat. He'd read in a book that photosynthesis consisted of CO_2 plus water making glucose" (p. 271).

Kamen made some targets for Ruben and Hassid who isolated the osazones from the plants to obtain the glucose. They found no activity in glucose! Where was the activity?

Ruben came to Kamen to ask him and tackle this problem with him. As we shall see later on, the first thing they did was a control experiment in the dark. They found an enormous fixation of CO_2.

"It petered out after about thirty minutes, but it was a reversible carboxylation" [28] (p. 262).

They then asked for advice from the Berkeley Department of Biochemistry, telling them about this fixation in the dark.

"We informed them that we had this activity going on in the dark and they said we were crazy. We then became aware of the fact that we seemed to run into some kind of dogma here. We had no prejudice at all, we had no knowledge of anything about biochemistry. That may have been an advantage" [28] (p. 262).

Van Niel, who was at the time residing at Hopkins Marine Station in Pacific Grove, Cal., had for several years developed a theory of CO_2 as a universal metabolic oxidizing agent.

"He was delighted to hear from us, and he said "of course, what you've got there is just another example of the effect of CO_2 as a universal metabolic oxidizing agent". We came under his influence immediately. We didn't have the preconditioning that would have made us wonder why the CO_2 was getting in — we thought it was natural that it should get in.

". . . Because of our particular feeling that it was a universal reaction, we did some work with liver slices. We got somebody to grind up some liver for us and we found fixation there too; we published that in 1940 in a paper in the PNAS [29] along with a lot of other things. We didn't pay much attention to the fixation in the mammalian tissue because we weren't mammalian chauvinists.

"It did seem at the time that we were so ignorant of the colossal importance of showing the fixation in mammalian tissue that we paid little attention to the fact that we had done it. We didn't demonstrate where it was. We just demonstrated that you couldn't get rid of it by boiling it with acid; that it was fixed in something more reduced than carbon dioxide. It seemed to us that it wasn't important at this point to show where it was. In fact, it might have been very difficult because of the necessary chemistry, and at that time also it wasn't very clear about the various reactions where it was possible. But Van Niel had pointed out that it had already been demonstrated in propionic acid bacteria (by Harland Wood); and we showed a little later a similar fixa-

References p. 138

tion in protozoa going from propionate to succinate, that is, carboxylation of propionate. So we thought, well it'll be a simple matter sometime to get around to liver if we feel like it. Meanwhile, Evans and Slotin published their experiment" [29a] (p. 262).

In a recent letter, Kamen remarks further:

"Harland Wood, in a letter to John Edsall and H.A. Krebs [9], remarks: "Frankly we at Iowa State College in Ames, Iowa were not very impressed by the fact that ^{11}CO$_2$ could be converted by yeast, *E. coli*, barley roots, ground rat liver, etc., to compounds in which the ^{11}C was not liberated by acid (Ruben and Kamen [29]). It was known that CO$_2$ could be reduced to formate by bacteria and could be converted to urea by animals and we felt that these investigators should have dug deeper and proved that ^{11}CO$_2$ was used to form carbon-carbon linkages and identified the compounds".

"This statement does not do justice to the work Ruben and I reported in 1940 [29]. We *did* show that carbon-carbon linkages were made, based on the general procedure of decarboxylation of labeled barium salts of the fixation products. We showed that the products of photosynthetic dark fixation of ^{11}CO$_2$ gave the expected yields of 50% labeled carbon in carbonate formed by this treatment, whereas the analogous fixation by yeast and the other heterotrophic systems yielded practically no labeled carbonate. Furthermore we never found appreciable fixation in any one-carbon compounds in these experiments, despite repeated intensive efforts. Moreover, while the dark fixation in algae was reversible, that in the heterotrophs was not. Chemically, one would be forced to conclude that linkages had been formed. As to why we did not proceed to identify the compounds, I think that is clear from the statements already alluded to in your account".

It is clear that if Ruben and Kamen first recognized a fixation of CO$_2$ in heterotrophs, it was the merit of Evans and Slotin to recognize the compound formed from pyruvate and its importance in gluconeogenesis.

6. Direct demonstration of CO$_2$ fixation in mammalian tissues

As stated above, Krebs recognized this as the missing link, which could only be demonstrated with the help of carbon isotope. This was accomplished by Hastings and his group at Harvard Medical School.

The background of this significant development has been narrated by Hastings [30—32] with admirable objectivity and exemplary modesty and it is of considerable historical interest from the point of view of methodology, as these experiments were started without reference to CO$_2$ fixation.

In the spring of 1939, in the house of the chemist J.B. Conant, then President of Harvard, a discussion took place at the luncheon table between Conant, G. Kistiakowsky and Hastings. Conant had published in the *Journal of Biological Chemistry* [32a] a paper on the mild oxidation by permanganate of acetaldehyde and lactic acid and had speculated that there might be a 2-carbon compound resulting from this mild oxidation of lactic acid, which could be a precursor of liver glycogen.

Conant was annoyed that this suggestion had never been criticized nor checked and he asked Hastings to prepare ^{11}C lactic acid (as the Harvard cyclotron had become available in 1939) and find out whether the radioactivity was present in glycogen.

To quote Hastings [32]:

"Of course, if we'd known what we were letting ourselves in for with a twenty minute half-life isotope, we probably would never have undertaken it at all. But this was the only objective we had, and when the President said this is what I want done, any discussion of how we were going to do it except make radioactive lactic acid and determine how it went to glycogen was furthest from our minds. The fixation of CO$_2$ may be a very important problem, but it certainly wasn't what we were thinking of at the time" (p. 254).

In October 1939, Birgit Vennesland [34], invited by Hastings, joined his laboratory in Boston. She had completed her Ph.D. thesis in the Department of Biochemistry of the University of Chicago under the direction of Martin Hanke [35] and had made the observation that the obligate anaerobe *Bacteroides vulgatus* depended on CO$_2$ for its growth. This had already been shown for different organisms [36]. She had a job in the department of Evans at Chicago and continued to take care of his laboratory when he was in the laboratory of Krebs in Sheffield. She was aware [34] of the work of Wood and Werkman as well as of the papers of Krebs. She had, in the autumn of 1939, received a fellowship from the International Federation of University Women to work with Meyerhof who was in Paris at the time. The political events prevented her from following her plans and she joined the team of Hastings [34].

In November 1939 at the first planning meeting in the house of Hastings, Birgit Vennesland recalls that she "suggested diffidently that experiments with ^{11}CO$_2$ were particularly important because there was a good chance that they may be positive. Nobody else thought the experiments would be positive but it was agreed that they were necessary as controls [34]" (p. 87).

According to Hastings [33]:

"In the course of the discussion, Dr. Vennesland suggested that the incorporation of the ^{11}CO$_2$ might be worth exploring. However, that suggestion was ignored, at the time" (p. 96).

Hastings [32] adds:

"I must confess that I thought it would be a waste of time and that we should concentrate on the metabolism of ^{11}C labelled lactic acid" (p. 254).

At this time, November 1939, neither Ruben and Kamen's, nor Evans and Slotin's results were known. Evans returned to Chicago from Sheffield in March 1940 and the paper by Evans and Slotin appeared late in 1940.

Referring to the statement of Hastings in the final mimeographed *Proceedings of the Conference on the Historical Developments of Bioenergetics* [32] which is enriched by consultation of the original protocols, we may state that reporting on their work on ^{11}C lactate on November 2, 1940 at the MIT conference, the authors reported that four experiments had been made with un-labelled lactate and H^{11}CO$_3$.

"We subsequently, states Hastings, did several more, including three experiments in which Birgit isolated and counted the glucosazones. I find no evidence that we undertook the ^{11}CO$_2$ experiments except as controls for our experiment with labelled lactate" (p. 255).

The experiments showed that about 20% of the lactate given to the rat was converted to CO$_2$. The experimenters did not expect to find radioactivity in glycogen, except coming from the lactate, but since CO$_2$ was produced it was decided to do experiments with the same amount of radioactivity as CO$_2$, as had been injected as lactate in the experiments. These experiments were difficult due to the short half life of ^{11}C and, considering the need of speed, a team of experts was gathered, including an organic chemist, R.D.

Cramer, in charge of the labelled lactic acid synthesis; A.K. Solomon, a physical expert with isotopes, in charge of preparation and measurement of ^{11}C and a biochemist, F. Klemperer.

To quote Hastings [32]:

"On March 2, 1940, we had our 7th attempted experiment with the radioactive lactic acid; this apparently went all right all the way through.

"On March 23 we had our first experiment with unlabelled lactic acid and ^{11}C labelled bicarbonate.

"On April 22, 1940, we had seven experiments with carbon ^{11}C labelled lactate that were successful and one control — the one that we did on March 23. This was reported at a meeting of the National Academy of Sciences on April 22, 1940, following a paper given by Kistiakowsky and Cramer on the synthesis of the radioactive lactic acid — because Cramer, an organic chemist, was responsible for the organic synthesis. The authors of our paper were Hastings, Kistiakowsky, Cramer, Klemperer, Solomon and Vennesland (John Buchanan had also joined our group and actively participated in our ^{11}C experiments from that time forward).

"Reference to the "control" experiment which was accomplished on March 23, 1940, was inserted at the last minute into the report which I delivered to the Academy in the following words. "In one control peritoneal injection of NaHCO$_3$ the carbon of which was radioactive; the glycogen, when isolated, had about the same small percentage of administered radioactivity as it had after radioactive lactate. More experiments on this point must be performed before its meaning becomes clear" (pp. 255—256).

In the summer of 1940, in Berkeley, Hastings had the opportunity of meeting Kamen and Ruben and discussing their problems. He showed his data and they showed their results which were about to be published in PNAS.

... "it was with a strong feeling of *déjà vu* that Ruben and I encountered the data shown us by Prof. Hastings when he and his colleagues from M.I.T. and Harvard visited us in the summer of 1940. We ... strongly encouraged Hastings to place firm belief in the Harvard findings that labelled lactate and CO$_2$ found their way into glycogen and that CO$_2$ incorporation was real [27] (p. 100).

This is also reported by Hastings [33]:

"In the summer of 1940, I had the opportunity to talk with Sam Ruben and Martin Kamen in Berkeley about their experiments on ^{11}CO$_2$ incorporation in plants, yeast cells and various cell preparations, and I told them about our experiences with ^{11}C labelled lactic acid and ^{11}CO$_2$ in rat liver glycogen".

"By fall, 1940, we had ten successful lactic acid experiments and we had done three more control experiments so we had a total of four experiments to report" (p. 96).

These data were incorporated in the report at the MIT Conference of November 2 and were briefly referred to in a paper by Conant, Cramer, Hastings, Klemperer, Solomon and Vennesland submitted to the *Journal of Biological Chemistry* in October 1940 * and which appeared in February 1941 [37]. A little later came Krebs' letter of November 5, 1940 [24], referred to above.

By that time the group of Hastings was convinced that CO$_2$ incorporation was real and experimented in that direction. Ruben and Kamen had not identified any compound resulting from CO$_2$ fixation. Then, writes B. Vennesland [34] to Krebs,

"we saw the note of Evans and Slotin. It was quite a November" (p. 88).

During 1940 and 1941 experiments were pursued in the laboratory of Hastings, in the line of what was a first tentative scheme of gluconeogenesis, and glycogen was found to contain significant amounts of the radioactive carbon introduced into the organism as bicarbonate (Solomon, Vennesland, Klemperer, Buchanan and Hastings [38]). They assumed a phosphorylation of a dicarboxylic acid formed from pyruvate and CO$_2$ and a conversion of this product into phosphopyruvate. They postulated that this phosphopyruvate went along the stages of a reversal of glycolysis, to glycogen, i.e. except for the step involving the formation of phosphopyruvate from pyruvate, the process of phosphopyruvate formation being the only step of the cycle found to be irreversible [39].

As Krebs [8] remarks:

"Subsequent work showed this to be quite a good prediction; though not fully correct, it was broadly correct" (p. 81).

---

* Commenting on the list of author's names, Hastings [32], p. 256 states:

"President Conant had been sent a copy of the manuscript because I had decided that, since he really initiated the research, he should take some of the responsibility also. Besides, with such a large number of authors I was hard put to know whose name should be first. If Conant joined us in the authorship, I could put them all alphabetically, and he rather reluctantly permitted me to do this."

Referring to this paper and a following one of 1942 [40], using glucose and $^{11}CO_2$ Hastings [32] writes that:

"It was in these papers, that Dr. Vennesland reported her success, in two instances out of several attempts, in obtaining and measuring the radioactivity of the osazones of the glucose from glycogen before the radioactivity of the ^{11}C had decayed. Until these data were obtained there was still the lingering possibility that finding radioactivity in our liver glycogen preparations was an experimental artifact rather than the participation of CO_2 in the reactions involved in the synthesis of glycogen.

"Had we been more attentive to Birgit Vennesland's suggestion in the Fall of 1939, our contribution of CO_2 fixation in mammalian tissues might have been made considerably earlier than it occurred" (p. 256).

Parallel to their experiments on CO_2 incorporation into rat glycogen in vivo, the members of the Hastings group were interested in CO_2 fixation by liver slices in vitro. These assays remained without success except in the use of rabbit liver slices and potassium instead of sodium in the incubation medium. Buchanan demonstrated in such experiments in vitro an incorporation of $^{11}CO_2$ in the glycogen of rabbit liver slices from pyruvate and bicarbonate, labelled with ^{11}C [41].

The work of the Hastings group in that field was interrupted by the war and could be resumed only when ^{14}C, after the war, became available.

Wood, Lifson and Lorber [41a] isolated the glycogen of rat liver following intraperitoneal administration of ^{13}C bicarbonate and feeding glucose by stomach tube. They determined the position of the labelled carbon in the glucose from the glycogen and found that carbon dioxide carbon is fixed in positions 3 and 4 of the glucose.

7. Phosphoenolpyruvate carboxykinase

Gluconeogenesis, a new formation of carbohydrate from non-carbohydrate, involves reactions of glycolysis in reverse and some additional reactions overcoming the energy barriers preventing a complete reversal of glycolysis.

The formation of phosphoenolpyruvate is a key step in the initiation of gluconeogenesis. Pyruvate could be converted to phosphoenolpyruvate by a reversal of the pyruvate-kinase reaction. That this reaction is essentially reversible was stated by Lardy and Ziegler [42].

Pyruvate + ATP → phosphoenolpyruvate + ADP

But, in 1954, Krebs [43] pointed out that energetically the pyruvate + ATP reaction is very unfavourable. This brought back the attention to the demonstration of an aerobic formation of phosphoenolpyruvate by tissue homogenates demonstrated, by Kalckar [44] in 1939.

That phosphoenolpyruvate could be formed by a dicarboxylic acid pathway involving malic enzyme (which led to confer a possible role to malate in gluconeogenesis) was suggested by enzyme studies (Ochoa, Mehler and Kornberg [45]; Utter and Kurahashi [46,47])

$$Pyruvate + CO_2 + NADPH + H^+ \rightarrow malate + NADP^+$$
$$Malate + NAD^+ \rightarrow oxaloacetate + NADH + H^+$$
$$Oxaloacetate + GTP \rightarrow phosphoenolpyruvate + GDP + CO_2$$

sum: Pyruvate + NADPH + NAD$^+$ + GTP
→ phosphoenolpyruvate + NADP$^+$ + NADH + GDP

As there was little direct experimental evidence for a direct carboxylation of pyruvate to oxaloacetate (Wood-Werkman reaction) in the animal liver, the theory of the direct carboxylation to oxaloacetate lost ground in favour of alternative formation of dicarboxylic acid via malate as illustrated above. Utter and Kurahashi [46] discovered phosphoenolpyruvate carboxykinase in avian liver and suggested its participation in aerobic formation of phosphoenolpyruvate in homogenates. Utter and Kurahashi recognized the catalyzed reaction as taking place between a nucleoside triphosphate and oxaloacetate, yielding reversibly phosphoenolpyruvate, CO$_2$ and the nucleoside diphosphate. This reaction was recognized as energetically more favourable. The systematic name given to the enzyme is [GTP:oxaloacetate carboxy-lyase (transphosphorylating); 4.1.1.32], catalysing the reaction

GTP + oxaloacetate = GDP + phosphoenolpyruvate + CO$_2$

That ITP can replace GTP in the reaction was demonstrated by Utter, Kurahashi and Rose [48].

8. Pyruvate carboxylase

The perspective was further modified by the isolation and partial purification, by Utter and Keech [49] of an enzyme from avian liver mitochondria, catalysing the formation of oxaloacetate from pyruvate and CO_2 in the presence of ATP, Mg^{2+} and acetyl-CoA

$$\text{Pyruvate} + CO_2 + \text{ATP} \xrightarrow[\text{acetyl-CoA}]{Mg^{2+}} \text{oxaloacetate} + \text{ADP} + P_i$$

Krampitz, Wood and C.H. Werkman [50] showed that crude extracts of *Micrococcus lysodeikticus* catalyse the exchange of $^{13}CO_2$ with the carboxyl group of oxaloacetate, but could not demonstrate a net synthesis of oxaloacetate from pyruvate and CO_2.

Kaltenbach and Kalnitsky [51] had reported the formation of small amounts of oxaloacetate from pyruvate. Later on, Woronick and Johnson [52] had reported that extracts from *Aspergillus niger* formed asparate and malate from pyruvate, CO_2 and ATP. The formation of these metabolic derivatives appeared to the authors as an indirect indication of the existence of a pyruvate carboxylation.

It was known that liver mitochondria can form phosphoenolpyruvate from pyruvate (Bartley [53], Bandursky and Lipmann [54]; Mendicino and Utter [55]). On the other hand neither pyruvate kinase, nor malic enzyme is present in significant amounts in mitochondria (Utter [56]). As the apparent precursor of phosphoenolpyruvate in liver mitochondria was oxaloacetate, Utter and Keech [57], sought a more direct pathway for the formation of oxaloacetate in liver mitochondria, in which they detected and from which they concentrated an enzyme they called pyruvate carboxylase and which carried out the reaction

$$\text{pyruvate} + \text{ATP} + CO_2 \xrightleftharpoons{\text{acetyl-CoA, } Mg^{2+}} \text{oxaloacetate} + \text{ADP} + P_i$$

This enzyme has received the systematic name [Pyruvate:carbondioxide ligase (ADP-forming); 6.4.1.1].

To quote Utter and Keech [57]:

"The pyruvate carboxylase reaction as described here is an elaborated version of the Wood-Werkman reaction postulated many years ago to account for dicarboxylic acid formation" (p. 2607).

Plate 197. Merton F. Utter.

As stated by Utter [58]:

"In view of the close association of this enzyme with phosphoenolpyruvate carboxykinase within the mitochondria and the favorable K_m values for the various reactants it is tempting to suggest that this enzyme performs the first step in the synthesis of phosphoenolpyruvate from pyruvate while phosphoenolpyruvate carboxykinase in the presence of GTP completes the process" (p. 339).

That oxaloacetate was the immediate precursor of phosphoenolpyruvate, was confirmed by Nordlie and Lardy [59].

The path from pyruvate to phosphoenolpyruvate in gluconeogenesis was thus reduced to two reactions involving a carboxylation of pyruvate, leading to oxaloacetic acid and a conversion of oxaloacetate to phosphoenolpyruvate by a decarboxylation and a phosphorylation (Fig. 6).

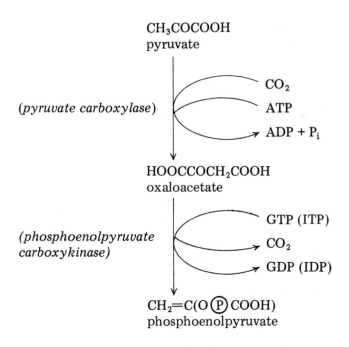

Fig. 6. From pyruvate to phosphoenolpyruvate.

9. Fructose diphosphatase

Another energy barrier opposed to a reversal of glycolysis is situated at the level of the glycolysis reaction catalysed by phosphofructokinase (see Chapter 21)

Fructose-1,6-diphosphate + ADP → fructose-6-phosphate + ATP

This endergonic reaction is bypassed in gluconeogenesis in the presence of fructose diphosphatase, an enzyme catalysing the *hydrolysis* of fructose-1,6-diphosphate to fructose-6-phosphate and phosphate. The enzyme was discovered by Gomori [60] in 1943. In the 1920s, in the course of various studies on phosphatases the term hexosediphosphatase occurs in the literature but in no case was a substrate sufficiently defined to justify the term. In studies carried out in an attempt to elucidate the mechanism of magnesium activation of phosphatase, Gomori found that

"organ extracts prepared in a certain way showed an extremely high activation by magnesium if hexose diphosphate was used as a substrate, whereas the activation was in the usual range if the substrate was glycerophosphate. Attempts at isolation of this highly activable enzyme resulted in extracts which showed a high splitting power towards hexosediphosphate but practically none towards glycerophosphate and phenyl phosphate" (p. 139).

The purified preparation of Gomori was obtained from rabbit liver, as well as by Pogell and McGilvery [61,62], by Mokrasch and McGilvery [63] and by Pogell [64]. In all cases only fructose-1,6-diphosphate was hydrolysed among a number of substrates. There was an exception. As observed by Byrne [65] the enzyme also hydrolyses sedoheptulose-1,7-biphosphate. It was confirmed at Ferrara (Italy) by Pontremoli, Traniello, Luppis and Wood [66] who crystallized the enzyme with the help of Horecker. The enzyme, commonly known as hexose diphosphatase has received the systematic name [D-fructose-1,6-biphosphate 1-phosphohydrolase; 3.1.3.11]. It catalyses the reaction

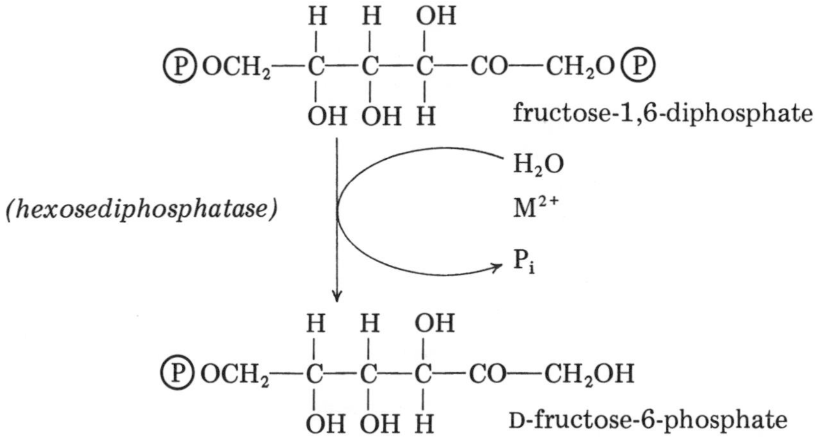

$$\text{(P)}\,OCH_2\!-\!\underset{\underset{OH}{|}}{\overset{\overset{H}{|}}{C}}\!-\!\underset{\underset{OH}{|}}{\overset{\overset{H}{|}}{C}}\!-\!\underset{\underset{H}{|}}{\overset{\overset{OH}{|}}{C}}\!-\!CO\!-\!CH_2O\,\text{(P)}$$

fructose-1,6-diphosphate

(hexosediphosphatase)

H_2O

M^{2+}

P_i

$$\text{(P)}\,OCH_2\!-\!\underset{\underset{OH}{|}}{\overset{\overset{H}{|}}{C}}\!-\!\underset{\underset{OH}{|}}{\overset{\overset{H}{|}}{C}}\!-\!\underset{\underset{H}{|}}{\overset{\overset{OH}{|}}{C}}\!-\!CO\!-\!CH_2OH$$

D-fructose-6-phosphate

10. From glucose-6-phosphate to glucose

The formation of glucose from glucose-6-phosphate involves hydrolysis in the presence of the specific glucose-6-phosphatase converting glucose-6-phosphate to glucose and phosphate. The substrate specificity of liver glucose-6-phosphatase was established by Crane [67]. The enzyme is also known to catalyze some trans-phosphorylations (Nordlie and Arion [68]).

Glucose-6-phosphatase has received the systematic name [D-glucose-6-phosphate phosphohydrolase; 3.1.3.9].

11. Recapitulation of intermediate stages in gluconeogenesis

In his Croonian Lecture of 1963, Sir Hans Krebs [1] has epitomized this subject as follows:

"It is now generally accepted that gluconeogenesis involves some reactions of glycolysis in reverse and some additional reactions which overcome the energy barriers preventing a direct reversal of glycolysis. . . . Energy barriers which obstruct the reversal of glycolysis occur between pyruvate and phosphopyruvate, between glucose-6-phosphate and glucose (see Krebs [43]; Horecker and Hiatt [69]) and between glucose-1-phosphate and glycogen. These barriers are circumvented by reactions special to gluconeogenesis. They are (1) the pyruvate carboxylase reaction converting pyruvate to oxaloacetate

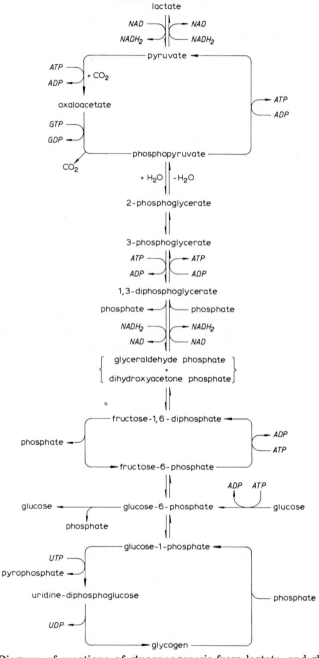

Fig. 7. Diagram of reactions of gluconeogenesis from lactate, and glycolysis (Krebs [1]).

by CO_2 fixation (Utter and Keech [49,57]) plus the phosphopyruvate carboxykinase reaction (Utter and Kurahashi [70]). (2) the fructose-1,6-diphosphatase which hydrolyses specifically the 1-phosphate ester bond of fructose-1,6-diphosphate; (3) the equally specific glucose-6-phosphatase which converts glucose-6-phosphate to glucose; (4) the uridine diphosphoglucose pathway of glycogen formation from glucose-1-phosphate".

The last item will be considered in the chapter of polysaccharide biosynthesis. Gluconeogenesis from lactate requires 6 ATP to form glucose and 7 ATP to form glycogen (GTP and UTP being equivalent to ATP).

REFERENCES

1   H.A. Krebs, Proc. Roy. Soc., Ser. B, 159 (1964) 545.
2   P.G. Stiles and G. Lusk, Am. J. Physiol., 9 (1903) 380.
3   L. Knopf, Arch. Exp. Pathol. Pharmacol., 49 (1903) 123.
4   A.J. Ringer and G. Lusk, Z. Physiol. Chem., 66 (1910) 106.
5   A.R. Mandel and G. Lusk, Am. J. Physiol., 16 (1906) 129.
6   A.I. Ringer, J. Biol. Chem., 17 (1914) 281.
7   H.D. Dakin and N.W. Janney, J. Biol. Chem., 15 (1913) 177.
8   H.A. Krebs, Mol. Cell. Biochem., 5 (1974) 79.
9   Letter of H.G. Wood to J. Edsall and H.A. Krebs, May 24, 1974, Mol.
     Cell. Biochem., 5 (1974) 91.
10   H.G. Wood and C.H. Werkman, Biochem. J., 30 (1936) 48; 30 (1936)
     618.
10a M.J. Johnson, W.H. Peterson and E.B. Fred, J. Biol. Chem., 91 (1931)
     569.
11   H.G. Wood and C.H. Werkman, Biochem. J., 32 (1938) 1262.
12   C.B. Van Niel, Annu. Rev. Biochem., 6 (1937) 308.
13   D.D. Woods, Biochem. J., 30 (1936) 515.
14   H.A. Barker, Arch. Microbiol., 7 (1936) 404.
15   A.S. Phelps, M.J. Johnson and W.H. Peterson, Biochem. J., 33 (1939)
     726.
16   S. Carson and S. Ruben, Proc. Natl. Acad. Sci. USA, 26 (1940) 726.
17   S.F. Carson, J.W. Foster, S. Ruben and M.D. Kamen, Science, 92
     (1940) 433.
18   H.G. Wood, C.H. Werkman, A. Hemingway and A.O. Nier, Proc. Soc.
     Exp. Biol. Med., 46 (1941) 313.
18a H.G. Wood, C.H. Werkman, A. Hemingway, A.O. Nier and C.G. Stuck-
     wisch, J. Am. Chem. Soc., 63 (1941) 2140.
18b P. Nahinsky and F. Ruben, J. Am. Chem. Soc., 63 (1941) 2275.
18c P. Nahinsky, C.N. Rice, F. Ruben and M.D. Kamen, J. Am. Chem. Soc.,
     64 (1942) 2299.
18d S.F. Carson, J.W. Foster, S. Ruben and H.A. Barker, Proc. Natl. Acad.
     Sci. USA, 27 (1941) 229.
19   H.G. Wood and C.H. Werkman, Biochem. J., 34 (1940) 7.
20   H.G. Wood, C.H. Werkman, A. Hemingway and A.O. Nier, J. Biol.
     Chem., 135 (1940) 789.
21   E.A. Evans Jr., Biochem. J., 34 (1940) 829.
22   E.A. Evans Jr., Letter to H.A. Krebs, December 28, 1973, Mol. Cell.
     Biochem., 5 (1974) 90.
23   H.A. Krebs and L.V. Eggleston, Biochem. J., 34 (1940) 1383.
24   Letter of H.A. Krebs to A.B. Hastings, November 5th, 1940, Mol. Cell.
     Biochem., 5 (1974) 83.
25   Letter of A.B. Hastings to H.A. Krebs, January 4, 1941, Mol. Cell. Bio-
     chem., 5 (1974) 84.
26   H.G. Wood, in J.F. Woessner and F. Huijing (Eds.), Miami Winter
     Symp., Vol. III, New York and London, 1972.

27   M.D. Kamen, Mol. Cell. Biochem., 5 (1974) 99.
28   M.D. Kamen, Proc. Conf. Histor. Devel. of Bioenerg., Oct. 11—13,
      1973, Boston 1975, Discussion, pp. 254—266.
29   S. Ruben and M.D. Kamen, Proc. Natl. Acad. Sci. USA, 26 (1940) 418.
29a   E.A. Evans Jr. and L. Slotin, J. Biol. Chem., 136 (1940) 301; 141
      (1941) 439.
30   A.B. Hastings, Mol. Cell. Biochem., 5 (1974) 85.
31   A.B. Hastings, Annu. Rev. Biochem., 39 (1970) 1.
32   A.B. Hastings, Proc. Conf. Histor. Devel. Bioenerg., Oct. 11—13, 1973,
      Boston 1975, Discussion, pp. 254—266.
32a   J.B. Conant and C.O. Tonberg, J. Biol. Chem., 88 (1930) 701.
33   A.B. Hastings, Mol. Cell. Biochem., 5 (1974) 95.
34   Letter of B. Vennesland to H.A. Krebs, April 16, 1973, Mol. Cell. Bio-
      chem., 5 (1974) 87.
35   B. Vennesland and M. Hanke, J. Bacteriol., 39 (1940) 139.
36   G.P. Gladstone, P. Fildes and G.M. Richardson, J. Exp. Pathol., 16
      (1935) 335.
37   J.B. Conant, R.D. Cramer, A.B. Hastings, F.W. Klemperer, A.K. Solo-
      mon and B. Vennesland, J. Biol. Chem., 137 (1941) 557.
38   A.K. Solomon, B. Vennesland, F.W. Klemperer, J.M. Buchanan and
      A.B. Hastings, J. Biol. Chem., 140 (1941) 171.
39   O. Meyerhof, P. Ohlmeyer, W. Gentner and W. Maier-Leibnitz, Bio-
      chem. Z., 298 (1938) 396.
40   B. Vennesland, A.K. Solomon, J.M. Buchanan and A.B. Hastings, J.
      Biol. Chem., 142 (1942) 379.
41   J.M. Buchanan, A.B. Hastings and F.B. Nesbett, J. Biol. Chem., 145
      (1942) 715.
41a   H.G. Wood, N. Lifson and V. Lorber, J. Biol. Chem., 159 (1945) 475.
42   H.A. Lardy and J.A. Ziegler, J. Biol. Chem., 159 (1945) 343.
43   H.A. Krebs, Johns Hopk. Hosp. Bull., 95 (1954) 19.
44   H. Kalckar, Biochem. J., 33 (1939) 631.
45   S. Ochoa, A.H. Mehler and A. Kornberg, J. Biol. Chem., 174 (1948)
      979.
46   M.F. Utter and K. Kurahashi, J. Biol. Chem., 207 (1954) 787.
47   M.F. Utter and K. Kurahashi, J. Am. Chem. Soc., 75 (1953) 758.
48   M.F. Utter, K. Kurahashi and I.A. Rose, J. Biol. Chem., 207 (1954)
      803.
49   M.F. Utter and D.B. Keech, J. Biol. Chem., 235 (1960) PC 17.
50   L.O. Krampitz, H.G. Wood and C.H. Werkman, J. Biol. Chem., 147
      (1943) 243.
51   J.P. Kaltenbach and G. Kalnitsky, J. Biol. Chem., 192 (1951) 629.
52   C.L. Woronick and M.J. Johnson, J. Biol. Chem., 235 (1960) 9.
53   W. Bartley, Biochem. J., 56 (1954) 387.
54   R.S. Bandurski and F. Lipmann, J. Biol. Chem., 219 (1956) 741.
55   J. Mendicino and M.F. Utter, J. Biol. Chem., 237 (1962) 1716.
56   M.F. Utter, Ann. N.Y. Acad. Sci., 72 (1959) 451.
57   M.F. Utter and D.B. Keech, J. Biol. Chem., 238 (1963) 2603.

58   M.F. Utter, in P.D. Boyer, H. Lardy and K. Myrbäck (Eds.), The Enzymes, 2nd ed., Vol. 5, New York and London, 1961, p. 319.
59   R.C. Nordlie and H.A. Lardy, J. Biol. Chem., 238 (1963) 2259.
60   G. Gomori, J. Biol. Chem., 148 (1943) 139.
61   B.M. Pogell and R.W. McGilvery, J. Biol. Chem., 197 (1952) 293.
62   B.M. Pogell and R.W. McGilvery, J. Biol. Chem., 208 (1954) 149.
63   L.C. Mokrasch and R.W. McGilvery, J. Biol. Chem., 221 (1956) 909.
64   B.M. Pogell, in R.W. McGilvery and B.M. Pogell (Eds.), Fructose-1,6-diphosphatase and its Role in Gluconeogenesis, American Institute of Biological Sciences, Washington D.C., 1961, p. 20.
65   W.L. Byrne, in R.W. McGilvery and B.M. Pogell (Eds.), Fructose-1,6-diphosphatase and its Role in Gluconeogenesis, American Institute of Biological Sciences, Washington D.C., 1961, p. 89.
66   S. Pontremoli, S. Traniello, B. Luppis and W.A. Wood, J. Biol. Chem., 240 (1965) 3459.
67   R.K. Crane, Biochim. Biophys. Acta, 17 (1955) 443.
68   R.C. Nordlie and W.J. Arion, J. Biol. Chem., 239 (1964) 1680.
69   B.L. Horecker and H. Hiatt, New Engl. J. of Med., 258 (1958) 177, 225.
70   M.F. Utter and K. Kurahashi, J. Biol. Chem., 207 (1954) 821.

Chapter 58

Biosynthesis of complex saccharides from monosaccharides

1. Introduction

This chapter is concerned with the biosynthesis, from monosaccharide derivatives, of such complex saccharides as oligosaccharides, glycosides and polysaccharides. We have recalled in Chapter 51 that the theory of saccharide formation by reversible zymo-hydrolysis was based by A. Croft Hill on an alleged promotion of a synthesis of maltose from glucose in the presence of maltase. One of the arguments against the theory of reversible zymo-hydrolysis was the realization in vitro of a synthesis of glycogen in the presence of a phosphorylase, which was not a reversal of the action of the hydrolytic enzyme.

But the phosphorylase theory was itself an aspect of the theory of reversible zymo-hydrolysis, in the particular case a reversed form of phosphorolysis and it eventually suffered the ill fate of the theory. For almost two decades (1939–1957), glycogen was nevertheless considered as biosynthesized according to the phosphorylase theory. The introduction of the concept of transglycosylation from sugar nucleotides led to renewed theory of the biosynthesis of complex saccharides. The present chapter is intended to explain the development of ideas on glycosylation reactions during the decade following 1950 *.

2. The concept of transglycosylation

In this process the glycosyl donor may be sugar phosphate, sugar

---

* For later developments, the reader is referred to Volume 17 of this Treatise.

Plate 198. William Zev Hassid.

nucleotide, oligosaccharide or polysaccharide. The acceptor may be alcohol, sugar phosphate, monosaccharide, polysaccharide, H_3PO_4. The product may be hexose phosphate, glycoside or di- to polysaccharide.

Doudoroff, Barker and Hassid [1] introduced the terms transglycosylase and transglycosylation instead of transglycosidase and transglycosidation after it was shown that sucrose can be formed in vitro by a transfer of D-glucose from α-D-glucose-1-P to D-fructose by the sucrose glycosyltransferase (EN 2.4.1.7) of the bacterium *Pseudomonas saccharophila*.

Transglycosylations correspond to the general equation

(glycosyl) → O—X + H—O—(acceptor)
(donor)

⇌ (glycosyl)—O—acceptor + H—O—X
(synthesized glycoside)   (deglycosylated donor)

X corresponds to a phosphate group (as in aldose-1-phosphates) or a nucleoside pyrophosphate (as in UDPglycosides) or a saccharide (sucrose, maltose, dextrins, polysaccharides). Three categories of enzymes catalyse the glycosyl transfer from these three categories of donors. These enzymes are the phosphorylases (X = phosphate group), the sugar nucleotide transglycosylates (X = a nucleoside pyrophosphate) and the transglycosylases proper (X = a saccharide) (literature on phosphorylase reactions, see Table 4, pp. 271–276 of Bernfeld [2]).

As pointed out by Hehre [3], it is not the glycoside group which is transferred in these reactions but the glycosyl group.

As stated by Hassid [4] (p. 316):

"In early classifications, the transferases that formed sugar esters were not considered to be in the same grouping as the transglycosylases that formed the true glycosidic linkages. However, since the sugar phosphates that participate in the enzymic formation of glycosyl compounds undergo cleavage between the carbon and oxygen (C—O—P) and not between the phosphorus and oxygen (C—O—P), the phosphorylases are classified as transglycosylases.

Scission of α-D-glucose-1-phosphate between the carbon and oxygen was shown to occur in the enzymic formation of glycogen by Cohn [5] through the use of ^{18}O isotope. Also the fact that D-ribose-1-phosphate could react with the imidazole nitrogen of hypoxanthine, in the presence of an enzyme from liver, to produce a C—N linkage, indicates scission of this pentose phosphate between C and O (Kalckar [6])."

Transglycosylation, as the mechanism of the biosynthesis of complex saccharides was first introduced in the form of the phosphorylase theory of glycogen and starch biosynthesis.

3. The phosphorylase theory of glycogen and starch biosynthesis (1939—1957)

We have, in Chapter 21, devoted a section to what Parnas [7] has called the phosphorolytic pathway of glycogen breakdown, and to phosphorylase and α-D-glucose-1-P (Cori, Colowick and Cori [8,9]).

Phosphorylase had been defined by Parnas and Baranowski [10] as catalysing the phosphorolytic breakdown of glycogen. The reaction was recognized as reversible by Cori, Schmidt and Cori [11] and the concept of the formation of complex saccharides through glycosyl transfer from an energy-rich donor was developed by Cori and Cori [12], by Hanes [13] and by others. The synthetic action of phosphorylase appeared limited to the catalysis of a chain growth reaction, consisting in a succession of steps in each of which a glucosyl radical from glucose-1-P was added to the C_4 hydroxyl of the non-reducing terminal unit of the polysaccharide. The reaction catalysed by phosphorylase is:

α-D-glucose-1-P + $[O\text{-}\alpha\text{-}D\text{-glucopyranosyl-}(1\text{—}4)\text{-}]_n$

$\rightleftharpoons [O\text{-}\alpha\text{-}D\text{-glucopyranosyl-}(1\text{—}4)]_{n+1}$ + inorganic P

The phosphorylases have the property of disrupting or biosynthesizing 1,4-α-glucoside linkages at the non-reducing end of the glycogen or starch chain. Cori and Cori represent the process and its reversal as follows:

α-D-glucose-1-P

Dialysed extracts of muscle, heart, liver, brain and yeast were shown by Cori and Cori [14] and Cori et al. [8,9] to contain a phosphorylating enzyme catalysing the reaction

$$glycogen + H_3PO_4 \rightarrow glucose\text{-}1\text{-}P \tag{1}$$

In the presence of this enzyme, glycogen is esterified with inorganic phosphate on atom 1.

Cori, Colowick and Cori [15] studied this reaction further.

As was stated in Chapter 21, Parnas [7], in a review article has pointed out that the reaction between glycogen and inorganic phosphate (not ATP) could be considered as an addition of one atom of hydrogen to the glycogen residue and of the phosphate residue to the esterified glucose unit and did not involve water. He called the process phosphorolysis.

As was retraced in Chapter 21, another enzyme present in the extracts catalyses the reaction

$$glucose\text{-}1\text{-}P \rightarrow glucose\text{-}6\text{-}P \tag{2}$$

This was shown by Cori, Colowick and Cori [16] to lead to a conversion of glucose-1-P to glucose-6-P.

Attempts to reverse reaction (2) having failed at that time, it became obvious that in order to proceed in the study of the phosphorylase reaction (1) it was necessary to separate the enzyme involved from the conversion enzyme (2). This was accomplished by adsorption with aluminium hydroxide, followed by elution with disodium phosphate, the result of two such treatments providing, after dialysis of the second elution to remove the inorganic phosphate a suitable solution of phosphorylase.

"When natural or synthetic 1-ester and 1 mM of adenylic acid are added to these dialyzed elutions, inorganic phosphate is liberated and a polysaccharide is formed" (Cori et al. [11], p. 464)

This polysaccharide was isolated by the methods in use for the preparation of glycogen from liver or muscle. This compound formed from glucose-1-P by muscle enzyme with addition of adenylic acid was found not to be destroyed by heating for 1 h in 20% NaOH at 100°C, to be insoluble in 50% alcohol in the presence of electrolytes and to show no reducing power. The rate of hydrolysis of the compound in 0.2 N HCl at 100°C is similar to that of glycogen and leads to a liberation of glucose. When added to muscle extract with inorganic phosphate and adenylic acid it is converted back to glucose-1-P. There is only one character which differentiated this product from glycogen: it gave a blue colour with iodine, though after a longer incubation, a purplish-brown colour is given.

Cori et al. [11] observed that without addition of adenylic acid the enzyme remains inactive and that neither inosinic acid or ATP can replace it. Galactose-1-P or mannose-1-P, as shown by Colowick [17], are not transformed into a polysaccharide and their phosphate group is not split off.

Cori et al. [11] observed a lag period. During this period the mixture remains clear and the iodine reaction remains negative. Often within 1 min, the mixture becomes opalescent, the iodine reaction becomes positive and inorganic phosphate is liberated. These striking phenomena were demonstrated by Cori at the meeting of April 12, 1939 of the Missouri branch of the Society for Experimental Biology and Medicine. The reaction (1), represents a reversible enzymatic equilibrium, with adenylic acid acting as coenzyme in both directions. The authors found the demonstration of it in the fact that if some conversion enzyme (catalysing reaction 2) is still present, the reaction is gradually reversed (this is expressed by a disappearance of inorganic phosphate).

These experiments showed the enzymatic synthesis, in vitro, of a high molecular polysaccharide from a phosphorylated monosaccharide and indicated that glucose-1-P was the substrate of glycogen biosynthesis in vitro, and that under these conditions the phosphorylase brought about glycogen biosynthesis or glycogenolysis. Nevertheless, Cori et al. vainly tried to obtain glycogen synthesis in perfused organs and in tissue slices. This was attributed by them to

"the sensitiveness of the mechanism by which glucose is phosphorylated, a process about which nothing is known beyond the fact that oxidative energy is necessary".

In yeast extracts, phosphorylation of glucose takes place anaerobically, linked with the oxidoreduction of cozymase.

Kiessling [18], by repeated fractionation with 0.3 saturated ammonium sulphate, obtained a fraction from yeast juice reversibly catalysing reaction (1). It contains the phosphorylase identified by Cori et al. [9]. The product formed from glucose-1-P is not distinguishable from glycogen. Furthermore, Kiessling showed that his phosphorylase preparation remained active after prolonged dialysis against 0.3 saturated ammonium sulphate, from which he concluded that adenylic acid is not required. Cori, Cori and Schmidt [19] compared the activity of phosphorylase preparations from the liver on the one hand and from skeletal, heart muscle or brain on the other, on the reaction

glycogen + H_3PO_4 ⇌ glucose-1-P.

When glucose-1-P is the substrate and the reaction is proceeding to the left, activity starts immediately with the liver phosphorylase while a lag period was observed with the phosphorylases of other origins. Cori and Cori [12] observed that while the lag period is short (5 to 45 min) in crude enzyme preparations, it is prolonged with further purification of the enzyme. In certain preparations a very long lag was observed, while when tested with glycogen and phosphate so that the reaction proceeded to the right, no lag appeared with the same preparation.

Cori and Cori noted that liver enzyme preparations always included some glycogen, while

"no or only doubtful traces of glycogen were found in the enzyme preparations from other tissues. It has been possible to abolish the lag period and to reactivate seemingly inactive enzymes by adding small amounts of glycogen . . . One may conclude that this enzyme which synthesizes a high molecular compound, glycogen, requires the presence of a minute amount of this compound in order to start activity" (pp. 397—398).

This was the origin of the concept of the role of primer. In 1940, Hanes [13] described the synthesis of starch by phosphorylase in the presence of small amounts of starch as primer. Cori, Swanson and Cori [20] explained the role of the primer as actually participating in the reaction. As stated by Hassid [4], these authors suggested that

"The primer is required because the enzyme is unable to cause a direct condensation of α-D-glucose-1-phosphate units, but acts as a medium for transferring D-glucose units from α-D-glucose-1-phosphate to the end of an already existing chain. The "priming" efficiency of a polysaccharide is a function of the number of nonaldehydic terminal D-glucose units" (p. 328).

Polysaccharides obtained by synthesis in vitro by phosphorylase resemble the amylose fraction of potato starch, a linear molecule possessing only 1,4-glycosidic linkages.

4. The unravelling of glucan structures

The history of this development is epitomized by Bernfeld [2] as follows (p. 291):

"That starch from most origins consists of two polysaccharides has been known for a long time (Samec and Blinc [21]) because of the different physico-chemical properties of these two components and because of their different behavior toward hydrolytic enzymes. They both have been known to be made up of D-glucopyranose units, which are linked together mostly by α-1,4-glucosidic bonds. The presence of a few α-1,6-glucosidic bonds in starch had also been established. The methylation procedure by Haworth and co-workers [22] yielded one nonreducing end-group for each 25 to 30 glucose residues in starch and amylopectin. Comparative end-group determinations of amylose and amylopectin revealed the unbranched nature of the former with only 0.3% end-groups (Meyer, Wertheim and Bernfeld [23]). Amylase was thus established to consist of chain-like molecules and to contain only α-1,4-glucosidic bonds. The results of end-group determinations, in conjunction with enzymatic studies and with investigations of the macromolecular behavior of amylose and amylopectin led to the advancement of the bush-like structure for amylopectin (Meyer and Bernfeld [24]) shown in Fig. 8 where the branching points are α-1,6-glucosidic bonds. A similar structure with a higher degree of branching has been suggested for glycogen (Meyer and Fuld [25]) and has been confirmed by Cori's group (Illingworth, Larner and Cori [26]; Larner, Illingworth, Cori and Cori [27]) using amylo-1,6-glucosidase as a specific enzyme to break the branching points in glycogen and amylopectin, after exposing them to the action of phosphorylase or β-amylase".

The multibranched model of glycogen and amylopectin structure suggested by Meyer has generally been accepted and preferred to the Haworth and Hirst laminated model (see literature in Hassid [4], p. 332).

It has been shown that starches consist of 20—30% of amylose, made up of linear chain molecules of several hundred D-glucose

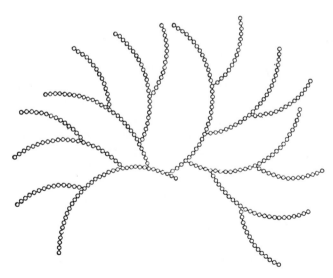

Fig. 8. Structure of amylopectin. Each circle represents a glucose unit. (From [24].)

residues joined through 1,4-α-linkages (Meyer and Gibbons [28]). The remaining part, amylopectin molecules, are highly branched, each consisting of thousands of D-glucose molecules. The branched structure of glycogen is similar to that of amylopectin but the outer branches are shorter (Meyer [29]).

5. Phosphorylase theory of sucrose biosynthesis

Several ways of accomplishing the chemical synthesis of sucrose from fructose and glucose derivatives have been described (literature in Hassid and Doudoroff [30]).

As was reported in Chapter 51, the biosynthesis of sucrose has been considered as taking place by a reversed inversion in the presence of the enzyme invertase. In 1931, Oparin and Kurssanov [31] described the synthesis of sucrose from inverted sugar in the presence of invertase and phosphatase, and of phosphate. This claim has not been substantiated by the later work of Lebedev and Dikanowa [32]. The notion that phosphorylation was an essential step in sucrose biosynthesis derived from the observation of large amounts of hexose phosphates in the leaves of such sucrose-producing plants as beets or peas (Burkard and Neuberg [33]; Hassid [34]; Kurssanov and Kriukova [35]). The importance of phosphorylation in the process of sucrose biosynthesis was also supported by Kriukova's [36] observation of an inhibition of sucrose biosynthesis in plants by iodoacetate, which inhibits phosphorylation. Inspired by the mechanism proposed by Cori and Cori for the biosynthesis of glycogen by phosphorylase, Doudoroff, Kaplan and Hassid [37] proposed a mechanism of sucrose biosynthesis based on the isolation from the bacterium *Pseudomonas saccharophila*, of a sucrose phosphorylase catalysing the breakdown of sucrose in the presence of phosphate, with the formation of glucose-1-P and D-fructose. This reaction was reversible and the same sucrose-phosphorylase catalysed the synthesis of sucrose from D-glucose-1-P and D-fructose.

6. Reactions catalysed by glucosyltransferase

The process just described can also be considered as a transglycosylation reaction in which the donor of glucose is α-D-glucose-phosphate and the acceptor is D-fructose, in the presence of sucrose glucosyltransferase, which is very specific towards the glucose moiety of the substrate but less towards the acceptor, which can be L-sorbose, D-xyloketose or L-araboketose, for instance (Hassid and Doudoroff [30]). In the bacteria provided with the transferase, sucrose can be formed by this pathway. But it is not the case in plants in which the transglycosylation from sugar nucleotide takes place. Maltose glucosyltransferase, identified by Fitting and Doudoroff [38] in the bacterium *Neisseria meningitidis*, catalyses the reversible reaction (Putman, Litt and Hassid [39]).

4-O-α-D-glucosyl-D-glucose + P_i
(maltose)

$\xrightleftharpoons{\text{maltose glucosyltransferase}}$ β-D-glucose-1-P + D-glucose

But in nature, maltose is mainly formed by the action of β-amylase on amylose or amylopectin.

A cellobiose glucosyltransferase has also been described in cellulolytic bacteria (Sih, Nelson and McBee [40]).

7. Introduction of the concept of a branching enzyme

This step took place in the context of the phosphorylase theory. Since natural glycogens and starches, in addition to 1,4-glycosidic linkages, contained 1,6-linkages, it was suggested that, in addition to the phosphorylase, a supplementary enzyme, which was eliminated during the preparation of phosphorylase, induced branching. Evidence for the existence of such branching enzymes was provided by Cori and Cori [41]. From heart, brain and liver, they isolated a "branching factor" synthesizing 1,6-linkages. They combined the action of this factor and of crystalline muscle phosphorylase and obtained glycogen. Soon after, Haworth, Peat and Bourne [42] isolated from potato juice an enzyme (Q-enzyme) which, in association with potato phosphorylase, produced amylopectin.

That the Q-enzyme was able to convert linear amylose to branched amylopectin without participation of inorganic phosphate was shown by Peat and his collaborators (Review: Peat [43]).

References p. 166

Plate 199. Luis Federico Leloir.

8. Sugar nucleotides and their function in the interconversions of sugars

The biosynthesis of sugars takes place, as shown in previous chapters, by photosynthesis, by gluconeogenesis and by other mechanisms. They can be converted into other sugars, but generally speaking these reactions take place at the level of sugar derivatives, such as sugar phosphates. In Chapter 21 we have considered the mechanism of formation of several of these phosphorylated derivatives.

Another type of derivative commonly formed is in the category of sugar nucleotides, a category which was discovered in 1950. They contain

"a sugar or a sugar derivative esterified by a glycosidic hydroxyl to the terminal phosphate residue of a nucleoside 5′-diphosphate" (Ginsburg [44], p. 35).

The first sugar nucleotide to be isolated was uridine diphosphate glucose (UDPG). It was identified by Caputto, Leloir, Cardini and Paladini [45] in a report from the Instituto de Investigaciones Bioquímicas, Fundacion Campomar, Buenos Aires, Argentina, published in 1950 and concerned with the utilization of D-galactose by yeast. Grant [46] published in 1935 the results of his studies on the phosphoric esters formed during the utilization of galactose by yeast adapted to galactose: he only found fructose diphosphate, glucose-6-phosphate and trehalose phosphate. After galactose feeding, phosphoric esters were investigated in the liver of rabbits by Kosterlitz [47] who found a non-reducing acid labile ester later identified by the same author [48] as galactose-1-P. It was found, by Caputto, Leloir and Trucco [49], and Trucco, Caputto, Leloir and Mittelman [50] as well as by Wilkinson [51], that the first step in the utilization of galactose by yeast was a phosphorylation product of galactose by ATP in the presence of galactokinase.

galactose + ATP → galactose-1-P + ADP

Caputto, Leloir, Trucco, Cardini and Paladini [52] showed that the enzymatic transformations leading from galactose-1-P to glucose-6-P were

galactose-1-P → glucose-1-P → glucose-6-P

     II   III

In the initial studies [53] it had been recognized that a thermo-stable factor was required in the overall reaction II and III, which led to the identification of glucose diphosphate as the coenzyme of reaction III [54].

Reaction II can be recognized as an inversion of C-4, catalysed by an enzyme which was called "galactowaldenase" in Leloir's laboratory. The purification of the coenzyme of this reaction was accomplished, as stated above, by Leloir and his colleagues. They observed that the coenzyme content of baker's yeast was doubled by a treatment with toluene. The yeast treated with toluene was extracted with 50% ethanol followed by precipitation with mercuric salts and dissolving in ammonium acetate, followed by reprecipitation and decomposition with H_2S in the cold. The absorption spectrum of the product showed a maximum absor-bance at 260 nm. After acid hydrolysis, uridine phosphate could be identified as a constituent of the coenzyme,

"by comparison with a known sample with respect to behavior during paper chromatography and to the spectra at different pH values" (p. 336).

The authors identified two phosphate groups, and a reducing sub-stance which they recognized, by several tests, as glucose.

Having identified uridine, two phosphate groups and glucose as components of the coenzyme, the authors determined several properties of the compound and concluded that the following formula agrees with these findings, a glucose-1-P molecule being attached to uridine 5′-phosphate forming a pyrophosphate link.

Uridine diphosphate glucose
UDPG

They called the compound uridine diphosphate glucose, the first sugar nucleotide to be isolated. The studies just referred to revealed the participation of the sugar nucleotide uridine diphosphate glucose in the conversion of galactose phosphate into glucose phosphate. This led Leloir [55] to investigate the

"possible changes brought about by the enzyme on the sugar moiety of the coenzyme" (p. 186)

which led to the detection of an enzyme that interconverts UDP-D glucose and UDP-D galactose, which was identified. In penicillin-treated *Staphylococcus aureus*, Park and Johnson [56] had recognized an accumulation of compounds of a similar nature. After the isolation of UDPG by Leloir and his colleagues, Park [57] returned to the penicillin-treated *Staphylococcus aureus* and identified among the accumulated compounds, UDP-*N*-acetylmuramic acid. UDP-*N*-acetylmuramic peptides were later recognized. UDPG was formed, as observed by Trucco [58], when ATP, UDP and glucose-1-P were incubated.

Kalckar [59] showed that the biosynthesis proceeded as follows

UTP + glucose-1-P ⇌ UDPG + pyrophosphate

in the presence of an enzyme which he called diphosphoglucose pyrophosphorylase.

This enzyme has been studied by several authors (Munch-Petersen, Kalckar, Cutolo and Smith [60]; Ginsburg [61]; Ganguli [62]; Turner and Turner [63]; Villar-Palasi and Larner [64]).

Besides UDPG, several uridine diphosphate nucleotides are known (literature in Bernfeld [2], p. 264). These nucleotides can be biosynthesized by reactions analogous to the formation of UDPG mentioned above, i.e. in general

UTP + aldose-1-P ⇌ UDP sugar + PP$_i$

But, as shown by Kalckar, Anderson and Isselbacher [65], UDP-sugars can also be formed according to the reaction

galactose-1-P + UDP glucose ⇌ glucose-1-P + UDP-galactose (*)

i.e. by incorporation of galactose-1-P followed by the exchange of the hexose-1-P moiety of uridine diphosphate for another hexose-1-P.

The enzyme catalysing reaction (*) was called galactose-1-phosphate uridyl transferase which has been identified by Kalckar, Braganca and Munch-Petersen [66] in galactose-adapted yeast. The uridine diphosphate sugars can, by epimerization, form UDP derivatives of other sugars by the inversion of a hydroxyl group of a UDP sugar.

When Leloir [55] described the reversible epimerization

UDP-glucose ⇌ UDP-galactose,

he revealed one of the functions of sugar nucleotides, their role in sugar conversions. This reaction has been further studied by a number of authors and the biological distribution of the enzyme involved (uridine diphospho-galactose-4-epimerase) has been reported (mammals, bacteria, yeast) (literature in Bernfeld [2], p. 265).

As stated by Hassid [4] (p. 314):

"Plant as well as animal tissues and microorganisms (Strominger, Maxwell, Axelrod and Kalckar [67]; Maxwell, Kalckar and Strominger [68]; Strominger and Mapson [69]; Smith, Mills, Bernheimer and Austrian [70]) contain an enzyme that oxidizes uridine diphosphate D-glucose to uridine diphosphate D-glucuronic acid; the latter can be epimerized to UDP-D-galacturonic acid or decarboxylated to uridine diphosphate pentose in plants".

9. The role of sugar nucleotides in transglycosylation

As stated above, the concept of a role of transglycosylation from a phosphate ester in the biosynthesis of carbohydrates was first introduced in the form of the phosphorylase theory of the biosynthesis of glycogen, starch and sucrose. Nevertheless, efforts were wasted in many laboratories in search of phosphorylases active in the biosynthesis of other sugar polymers.

Another form of transglycosylation (from sugar nucleotides) was reported in 1953 by Dutton and Storey [71–73], with the synthesis of a glycoside, and by Leloir and Cabib [74] with the synthesis of trehalose phosphate.

(a) The formation of glycosidic bonds by transglycosylation

The glycosidic linkage is defined as an acetal bond between the aldehyde or keto group of a sugar and a hydroxy group, for instance in di-, oligo- and polysaccharides and in glycosides (a sugar linked with an alcohol or a phenol). The role of sugar nucleotides as glycosyl donors was first recognized by Dutton and Storey [71—73], with the synthesis of the glycoside β-*o*-aminophenol glucuronide

> UDP-D-glucuronic acid + *o*-aminophenol → *o*-aminophenyl
> β-D-glucuronic acid

Isselbacher and Axelrod [75]; Dutton [76]; Grodsky and Carbone [77] have shown that a number of other acceptors, besides *o*-aminophenol take part in similar reactions.

In plants, generally, phenolic compounds are combined in the form of glycosides, with glucose.

The glycoside arbutin is for instance synthesized from hydroquinone with the participation of UDP-D-glucose as glycosyl donor in the presence of an enzyme preparation from wheat germ (Yamaha and Cardini [78]).

> UDP-D-glucose + hydroquinone → UDP + hydroquinone
> β-D-glucopyranoside (arbutin)

A large variety of reactions involved in the formation of glycosides have been described (see Hassid [4]).

(b) UDPG as glucose donor in the synthesis of trehalose phosphate

This was first demonstrated by the discovery of Leloir and Cabib [74] of an enzyme which catalyses the formation of trehalose phosphate as follows

> UDPG + glucose-6-P → UDP + trehalose phosphate

This enzyme has been purified 15- to 20-fold from brewer's yeast, by Cabib and Leloir [79]. The disaccharide obtained in the reaction was identified as trehalose phosphate by comparison of its properties with those of the pure compound.

(c) UDPG as glucose donor in the synthesis of sucrose
and of sucrose phosphate

As stated above, it had been believed that sucrose was biosynthesized through transglycosylation by a phosphorylase. The enzyme involved had been called sucrose phosphorylase, catalysing the reversible reaction

$$\text{glucose-1-P} + \text{fructose} \rightleftharpoons \text{sucrose} + P_i$$

But the reaction was shown to be displaced towards the phosphorolysis of sucrose.

In many higher plants, which are the main producers of sucrose in nature, a mechanism utilizing UDPG as glycosyl donor and β-D-fructose as acceptor has been described (Buchanan [80]; Cardini, Leloir and Chiriboga [81]; Burma and Mortimer [82]; Leloir and Cardini [83]).

$$\text{UDPG} + \text{fructose} \rightleftharpoons \text{UDP} + \text{sucrose}$$

Another plant enzyme described by Leloir and Cardini [83] can lead to sucrose biosynthesis by the reaction

$$\text{UDPG} + \text{fructose-6-P} \rightleftharpoons \text{UDP} + \text{sucrose phosphate}$$

The reaction is pulled towards biosynthesis by adding the reaction

$$\text{sucrose phosphate} \rightarrow \text{sucrose} + P_i$$

(d) UDPG as glucose donor in the synthesis of cellulose in microorganisms

An insoluble enzyme which biosynthesizes cellulose from UDPG was recognized by Glaser [84,85] in the microorganism *Acetobacter xylinum*. But in plants, cellulose biosynthesis appears to take place by another mechanism, from guanosine diphosphate D-glucose (GDP-glucose). Barber and Hassid [86] have detected an enzyme (pyrophosphorylase) in peas forming GDP-glucose from GTP (guanosine triphosphate) and α-D-glucose-1-P.

The formation of cellulose in plants is supposed to begin with a formation of GDP-glucose

$$\text{GTP} + \alpha\text{-D-glucose-1-P} \xrightleftharpoons{\text{pyrophosphorylase}} \text{GDP-D-glucose} + PP_i$$

Cellulose could then be formed by a transferase catalysing repetitive D-glucosyl transfers to an acceptor (see Hassid [4], p. 364)

$$n(\text{GDP-D-glucose}) + \text{acceptor} \rightarrow \text{acceptor-}(\beta\text{-1,4-D-glucose})n$$
$$+ n\text{GDP (cellulose)}$$

But the final answer is not yet known.

(e) UDPG as glucose donor in the biosynthesis of glycogen

Leloir and Cardini [87] incubated UDPG with a liver enzyme and a small amount of glycogen. They observed that approximately equal amounts of UDP and of glycogen were formed. Such changes could only be obtained if the liver preparation was freed of amylase.

As stated by the authors:

"Other preparations obtained by ammonium sulfate precipitation contained amylase and therefore lost their glycogen. With such enzyme no UDP formation took place unless a primer was added. . . . glycogen and soluble starch acted as primers whereas glucose and maltose were ineffective. Several mono-, di- and oligosaccharides and hexose phosphates were tested with negative results. Treatment of glycogen with α-amylase destroyed its priming capacity. It can be concluded that UDPG acts directly as a glucose donor to glycogen and that the reaction is thus similar to polysaccharide formation from glucose-1-phosphate with animal phosphorylase which requires a primer of high molecular weight. The enzyme was found in the soluble fraction of liver and became very unstable after purification".

This work of Leloir and Cardini [87] marked the end of the period during which glycogen was believed to be synthesized by phosphorylase. That this enzyme is not responsible for the catalysis of the α-1,4-D-glucose-linked glycogen chain from α-D-glucose-1-P had been repeatedly suggested (literature in Manners [88]). Another kind of argument against the phosphorylase theory came from pharmacological experiments, when it was observed that agents increasing phosphorylase activity such as epinephrine or glucagon produced an increase of glycogen breakdown and conversely such agents as high potassium concentration, known as lowering the phosphorylase activity, increased glycogen concentration (Sutherland and Cori [89]; Cahill, Ashmore, Zottu and Hastings [90]). Mommaerts, Illingworth, Pearson, Guillory and Searaydarian [91] later showed that phosphorylase is not detectable in the muscles of the patients suffering from McArdle's disease. Nevertheless these muscles contain normal, or slightly increased amounts of glycogen. From 1957 on it was admitted that the role of phosphorylase is limited to glycogen degradation and that it plays no part, in vivo, in the biosynthesis of glycogen.

When reporting on the subject in the plenary session of the 6th

International Congress of Biochemistry, New York 1964, Leloir [92] traced the present picture of the biosynthesis of glycogen from UDPG in the presence of glycogen synthase. The synthesis begins with a transfer of D-glucose units from UDPG by the enzyme UDP-D-glucose—glycogen glucosyltransferase (EN 2.4.1.11), forming α-1,4-linked D-glucose chains. When the chain becomes about 10 units long, portions are transferred forming α-1,6-D-glucose linkages by a branching enzyme, α-1,4-glucan: α-1,4-glucan-6-glucosyltransferase. Leloir [93], pointing to the thermodynamic superiority of the synthetase reaction as favouring the formation of α-1,4-glycosidic bonds much more than the phosphorylase reaction, suggests that it may be due to the fact that the synthetase reaction liberates a secondary acid group in the form of UDP while the phosphorylase reaction liberates a tertiary group in the form of orthophosphate. At neutral pH, secondary groups, contrary to tertiary groups, are nearly completely ionized. Thus the driving force of reactions involving UDP-linked sugars is attributed by Leloir to the free energy of dissociation of the secondary acid group liberated.

Muscle glycogen synthetases have been shown to exist in two interconvertible forms. One is dependent (D form) for activity on glucose-6-P while the other (I) is independent of it. The history of this development has been retraced by E. Helmreich in Chapter II of volume 17 of this Treatise (p. 20—22).

(f) UDPG as glucose donor in the biosynthesis of starch

The phosphorylase theory of starch biosynthesis has followed the same fate as the phosphorylase theory of glycogen formation. This issue was foreshadowed by a series of contrary reports before the theory came to grief. Ewart, Siminovitch and Briggs [94] remarked, for instance, that the phosphate level in plant cells is such that the direction of the phosphorylase reaction should be towards starch breakdown. Stocking [95] had already remarked that starch is biosynthesized in the plastids while phosphorylase is found in the cytoplasm.

Leloir, de Fekete and Cardini [96], inspired by the discovery of the enzyme synthesizing glycogen from UDPG found a similar enzyme catalysing the biosynthesis of starch from UDPG in plants.

(g) UDP-galactose as a galactose donor to glucose in lactose
biosynthesis

With tissue slices from guinea pig mammary glands, Grant [97] demonstrated a biosynthesis of lactose from D-glucose. This observation was confirmed (Malpress and Morrison [98]; Heyworth and Bacon [99]). The first experimenters who obtained a small amount of lactose from D-glucose with a homogenate from lactating mammary glands of guinea pigs were Reithel, Horowitz, Davidson and Kittinger [100]. Gander, Petersen and Boyer [101, 102], using homogenates of the bovine udder, reported the production of lactose phosphate from UDPgalactose and glucose-1-P.

The authors visualized as follows the steps of the pathway, involving the formation of lactose-1-P as intermediate:

$$2\text{-D-Glucose} \xrightarrow[\text{ATP}]{\text{hexokinase}} 2\text{-glucose-6-P}$$

$$\xrightarrow{\text{phosphoglucomutase}} 2\text{-}\alpha\text{-D-glucose-1-P}$$

$$\alpha\text{-D-Glucose-1-P} + \text{UTP} \xrightarrow{\text{pyrophosphorylase}} \text{UDP-D-glucose}$$
$$+ \text{ pyrophosphate}$$

$$\text{UDP-D-glucose} \xrightarrow{\text{UDP-D-galactose 4-epimerase}} \text{UDP-D-galactose}$$

$$\text{UDP-D-galactose} + \alpha\text{-D-glucose-1-P}$$

$$\xrightarrow{\text{galactosyltransferase}} \text{lactose-1-P} + \text{UDP}$$

$$\text{Lactose-1-P} \xrightarrow{\text{phosphomonoesterase}} \text{lactose} + \text{phosphate}$$

The authors erroneously reported the existence of a galactosyltransferase catalysing the transfer of D-galactose from UDP-D-galactose to α-D-glucose-1-P. Hexokinase and phosphoglucomutase (Kittinger and Reithel [103]) as well as UDP-D-galactose-4-epimerase (Caputto and Trucco [104]) had been shown to be present in mammary tissues. This scheme proposing UDP-D-galactose as a D-galactose donor appeared to be similar to the role of UDP-D-glucose as glucosyl donor in the formation of disaccharides in plants and microorganisms (see above). The scheme proposed lactose-1-P as the penultimate step. This was logical since trehalose phosphate and sucrose phosphate are formed in the course of the synthesis of trehalose and of sucrose (see above). Further, McGeown and Malpress [105] had reported small amounts of lactose phosphate in milk. But results obtained in whole animals and with perfused mammary glands did not agree with the scheme of Gander et al. [102]. That blood D-glucose was the main precursor of both hexose moieties of lactose was shown by Baxter, Kleiber and Black [106] and by Kleiber, Black, Brown, Baxter, Luick and Stadtman [107], but in experiments in which labelled acetate was used as precursor, Schambye, Wood and Kleiber [108]; Wood, Schambye and Peeters [109] and Wood, Siu and Schambye [110] showed that the D-glucose and D-galactose components of the isolated lactose were not equally labelled and did not have the same ^{14}C distribution pattern, which would be the case if lactose were biosynthesized by the mechanism proposed by Gander et al. [102].

As stated by Wood, Joffe, Gillespie, Hansen and Hardenbrook [111],

"Wood et al. [110] have injected acetate-1-^{14}C unilaterally into the pudic artery which supplies blood to one half of the cow's udder and investigated separately the milk from the two sides of the udder [109], whereas that from the noninjected side resembled the lactose obtained when acetate-1-^{14}C was administered intravenously to the cow [108]. Thus the milk from the injected side reflected mainly the metabolism of the udder per se whereas that from the noninjected side reflected the composite metabolism of the whole animal. On the injected side the distribution of ^{14}C in the six carbons of the galactose differed greatly from that of the glucose and it was concluded that the glucose and galactose moieties arise from different precursors" [108–110] (p. 1264).

Wood et al. [111] studied the distribution of ^{14}C in the six carbons of the blood glucose and in the glucose and galactose moieties of lactose. The results showed that the carbon chain of glucose is converted intact to the glucose moiety of lactose and therefore that this glucose moiety derives from blood glucose. On the other hand, in the injected side, the galactose moiety of lactose had a much higher activity than blood glucose. The distribution pattern of ^{14}C was entirely different. The authors proposed that the asymmetrical distribution patterns of ^{14}C found in the galactose moiety may be caused by the transaldolase exchange reaction taking place in the pentose phosphate cycle and by a slow trioseisomerase reaction.

Wood, Gillespie, Joffe, Hansen and Hardenbrook [112] accomplished a unilateral injection of [2,6-^{14}C]glucose and [6-^{14}C]-glucose into the pudic artery of a cow and degraded the glucose and the galactose of the lactose. They found that

"with glucose-2,6-^{14}C the glucose moiety was almost exclusively and equally labelled in C-2 and C-6. This finding is in accord with the proposal that glucose is a direct precursor of the glucose moiety. The galactose was labelled to the highest degree in C-2 and C-6 but also contained considerable activity in C-1, C-3 and C-5. With glucose-^{14}C both the glucose and galactose were almost exclusively labelled in C-6. Thus C-2 of glucose was randomized to C-1, C-3 and C-5 of galactose, but there was little randomization of C-6 of glucose. The pentose cycle is considered to be the mechanism whereby C-2 is randomized into C-1 and C-3. It is proposed that the C-5 labelling occurs by way of the aldose and triosephosphate isomerase reactions, forming glyceraldehyde-2,3-^{14}C phosphate which then exchanges by means of the transaldolase reaction with fructose-6-phosphate" (p. 1277).

The concept which results from these experiments is that blood glucose is directly used while it is transformed into the D-galactosyl moiety by hexose phosphate intermediates.

Watkins and Hassid [113] examined homogenates of guinea pig mammary glands with a variety of ^{14}C-labelled substrates and

"although lactose formation could be demonstrated in some of the mixtures, the amounts of radioactivity incorporated were very small. The supernatant solutions remaining after removal of most of the cellular fragments also gave only slight indication of lactose synthesis. Concentrated suspensions of particles isolated from the homogenates were, however, found to give active synthesis of lactose. Similar results were obtained with particulate preparations from both guinea pig and cow mammary glands. No evidence was obtained for the formation of lactose-1-phosphate with either particulate or cell-free preparations from either guinea pig or bovine mammary glands" (p. 1438).

With their particulate enzyme preparation, Watkins and Hassid observed the formation of radioactive lactose in mixtures containing either [^{14}C]UDP-D-galactose or [^{14}C]UDP-D-glucose. But no labelled lactose was detected after incubating the particulate preparation for 1 h in the absence of a nucleotide sugar, with either [^{14}C]D-glucose, [^{14}C]D-galactose, [^{14}C]α-D-galactose-1-P.

References p. 166

The lactose formed in a mixture containing $[^{14}C]$UDP-D-glucose and unlabelled D-glucose was labelled in the D-galactose moiety. From these data the authors concluded that UDP-D-galactose was the donor of the galactose moiety of lactose.

That D-glucose is the acceptor was suggested by the fact that the substrate mixture which gave the most active lactose synthesis was one containing $[^{14}C]$UDP-D-galactose and D-glucose, and the authors proposed that the final stage in the formation of lactose involves the transfer of D-galactose from UDP-D-galactose to D-glucose according to the reaction

$$\text{UDP-D-galactose + D-glucose} \xrightarrow{\text{galactosyl transferase}} \text{lactose + UDP}$$

It was not possible to solubilize the enzyme from this preparation, but Babad and Hassid [114,115] later found the enzyme, lactose synthase, in milk.

The series of enzymes involved in the conversion of D-glucose into UDP-D-galactose have been found in the mammary gland (Watkins and Hassid [113]; Smith and Mills [116]; Maxwell, Kalckar and Burton [117]; Kittinger and Reithel [103].

(h) Glycosyl transfer from UDPAG in chitin biosynthesis

Chitin is a structural polysaccharide which is found in the exoskeleton of arthropods, in the shells of molluscs and in the cell walls of fungi. It is a polymer of N-acetyl-D-glucosamine units, combined through β-(1 → 4)-linkages.

A particulate preparation from *Neurospora crassa*, prepared by Glaser and Brown [118], catalysed the synthesis of an insoluble polysaccharide from the sugar nucleotide uridine diphosphate-N-acetyl-D-glucosamine (UDPAG).

This transglycosylase catalyses the reaction

$$\text{UDPAG} + (AG)_n \rightleftharpoons \text{UDP} + (AG)_{n+1}$$

10. The enzymes of the biosynthesis of complex saccharides from sugar nucleotides

In photosynthesis carbon may be drained off to enter molecules of pyruvate and be utilized for the formation of other compounds such as fatty acids (see Chapter 59) and amino acids (see Chapters 64—74).

According to present knowledge, glucose-6-P formed in photosynthesis (Chapter 56) or derived from gluconeogenesis (Chapter 57) may contribute to the formation of a variety of sugar phosphates and sugar nucleotide derivatives which contribute to the biosynthesis of complex saccharides. After the theories of reversible zymohydrolysis and the theory of biosynthesis of complex saccharides by phosphorylase were both discarded, they were replaced by the theory involving the participation of different sugar phosphates and sugar nucleotides. The enzymology of the pathways now considered by the prevailing theory as forming, in nature, the complex saccharides from sugar nucleotide precursors, was clarified during the period following 1955. At that time, it was long known that phosphoglucomutase catalysed the conversion of D-glucose-6-P to α-D-glucose-1-P. A phosphorylase found in plants, yeast and animals was recognized by Kalckar [119] as catalysing the formation of uridine diphosphate-D-glucose from α-D-glucose-1-P and uridine triphosphate. That the interconversion of uridine diphosphate-D-glucose and uridine diphosphate-D-galactose was catalysed by 4-epimerases present in the generality of cells was demonstrated (Kalckar [119]; Neufeld, Ginsburg, Putman, Fanshier and Hassid [120]).

As stated by Hassid [4] concerning the identification of transglycosylases:

"Plant and animal tissues have been shown to contain transglycosylases capable of catalyzing the transfer of D-glucose, D-galactose, N-acetyl-D-glucosamine or other monosaccharides from sugar nucleotides containing various bases (uridine, guanine, thymine, adenine, cytidine) to sugar acceptors, forming complex saccharides such as sucrose (Cardini et al. [81]) sucrose phosphate (Leloir and Cardini [121]); trehalose (Bean and Hassid [122]); lactose (Watkins and Hassid [113]; Babad and Hassid [115]); raffinose (Pridham and Hassid [123]); cellulose (Elbein, Barber and Hassid [124]); Barber, Elbein and Hassid [125]); β-1,3-glucan (Feingold, Neufeld and Hassid [126]); starch (Leloir et al. [96]; Recondo and Leloir [127]); glycogen (Leloir [92]) and pectin (Villemez, Lin and Hassid [128]). There is good reason to believe that most of the other numerous complex saccharides in nature are formed from sugar nucleotide precursors" (p. 314).

REFERENCES

1 M. Doudoroff, H.A. Barker and W.Z. Hassid, J. Biol. Chem., 168 (1947) 725.
2 P. Bernfeld, in P. Bernfeld (Ed.), Biogenesis of Natural Compounds, Oxford, 1963, 233.
3 E.J. Hehre, Adv. Enzymol., 11 (1951) 297.
4 W.Z. Hassid, in D.M. Greenberg (Ed.), Metabolic Pathways, Vol. I, New York and London, 1967, p. 307.
5 M. Cohn, J. Biol. Chem., 180 (1949) 771.
6 H.M. Kalckar, J. Biol. Chem., 167 (1947) 477.
7 J.K.L. Von Parnas, Erg. Enzymf., 6 (1937) 57.
8 C.F. Cori, S.P. Colowick and G.T. Cori, J. Biol. Chem., 121 (1937) 465.
9 G.T. Cori, S.P. Colowick and C.F. Cori, J. Biol. Chem., 123 (1938) 375.
10 J.K. Parnas and T. Baranowski, C. R. Soc. Biol., 120 (1935) 307.
11 C.F. Cori, G. Schmidt and G.T. Cori, Science, 89 (1939) 464.
12 G.T. Cori and C.F. Cori, J. Biol. Chem., 131 (1939) 397.
13 C.S. Hanes, Proc. Roy. Soc., Ser. B., 128 (1939—1940) 421.
14 G.T. Cori and C.F. Cori, Proc. Soc. Exp. Biol. Med., 36 (1937) 119.
15 G.T. Cori, S.P. Colowick and C.F. Cori, J. Biol. Chem., 123 (1938) 381.
16 G.T. Cori, S.P. Colowick and C.F. Cori, J. Biol. Chem., 124 (1938) 543.
17 S.P. Colowick, J. Biol. Chem., 124 (1938) 557.
18 W. Kiessling, Naturwissenschaften, 27 (1939) 129.
19 G.T. Cori, C.F. Cori and G. Schmidt, J. Biol. Chem., 129 (1939) 629.
20 G.T. Cori, M.A. Swanson and C.F. Cori, Fed. Proc., 4 (1945) 234.
21 M. Samec and M. Blinc, Kolloid Beih., 47 (1936) 371.
22 W.N. Haworth, E.L. Hirst and F.A. Isherwood, J. Chem. Soc., (1937) 577.
23 K.H. Meyer, M. Wertheim and P. Bernfeld, Helv. Chim. Acta, 23 (1940) 865.
24 K.H. Meyer and P. Bernfeld, Helv. Chim. Acta, 23 (1940) 875.
25 K.H. Meyer and M. Fuld, Helv. Chim. Acta, 24 (1941) 375.
26 B. Illingworth, J. Larner and G.T. Cori, J. Biol. Chem., 199 (1952) 631.
27 J. Larner, B. Illingworth, G.T. Cori and C.F. Cori, J. Biol. Chem., 199 (1952) 641.
28 K.H. Meyer and G.C. Gibbons, Adv. Enzymol., 12 (1951) 341.
29 K.H. Meyer, Natural and Synthetic High Polymers, 2nd ed., New York, 1950, p. 468.
30 W.Z. Hassid and M. Doudoroff, Adv. Carbohydr. Chem., 5 (1950) 29.
31 A. Oparin and A. Kurssanow, Biochem. Z., 239 (1931) 1.
32 A. Lebedew and A. Dikanowa, Z. Physiol. Chem., 231 (1935) 271.
33 J. Burkard and C. Neuberg, Biochem. Z., 270 (1934) 229.
34 W.Z. Hassid, Plant Physiol., 13 (1938) 641.
35 A. Kurssanov and N. Kriukova, Biokhimiya, 4 (1939) 229.
36 N. Kriukova, Biokhimiya, 5 (1940) 574.
37 M. Doudoroff, N. Kaplan and W.Z. Hassid, J. Biol. Chem., 148 (1943) 67.

38 C. Fitting and M. Doudoroff, J. Biol. Chem., 199 (1952) 153.
39 E.W. Putman, C.F. Litt and W.Z. Hassid, J. Am. Chem. Soc., 77 (1955) 4351.
40 C.J. Sih, N.M. Nelson and R.H. McBee, Science, 126 (1957) 1116.
41 G.T. Cori and C.F. Cori, J. Biol. Chem., 151 (1943) 57.
42 W.N. Haworth, S. Peat and E.J. Bourne, Nature (London), 154 (1944) 236.
43 S. Peat, Adv. Enzymol., 11 (1951) 339.
44 V. Ginsburg, Adv. Enzymol., 26 (1964) 35.
45 R. Caputto, L.F. Leloir, C.E. Cardini and A.C. Paladini, J. Biol. Chem., 184 (1950) 333.
46 G.A. Grant, Biochem. J., 29 (1935) 1661.
47 H.W. Kosterlitz, Biochem. J., 31 (1937) 2217.
48 H.W. Kosterlitz, Biochem. J., 37 (1943) 318.
49 R. Caputto, L.F. Leloir and R.E. Trucco, Enzymologia, 12 (1948) 350.
50 R.E. Trucco, R. Caputto, L.F. Leloir and N. Mittelman, Arch. Biochem., 18 (1948) 137.
51 J.F. Wilkinson, Biochem. J., 44 (1949) 460.
52 R. Caputto, L.F. Leloir, R.E. Trucco, C.E. Cardini and A.C. Paladini, J. Biol. Chem., 179 (1949) 497.
53 R. Caputto, L.F. Leloir, R.E. Trucco, C.E. Cardini and A. Paladini, Arch. Biochem., 18 (1948) 201.
54 C.E. Cardini, A.C. Paladini, R. Caputto, L.F. Leloir and R.E. Trucco, Arch. Biochem., 22 (1949) 87.
55 L.F. Leloir, Arch. Biochem. Biophys., 33 (1951) 186.
56 J.T. Park and M.J. Johnson, J. Biol. Chem., 179 (1949) 585.
57 J.T. Park, J. Biol. Chem., 194 (1952) 877, 885, 897.
58 R.E. Trucco, Arch. Biochem. Biophys., 34 (1951) 482.
59 H.M. Kalckar, Science, 125 (1957) 105.
60 A. Munch-Petersen, H.M. Kalckar, E. Cutolo and E.E.B. Smith, Nature (London), 172 (1953) 1036.
61 V. Ginsburg, J. Biol. Chem., 232 (1958) 55.
62 N.C. Ganguli, J. Biol. Chem., 232 (1958) 337.
63 D.H. Turner and J.F. Turner, Biochem. J., 69 (1958) 448.
64 C. Villar-Palasi and J. Larner, Arch. Biochem. Biophys., 86 (1960) 61.
65 H.M. Kalckar, P.E. Anderson and K.J. Isselbacher, Proc. Natl. Acad. Sci. U.S.A., 42 (1956) 49.
66 H.M. Kalckar, B. Braganca and A. Munch-Petersen, Nature (London), 172 (1953) 1038.
67 J.L. Strominger, E.S. Maxwell, J. Axelrod and H.M. Kalckar, J. Biol. Chem., 224 (1957) 79.
68 E.S. Maxwell, H.M. Kalckar and J.L. Strominger, Arch. Biochem. Biophys., 65 (1956) 2.
69 J.L. Strominger and L.W. Mapson, Biochem. J., 66 (1957) 567.
70 E.E.B. Smith, G.T. Mills, H.P. Bernheimer and R. Austrian, Biochim. Biophys. Acta, 28 (1958) 211.
71 G.J. Dutton and I.D.E. Storey, Biochem. J., 53 (1953) XXXVII.
72 G.J. Dutton and I.D.E. Storey, Biochem. J., 57 (1954) 275.

73 I.D.E. Storey and G.J. Dutton, Biochem. J., 59 (1955) 279.
74 L.F. Leloir and E. Cabib, J. Am. Chem. Soc., 75 (1953) 5445.
75 K.J. Isselbacher and J. Axelrod, J. Am. Chem. Soc., 77 (1955) 1070.
76 G.J. Dutton, Biochem. J., 64 (1956) 693.
77 G.M. Grodsky and J.V. Carbone, J. Biol. Chem., 226 (1957) 449.
78 T. Yamaha and C.E. Cardini, Arch. Biochem. Biophys., 86 (1960) 127.
79 E. Cabib and L.F. Leloir, J. Biol. Chem., 231 (1958) 259.
80 J.G. Buchanan, Arch. Biochem. Biophys., 44 (1953) 140.
81 C.E. Cardini, L.F. Leloir and J. Chiriboga, J. Biol. Chem., 214 (1955) 149.
82 D.P. Burma and D.C. Mortimer, Arch. Biochem. Biophys., 62 (1956) 16.
83 L.F. Leloir and C.E. Cardini, J. Biol. Chem., 214 (1955) 157.
84 L. Glaser, J. Biol. Chem., 232 (1957) 627.
85 L. Glaser, Biochim. Biophys. Acta, 25 (1957) 436.
86 G.A. Barber and W.Z. Hassid, Biochim. Biophys. Acta, 86 (1964) 397.
87 L.F. Leloir and C.E. Cardini, J. Am. Chem. Soc., 79 (1957) 6340.
88 D.J. Manners, Adv. Carb. Chem., 17 (1962) 371.
89 E.W. Sutherland and C.F. Cori, J. Biol. Chem., 188 (1951) 531.
90 G.F. Cahill Jr., J. Ashmore, S. Zottu and A.B. Hastings, J. Biol. Chem., 224 (1957) 237.
91 W.H.F.M. Mommaerts, B. Illingworth, C.M. Pearson, R.J. Guillory and K. Searaydarian, Proc. Natl. Acad. Sci. U.S.A., 45 (1959) 791.
92 L.F. Leloir, Proc. 6th Int. Cong. Biochem., New-York, 1964, p. 15.
93 L.F. Leloir, in W.J. Whelan and M.P. Cameron (Eds.), Ciba Foundation Symposium on Control of Glycogen Metabolism, Boston, 1964, pp. 68 and 82.
94 M.H. Ewart, D. Siminovitch and D.R. Briggs, Plant Physiol., 29 (1954) 407.
95 C.R. Stocking, Am. J. Bot., 39 (1952) 283.
96 L.F. Leloir, M.A.R. de Fekete and C.E. Cardini, J. Biol. Chem., 236 (1961) 636.
97 G.A. Grant, Biochem. J., 29 (1935) 1905.
98 F.H. Malpress and A.B. Morrison, Biochem. J., 46 (1950) 307.
99 R. Heyworth and J.S.D. Bacon, Biochem. J., 61 (1955) 225.
100 F.J. Reithel, M.G. Horowitz, H.M. Davidson and G.W. Kittinger, J. Biol. Chem., 194 (1952) 839.
101 J.E. Gander, W.E. Petersen and P.D. Boyer, Arch. Biochem. Biophys., 60 (1956) 259.
102 J.E. Gander, W.E. Petersen and P.D. Boyer, Arch. Biochem. Biophys., 69 (1957) 85.
103 G.W. Kittinger and F.J. Reithel, J. Biol. Chem., 205 (1953) 527.
104 R. Caputto and R.E. Trucco, Nature (London), 169 (1952) 1061.
105 M.G. McGeown and F.H. Malpress, Biochem. J., 52 (1952) 606.
106 C.F. Baxter, M. Kleiber and A.L. Black, Biochim. Biophys. Acta, 21 (1956) 277.
107 M. Kleiber, A.L. Black, M.A. Brown, C.F. Baxter, J.R. Luick and F.H. Stadtman, Biochim. Biophys. Acta, 17 (1955) 252.
108 P. Schambye, H.G. Wood and M. Kleiber, J. Biol. Chem., 226 (1957) 1011.

109   H.G. Wood, P. Schambye and G.J. Peeters, J. Biol. Chem., 226 (1957) 1023.
110   H.G. Wood, P. Siu and P. Schambye, Arch. Biochem. Biophys., 69 (1957) 390.
111   H.G. Wood, S. Joffe, R. Gillespie, R.G. Hansen and H. Hardenbrook, J. Biol. Chem., 233 (1958) 1264.
112   H.G. Wood, R. Gillespie, S. Joffe, R.G. Hansen and H. Hardenbrook, J. Biol. Chem., 233 (1958) 1271.
113   W.M. Watkins and W.Z. Hassid, J. Biol. Chem., 237 (1962) 1432.
114   H. Babad and W.Z. Hassid, J. Biol. Chem., 239 (1964) PC 946.
115   H. Babad and W.Z. Hassid, J. Biol. Chem., 241 (1966) 2672.
116   E.E.B. Smith and G.T. Mills, Biochim. Biophys. Acta, 18 (1960) 152.
117   E.S. Maxwell, H.M. Kalckar and R.M. Burton, Biochim. Biophys. Acta, 18 (1955) 444.
118   L. Glaser and D.H. Brown, J. Biol. Chem., 228 (1957) 729.
119   H.M. Kalckar, in W.D. McElroy and B. Glass (Eds.), Symposium on the Mechanism of Enzyme Action, Baltimore, 1954, p. 675.
120   E.F. Neufeld, V. Ginsburg, E.W. Putman, D. Fanshier and W.Z. Hassid, Arch. Biochem. Biophys., 69 (1957) 602.
121   L.F. Leloir and C.E. Cardini, J. Biol. Chem., 214 (1955) 157.
122   R.C. Bean and W.Z. Hassid, J. Am. Chem. Soc., 77 (1955) 5739.
123   J.B. Pridham and W.Z. Hassid, Plant Physiol., 40 (1965) 984.
124   A.D. Elbein, G.A. Barber and W.Z. Hassid, J. Am. Chem. Soc., 86 (1964) 309.
125   G.A. Barber, A.D. Elbein and W.Z. Hassid, J. Biol. Chem., 239 (1964) 4056.
126   D.S. Feingold, E.F. Neufeld and W.Z. Hassid, J. Biol. Chem., 233 (1958) 783.
127   E. Recondo and L.F. Leloir, Biochem. Biophys. Res. Commun., 6 (1961) 85.
128   C.L. Villemez, T.S. Lin and W.Z. Hassid, Proc. Natl. Acad. Sci. U.S.A., 54 (1965) 1626.

Chapter 59

Biosynthesis of fatty acids and glycerides

1. Fatty acid biosynthesis considered as the reversal of β-oxidation

The controversies concerning fat biosynthesis in animals were narrated in Chapter 42. We have also, in Chapter 50 (Section 4) retraced the general recognition of the conversion of carbohydrates; and indirectly of proteins, into fats, and the suggestion of Magnus-Levy who considered acetaldehyde as the intermediate between carbohydrates and fats. In the same line, Smedley-MacLean and Hoffert (1926) observed that yeast cells receiving acetate as the only source of carbon, accumulated lipids. At the time the mechanism of fatty acid-β-oxidation was recognized (see Chapter 34), and it was assumed, in accordance with the concept of "reversible zymo-hydrolysis", which had considered the biosynthesis of glycerides as an example (see Chapter 51), that synthesis of fatty acids was merely the reversal of β-oxidation.

At the very beginning of the application of isotopic methods, it was conceived that the fatty acids were biosynthesized by the head and tail condensation of acetate units (for a review of the experimentation with labelled acetate, see Bloch [1–3]). Furthermore, the experimentation was interpreted as confirming the concept according to which, in the formation of long chain amino acids, every carbon of the chain originated from acetate carbon (Rittenberg and Bloch [4]). Stadtman and Barker [5] published in 1949 experiments on the anaerobic synthesis of butyrate from ethanol by extracts of *Clostridium kluyveri*.

As stated by Green and Gibson [6]:

"Since the same extract could carry out either the anaerobic synthesis of

Plate 200. Samuel Gurin.

fatty acids from ethanol (by way of acetyl phosphate) or the oxidation of the same fatty acids to acetate (in the form of acetyl phosphate) the case for assuming the identity of the oxidative synthetic process seemed overwhelming" (p. 331).

After 1953, when an enzymological definition of the cycle of fatty acid oxidation was at hand (see Chapter 34), it was commonly stated in textbooks that fatty acid biosynthesis was the reversal of β-oxidation (see Stansly and Beinert [7]; Seubert, Freull and Lynen [8]). Acetyl-CoA was at the time considered as the form of acetate involved (Lipmann [9]; Lynen and Reichert [10]). It was nevertheless observed that while acetyl-CoA, in presence of the enzymes of fatty acid oxidation, was converted into butyryl-CoA, no higher fatty acid could be obtained under such conditions (Brady and Gurin [11]).

2. Attempts at biosynthesis of fatty acids in the presence of β-oxidation enzymes and coenzymes

In 1953, Stansly and Beinert [7] attempted to convert acetyl-CoA to higher fatty acyl derivatives of CoA after adding NADH, a reduced dye and the series of purified enzymes of the fatty oxidation cycle. As the authors were unable to demonstrate the formation of any fatty acyl-CoA higher than butyryl-CoA, their negative results cast doubt on the validity of the concept accepted by many workers (Lynen [12]) according to which the biosynthesis of fatty acids occurs by the reversal of β-oxidation in the presence of the enzymes and coenzymes involved in this catabolic pathway.

3. Biosynthesis of fatty acids in cell-free preparations

This was accomplished for the first time by Stadtman and Barker [5] in 1949. These authors were able to demonstrate the conversion of labelled acetate into fatty acids by a water soluble enzyme preparation obtained from *Clostridium kluyveri*. But these experiments only led to formation of short fatty acids.

Later on, the formation of long-chain fatty acids from acetate in homogenates and in particle-free extracts of pigeon liver ("high speed supernatant" left after centrifugation at $100\,000\,g$ for

30 min) was reported by Brady and Gurin [11] and by Van Baalen and Gurin [13]. The authors showed that the system incorporated labelled acetate predominantly into fatty acids rather than into glycerides. They were able to demonstrate by treatment of the extracts with charcoal, a requirement for ATP, NAD and CoA.

Popjàk and Tietz [14], using soluble preparations of mammary gland of lactating rats or rabbits obtained similar results. They reported that a soluble extract of mammary gland would synthesize fatty acids in the presence of added ATP and α-ketoglutarate. After treatment with Dowex-I ion exchange resin, a requirement for CoA and NAD was revealed.

But the rupture with the concept of a reversal of β-oxidation was accomplished in 1957 when the group of the Institute for Enzyme Research of the University of Wisconsin in Madison, Wis. reported on the purification of enzymes from pigeon liver,

"that carried out the de novo synthesis of long chain fatty acids from acetyl-CoA in the absence of any of the enzymes of the β-oxidation sequence" (Green and Allman [15], p. 38).

Wakil, Porter and Gibson [16] described the preparation and purification of a soluble enzyme system from pigeon liver which actively synthesized long-chain fatty acids from acetate. The authors prepared four enzyme fractions which only on recombination catalysed fatty acid biosynthesis.

The cofactor requirements were defined in a second paper from the same laboratory (Porter, Wakil, Tietz, Jacob and Gibson [17]). With the most purified fractions, the authors observed that the following cofactors are necessary: ATP, CoA, NADH, NADPH, Mn^{2+}, isocitrate and a suitable sulfhydryl compound such as glutathione or cysteine. Glucose-1-P, which was needed in crude systems can be replaced by NADH in more purified systems. Porter and Tietz [18] inquired about the nature of the fatty acids biosynthesized in the purified system. This topic had a previous history which has been retraced by Porter and Tietz referring to a series of different soluble systems.

"Stadtman and Barker [5], employing a soluble extract of *Clostridium kluyveri*, identified the products of synthesis in this system as butyric and caproic acids by partition chromatography and Duclaux distillation. Brady and Gurin [19] separated the long-chain fatty acids ($>C_{10}$) synthesized by a soluble pigeon liver preparation by the copper-lime precipitation method and they showed that the acids formed were unesterified. On decarboxylation of the fatty acids (synthesized from ^{14}C-1-acetate) the content of ^{14}C in the terminal carbon atom of the fatty acids was found to be only slightly above that expected on the basis of successive condensations of acetate units. Popjàk and Tietz [14] separated the fatty acids synthesized from ^{14}C-acetate by homogenates of mammary gland by employing reversed-phase chromatography. Even-numbered, n-saturated acids from C_6 to C_{18} were identified, as well as oleic acid. Most of the radioactivity was distributed among the C_6, C_{14}, C_{16} and oleic acids. Later Hele and Popjàk [20], employing a purified, soluble enzyme system from mammary gland, identified the hydroxamates of β-hydroxybutyrate, butyrate, β-hydroxyoctanoate, and octanoate, after treating the incubation mixture with hydroxylamine" (p. 42).

Porter and Tietz observed that the principal fatty acid synthesized by their pigeon liver system was palmitic acid. Smaller portions of myristic, lauric and decanoic acid were also formed. Short-chain fatty acids (butyrate, hexanoate, etc.) neither accumulated during the synthesis, nor were they incorporated into the longer-chain acids.

Tietz [21] observed that the enzyme system prepared from chicken has essentially the same properties as that of the pigeon system.

4. Absolute requirement for bicarbonate.
Malonyl-CoA as intermediate

Gibson, Titchener and Wakil [22] reported on progress in further purification of the enzyme fractions. They substituted acetyl-CoA for acetate and CoA (thus eliminating the requirement for acetic thiokinase). After these studies the requirement for the biosynthesis of palmitate narrowed down to two protein fractions: these two fractions contained very little or none of the enzymes of the fatty acid oxidation sequence. Furthermore, NADPH alone could provide electrons for the reductive steps. The synthesis of long-chain fatty acids from acetyl-CoA was catalysed by two highly purified protein fractions in the presence of ATP, Mn^{2+}, HCO_3^- and NADPH. Though bicarbonate is absolutely required, $H^{14}CO_3^-$ does not become incorporated into long-chain fatty acids during active synthesis from unlabelled acetyl-CoA.

This fact, together with the knowledge that acetaldehyde is a

Plate 201. Salih J. Wakil.

better precursor than acetate (Brady and Gurin [23]) and that the addition of malonate causes a marked enhancement of fatty acid synthesis (Popjàk and Tietz [24]) suggested the hypothesis that the requirement for bicarbonate (not incorporated) was linked with a biosynthesis of malonyl-CoA from acetyl-CoA, malonyl-CoA playing the role of C_2 donor in the elongation of fatty acids (Wakil [25]). The role of bicarbonate was thus made clear as required for synthesis of malonyl-CoA, the C_2 donor from acetyl-CoA, the most effective acyl-primer.

The part played by malonyl-CoA as an intermediate was also strongly suggested in a paper of Brady [26]. Brady epitomizes as follows his views of the biosynthesis of fatty acids:

"The results of the present experiments are consistent with a mechanism in which the formation of long-chain fatty acids occurs by the successive condensation of aliphatic aldehydes with malonyl CoA in the presence of Mn^{2+} ions. The aliphatic aldehydes are produced by the reduction of the respective thiol ester derivatives of coenzyme A in the presence of NADPH. The aldehydes condense with the activated methylene carbon atom of malonyl CoA, and the free carboxyl carbon atom of malonyl CoA is displaced in the course of the reaction. The product of the condensation reaction would be a β-hydroxy derivative of Coenzyme A which could be dehydrated and then reduced with a second molecule of NADPH. Acetaldehyde may perform a dual role in the system employed in the present investigation by initiating the condensing reaction with malonyl CoA as well as being a substrate for the generation of NADPH for the saturation of ethylenic derivatives and reduction of thiol esters. The necessity of ATP for the conversion of acetyl CoA to fatty acids may be due to the requirement of this nucleotide for the activation of carbon dioxide. Malonyl CoA may be formed from acetyl CoA, ATP and CO_2 in a reaction resembling the carboxylation of propionyl CoA to form methylmalonyl CoA" (pp. 996—997).

Therefore, by 1958, after the contributions by Wakil and by Brady, evidence was available which strongly suggested that malonyl-CoA was an intermediate of fatty acid biosynthesis.

As stated above, the fatty acid synthesizing system had been divided in two fractions R_1 and R_2, which were submitted to more purification attempts (see Strickland [27], p. 89, for literature).

Biotin had been shown to be concentrated in the R_1 fraction and to persist with this fraction during purification attempts. In 1959, Wakil and Ganguly [28,29] showed that the first step in fatty acid biosynthesis is a carboxylation of acetyl-CoA to malonyl-CoA by the biotin-rich R_1 fraction in the presence of ATP and Mn^{2+}. The malonyl-CoA was then converted to palmitate by the R_2 fraction and NADPH. Thus the two fractions serve different functions. One catalyses the carboxylation of acetyl-CoA while the other catalyses the reductive condensation of malonyl-CoA units to long-chain fatty acids with NADPH as reducing agent.

It was also shown by Ganguly [30] that malonyl-CoA serves as a substrate in a wide variety of tissues.

Arguments leading to the conclusion that palmitate is formed from 1 mole of acetyl-CoA and 7 moles of malonyl-CoA were presented by Brady, Bradley and Trams [31].

The problem of the introduction of malonyl-CoA in the biosynthetic pathway was the subject of different schemes presented by Wakil and Ganguly [28], by Brady [31] and by Lynen [32]. Lynen [32] used a partially purified preparation from yeast. He presented evidence that the malonyl-CoA was formed through fixation of CO_2 to acetyl-CoA. He suggested that malonyl-CoA condenses with acetyl-CoA.

In the scheme of Wakil and Ganguly [29], the postulated aceto-malonyl-CoA was believed to undergo reduction, dehydration and reduction before the advent of decarboxylation.

Controversies concerning the question of the fate of malonyl-CoA have been reflected in several general reviews (Green and Gibson [6], Strickland [27]).

When the roadblock was removed by the discovery of the acyl carrier protein, the final resolution of the mechanism of fatty acid biosynthesis went to its completion.

5. Acetyl-CoA carboxylase

As stated above, Wakil and Ganguly [29] showed in 1959 that the first step in fatty acid biosynthesis was a carboxylation of acetyl-CoA to malonyl-CoA by the biotin-rich R_1 fraction in the presence of ATP. The carboxylase involved was shown to be virtually specific for acetyl-CoA (Waite and Wakil [33]). The product of the carboxylation reaction was identified as malonyl-CoA by Gibson, Titchener and Wakil [34]. It was the first biotin enzyme to be identified (see Wakil in Chapter II of Volume 18S of this Treatise, where subsequent studies on the enzyme are reported).

The enzyme has received the systematic name [Acetyl-CoA: carbon dioxide ligase (ADP-forming); 6.4.1.2].

6. Fatty acid synthase

The first overall step is the formation of malonyl-CoA and the second overall step its conversion to palmitic acid. The overall reaction can be stated as follows

$$CH_3COS \; CoA + 7 \; HOOCCH_2COS \; CoA + 14 \; NADPH + 14 \; H^+$$

$$\rightarrow CH_3CH_2(CH_2CH_2)_6CH_2COOH + 7 \; CO_2 + 14 \; NADP^+$$

$$+ \; 8 \; CoA + 6 \; H_2O$$

It is catalysed by a complex of enzymes. This multienzyme complex has been isolated from liver and from yeast (Bressler and Wakil [35]; Lynen [36]; Hsu, Wasson and Porter [37]). Lynen has postulated a stepwise sequence of reactions with protein-bound acyl derivatives. He has focussed his attention on the fatty acid synthase of yeast. As the overall synthesis is susceptible to sulfhydryl reagents, Lynen has postulated that the acyl groups of acetyl-CoA and malonyl-CoA are transferred to SH groups in the synthase complex. The component steps in the biosynthesis pathway were considered as involving the transfer of the acyl groups from the SH of one enzyme to the SH of the next one in the complex. An argument for this theory was found in the demonstration that malonyl-CoA forms a malonyl synthase complex and that acetyl-CoA forms an acetyl synthase complex. Both reactions involved a release of CoA.

As stated by Green and Allman [15]:

"When the synthase is incubated with both malonyl-CoA and acetyl-CoA in the absence of NADPH, an acetoacetyl derivative of the synthase is formed with the concomitant release of CO_2. In turn, the acetoacetyl derivative of the synthase, after separation from excess malonyl- and acetyl-CoA, can be reduced by NADPH to the butyryl derivative.

"The formation of an acetoacetyl-enzyme in the pigeon liver system was also demonstrated by Brodie, Wasson and Porter [38]. This reduction has been shown to proceed stepwise via the β-hydroxybutyryl and trans-$\Delta^{2,3}$-enoyl buturyl derivatives. In principle the steps are identical with those for β-oxidation. The important difference is the β-hydroxy derivative in synthesis is of the D-(—) series, whereas the β-hydroxy derivative in β-oxidation is of the L(+) series. From these demonstrations of the stepwise conversion of aceto-acetyl-synthase to butyryl-synthase with NADPH as reductant, Lynen postulated the existence of two reductases (β-ketoacyl reductase), one hydrase (enoyl hydrase) and one condensing enzyme (malonyl-acetyl condensing enzyme)" (p. 44).

This scheme of Lynen was the first indication of the stepwise character of fatty acid biosynthesis. It must be recognized that the evidence in favour of it was of indirect nature and that the specificity of the enzymes involved appeared different when they were studied in their biological context.

A number of authors confirmed for other systems the concerted condensation-decarboxylation demonstrated for yeast, in the absence of NADPH, by Lynen (literature in Green and Allmann [15]).

7. Discovery of acyl carrier protein

There are several forms of the system of fatty acid synthetases according to the species considered (see Wakil in Vol. 18S). In E. coli for instance, the system is easily resolved into its component enzymes, as demonstrated by Alberts, Goldman and Vagelos [39]; Lennarz, Light and Bloch [40]; Wakil, Pugh and Sauer [41]. Independently it was recognized by Wakil et al. [41] and by Majerus, Alberts and Vagelos [42] that in the E. coli system, the acyl intermediates are bound via a thioester linkage to a low molecular protein. By agreement between K. Bloch, P. Stumpf and S.J. Wakil, this protein was called acyl carrier protein (ACP).

To quote Alberts, Majerus, Talamo and Vagelos [43]:

"The involvement of ACP as the acyl carrier in fatty acid biosynthesis was established earlier with the demonstration of the formation of acetoacetyl-ACP and the conversion of this compound to long-chain fatty acids (Goldman, Alberts and Vagelos [44]). The intermediate steps in the condensation reaction include the transfer of the acetyl and malonyl moieties of acetyl-CoA and malonyl CoA to ACP to form acetyl-ACP and malonyl-ACP respectively. Acetyl-ACP and malonyl-ACP are the condensing units in the biosynthesis of fatty acids. Malonyl-CoA cannot substitute for malonyl-ACP in the condensation reaction. Thus, when malonyl-CoA was incubated with the malonyl transacylase-condensing fraction and chemically prepared acetyl-ACP (where the sulfhydryl groups were totally acylated), condensation did not occur until free ACP was added. With the addition of ACP, the transfer of the malonyl group from CoA to ACP occurred rapidly, forming malonyl-ACP which then condensed with acetyl-ACP to form acetoacetyl-ACP" (p. 1570).

Other proofs of the involvement of acyl-ACP rather than acyl-CoA derivatives are provided by the comparison of the relative reactivity of the various substrates with the enzymes concerned. For instance acetoacetyl-ACP is reduced sixty times faster by β-keto-acyl-ACP reductase than acetoacetyl-CoA.

The intermediates of the fatty acid biosynthesis pathway have been recognized as attached to the thiol group of the 4'-phospho-panthoteine prosthetic group. The acetyl and malonyl groups are transferred from their CoA derivative to ACP and all the ensuing intermediates remain bound to ACP until the palmityl-ACP is hydrolysed to palmitic acid and ACP. ACP from *E. coli* has been purified to homogeneity by Vanaman, Wakil and Hill [45] and its chemistry extensively studied (literature in Wakil, Vol. 18S).

8. Transacylases

The acyl groups of acetyl-CoA are transferred from CoA to ACP in the presence of acetyl-CoA-ACP transacylase, which has been isolated from *E. coli* by Alberts et al. [43] and by Williamson and Wakil [46]. On the other hand, the malonyl-CoA-ACP transacylase, isolated by Williamson and Wakil and by Alberts et al. catalyses the specific transfer of the malonate group from CoA to ACP. The reactions just described correspond to the following

$$CH_3COS\text{-}CoA + ACP\text{-}SH \rightleftharpoons CH_3COS\text{-}ACP + CoA\text{-}SH \qquad (1)$$

$$HOOCCH_2COS\text{-}CoA + ACP\text{-}SH \rightleftharpoons HOOCCH_2COS\text{-}ACP$$
$$+ CoA\text{-}SH \qquad (2)$$

9. Acyl-malonyl-ACP condensing enzyme

From *E. coli*, an enzyme catalysing the condensation of acetyl-ACP (or higher homologues) with malonyl-ACP to form aceto-acetyl-ACP (or higher homologues) and CO_2 was isolated by

References p. 190

Toomey and Wakil [47] and by Alberts, Majerus and Vagelos
[48]. This enzyme catalyses the reaction

$$CH_3COS\text{-}ACP + HOOCCH_2COS\text{-}ACP$$

$$\rightarrow CH_3COCH_2COS\text{-}ACP + CO_2 + ACP\text{-}SH \qquad (3)$$

The enzyme is called acyl-malonyl-ACP condensing enzyme. It has
an absolute specificity for the ACP derivatives (on further studies
on the enzyme, see Wakil in Vol. 18S).

10. β-Ketoacyl-ACP reductase

The acetoacetyl-ACP produced as stated in Section 9 is subsequently reduced to β-hydroxybutyryl-ACP, dehydrated to crotonyl-ACP and reduced to butyryl-ACP. The first of these reactions is

$$CH_3COCH_2COS\text{-}ACP + NADPH + H^+$$

$$\rightleftharpoons D(-)\text{-}CH_3CHOH\ CH_2COS\text{-}ACP + NADP^+ \qquad (4)$$

The enzyme catalysing this reaction, called β-ketoacyl-ACP
reductase has been isolated from *E. coli* and purified by Alberts et
al. [43] and by Toomey and Wakil [49]. Later studies on this
enzyme are reported in the Chapter by Wakil in Vol. 18S.

It should be noted that the product of reaction (4) is the
D-isomer of β-hydroxyacyl-ACP. This is the optical antipode of the
hydroxyacyl-CoA intermediate in β-oxidation.

11. Enoyl-ACP hydrase

The formation of β-hydroxybutyryl-ACP is followed by an elimination of water leading to the formation of a *trans*-enoyl-ACP
derivative.

The enzyme involved, called enoyl-ACP hydrase (see Majerus,
Alberts and Vagelos [50]) acts specifically on hydroxyacyl esters
of ACP of the D-series (in opposition to the less specific crotonase,
the corresponding hydrase of the β-oxidation sequence).

12. Enoyl-ACP reductase

This enzyme catalyses the reduction of *trans*-α,β-unsaturated acyl-ACP (crotonyl-ACP) to the saturated derivative (butyryl-ACP) (see Wakil in Vol. 18S for the properties and forms of this enzyme).

The pathway from acetyl-CoA to butyryl-ACP is shown in Fig. 9 (in *E. coli*, yeast and, extra-mitochondrially, in mammals). Starting from acetyl-ACP, condensing with malonyl-ACP providing a C_2 unit, the pathway of fatty acid biosynthesis, as shown in Fig. 9 continues through a reduction of the β-ketoacyl-ACP by NADPH, a dehydration of the β-hydroxyacyl-ACP and finally a reduction of the *trans*-2-enoyl-ACP to the saturated acyl-ACP.

As shown in Fig. 10 the cycle begins again with the condensation of the newly formed acyl-ACP with malonyl-ACP, leading from butyryl-ACP to hexanoyl-ACP or, generally speaking from *n*-acyl-ACP to *n + 2* acyl-ACP.

13. Elongation of fatty acids in animal tissues

The system described in Fig. 9 and 10 is the classical system of de novo biosynthesis in animal systems, leading to palmitic acid (80%) and stearic acid (20%).

(a) Mitochondrial elongation system

In 1953, Seubert, Freull and Lynen [51] showed that octanoyl-CoA as well as capryl-CoA could be synthesized from hexanyl-CoA and NADPH by rat liver mitochondria. Wakil [52,54], Harlan and Wakil [53,55], Holloway and Wakil [56] have accomplished the elongation with soluble mitochondrial extracts and proved the requirement for NADPH, NADH, acetyl-CoA, acetyl-CoA primer (C8 to 18).

The mechanism appears as a reversed β-oxidation, except that the acyl dehydrogenase is replaced by enoyl-CoA reductase (see Wakil in Vol. 18S).

(b) Microsomal elongation system

That microsomes are able to convert fatty acyl-CoA derivatives to longer chains in the presence of malonyl-CoA and NADPH has

References p. 190

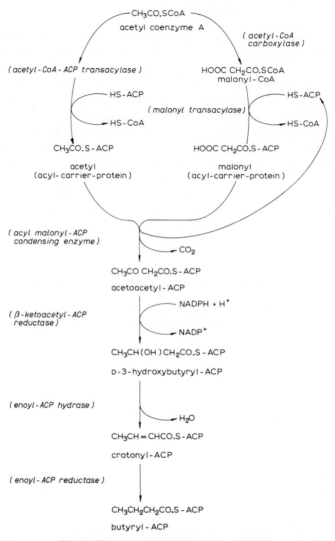

Fig. 9. From acetyl-CoA to butyryl-ACP.

been demonstrated by several authors (Stoffel and Ach [57]; Nutgeren [58]; Mohrhauer, Christiansen, Gan, Deubig and Holman [59]) (on the best available evidence concerning the mechanism involved, see Wakil in Vol. 18S).

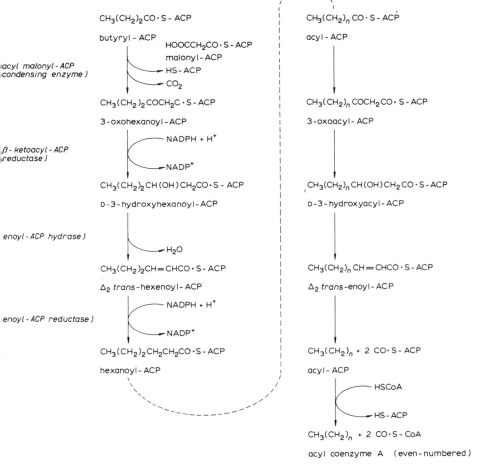

Fig. 10. Fatty acid biosynthesis.

14. Desaturation of fatty acids

That a saturated fatty acid can be converted to the corresponding unsaturated acid in the presence of reduced pyridine nucleotide and oxygen was observed by Bloomfield and Bloch [60] in 1960. The enzymatic (desaturase) system involved was located in microsomes by Bernard, von Bülow-Köster and Wagner [61]. More

References p. 190

recent developments concerning the desaturating capacity of microsomes are retraced by Wakil in Vol. 18S.

15. Biosynthesis of triglycerides

That the glycerol moiety involved in triglyceride biosynthesis mainly derives from the specific isomer L-α-glycerol phosphate was demonstrated by Weiss and Kennedy [62], by Buell and Reiser [63] and by Weiss, Kennedy and Kiyasu [64].

L-α-Glycerol phosphate may be formed from dihydroxyacetone phosphate, an intermediate of the glycolysis pathway by the action of glycerol-3-phosphate dehydrogenase. This enzyme has been crystallized from muscle by Baranowski [65]. But the L-α-glycerol phosphate may also derive from the phosphorylation of glycerol in the presence of ATP and of glycerol kinase. First partially purified by Bublitz and Kennedy [66], the enzyme has been crystallized by Wieland and Suyter [67]. It has received the systematic name [ATP : glycerol-3-phosphotransferase; 2.7.1.30].

Lands and Hart [68] and Lands [69] have studied the reaction of L-α-glycerol phosphate with acyl-CoA using guinea pig microsomes to form diacyl glycerol phosphate (L-α-phosphatidic acid) from α-glycerol phosphate and the CoA esters of stearate and linoleate. As the intermediate lysophosphatidic acid (mono-acylglycerol phosphate) is very rapidly acylated to phosphatidic acid, it does not accumulate. The successive acetylations are catalysed by glycerol phosphate acyltransferase which has received the systematic name [Acyl-CoA : Sn-glycerol-3-phosphate O-acyltransferase; 2.3.1.15].

In 1957, Smith, Weiss and Kennedy [71] have discovered in the liver an enzyme removing phosphate from phosphatidic acid to form D-α,β-diglyceride. This suggested a role of phosphatidic acid in lipid biosynthesis. The enzyme phosphatidate phosphatase catalyses the reaction

an L-α-phosphatidate + H_2O = a D-2,3 (or L-1,2) diglyceride

+ orthophosphate

It has received the systematic name [L-α-Phosphatidate phosphohydrolase; 3.1.3.4].

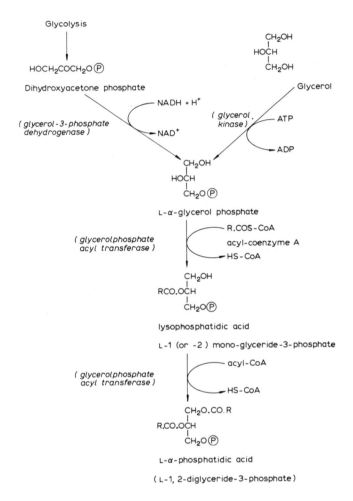

Fig. 11. Biosynthesis of phosphatidic acid (partly after Dagley and Nicholson [70]).

Weiss and Kennedy [62] showed that D-α,β-diglyceride resulting from the dephosphorylation of phosphatidic acid was a precursor of triglycerides. Kennedy and Weiss [72], during the same year, showed that it was also the precursor of the phosphoglycerides, phosphatidyl choline and phosphatidyl ethanolamine.

References p. 190

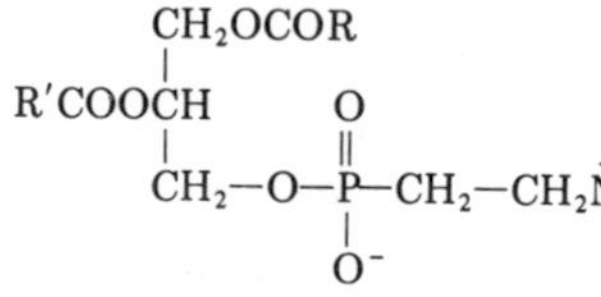

$$CH_2OCOR$$
$$R'COOCH \quad O$$
$$CH_2-O-\overset{\overset{\parallel}{}}{P}-CH_2-CH_2\overset{+}{N}(CH_3)_3$$
$$O^-$$

L-α-lecithin
(L-α-phosphatidyl choline)

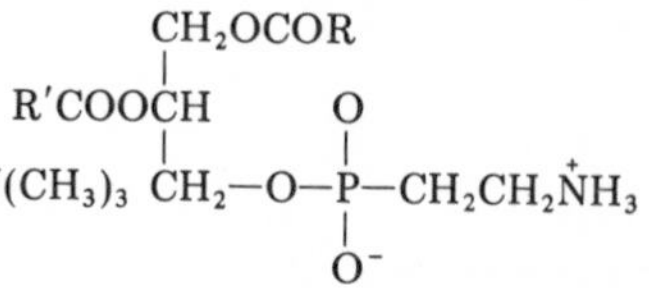

$$CH_2OCOR$$
$$R'COOCH \quad O$$
$$CH_2-O-\overset{\parallel}{P}-CH_2CH_2\overset{+}{N}H_3$$
$$O^-$$

L-α-phosphatidyl ethanolamine

After showing the importance of phosphatidic acid as a source of α,β-diglyceride in the biosynthesis of phosphoglycerides, Weiss and Kennedy [62] investigated the possibility that these diglycerides may also function in triglyceride synthesis. They showed the net synthesis of triglyceride by a particulate enzyme of chicken liver after adding palmityl-CoA and α,β-diglyceride.

The enzyme involved, diacylglycerol acyltransferase (also called diglyceride acyltransferase), catalyses the reaction

acyl-CoA + 1,2-diacylglycerol = CoA + triacylglycerol

(Goldman and Vagelos [73]; Weiss et al. [64]). It has received the

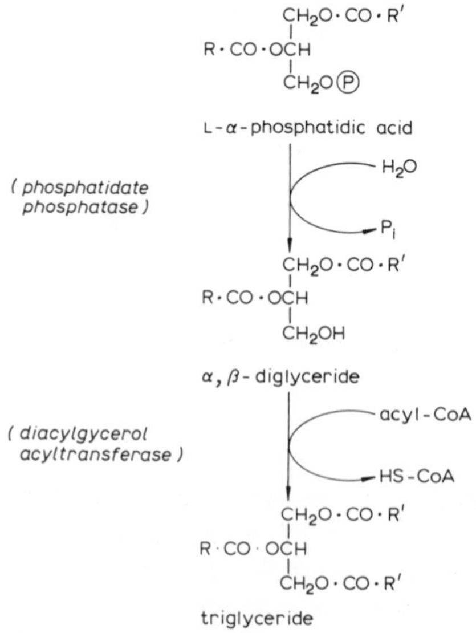

$$CH_2O \cdot CO \cdot R'$$
$$R \cdot CO \cdot OCH$$
$$CH_2O\,\textcircled{P}$$

L-α-phosphatidic acid

(phosphatidate
phosphatase)

$$\quad - H_2O$$
$$\quad - P_i$$

$$CH_2O \cdot CO \cdot R'$$
$$R \cdot CO \cdot OCH$$
$$CH_2OH$$

α,β- diglyceride

(diacylglycerol
acyltransferase)

$$\quad - acyl-CoA$$
$$\quad - HS-CoA$$

$$CH_2O \cdot CO \cdot R'$$
$$R \cdot CO \cdot OCH$$
$$CH_2O \cdot CO \cdot R'$$

triglyceride

Fig. 12. From phosphatidate to triglyceride.

systematic name [Acyl-CoA : 1,2 diacylglycerol O-acyltransferase; 2.3.1.20].

Triglyceride synthesis by the glycerol-phosphate pathway described above has been found to obtain in adipose tissue (see Chapter II, by G.V. Marinetti, in Vol. 18 of this Treatise).

In the intestinal lumen, triglycerides are hydrolysed and resynthesis of triglycerides takes place during the passage from lumen to the lymphatic vessels. To quote Marinetti,

"It was first assumed that the synthesis of triglycerides occurred by the glycerol phosphate pathway which prevailed in liver. However, Clark and Hübscher [74,75] suggested another pathway in which monoglycerides were acylated directly to triglycerides without going through glycerol phosphate and phosphatidic acid as intermediates. This pathway, now coined the "monoglyceride path" was substantiated by Senior and Isselbacher [76] and by Johnston and Brown [77]. Further studies have shown that in intestine the monoglyceride pathway is more important than the glycerol-phosphate pathway for resynthesis of triglycerides" (p. 137). (For more recent developments see Marinetti.)

REFERENCES

1   K. Bloch, Physiol. Rev., 27 (1947) 574.
2   K. Bloch, Cold Spring Harbor Symp., 13 (1948) 29.
3   K. Bloch, Annu. Rev. Biochem., 21 (1952) 273.
4   D. Rittenberg and K. Bloch, J. Biol. Chem., 154 (1944) 311.
5   E.R. Stadtman and H.A. Barker, J. Biol. Chem., 180 (1949) 1085, 1095, 1117.
6   D.D. Green and D.M. Gibson, in D.M. Greenberg (Ed.), Metabolic Pathways, 2nd ed. of Chemical Pathways of Metabolism, Vol. I, 1960, p. 301.
7   P.G. Stansly and H. Beinert, Biochim. Biophys. Acta, 11 (1953) 600.
8   W. Seubert, F. Freull and F. Lynen, Angew. Chem., 69 (1957) 359.
9   F. Lipmann, Harvey Lect., 44 (1950) 99.
10 F. Lynen and E. Reichert, Angew. Chem., 63 (1951) 47.
11 R.O. Brady and S. Gurin, J. Biol. Chem., 187 (1950) 589.
12 F. Lynen, Fed. Proc., 12 (1953) 683.
13 J. Van Baalen and S. Gurin, J. Biol. Chem., 205 (1953) 303.
14 G. Popjàk and A. Tietz, Biochem. J., 56 (1954) 46.
15 D.E. Green and D.W. Allman, in D.M. Greenberg (Ed.), Metabolic Pathways, 3rd ed., Vol. III, 1969, p. 37.
16 S.J. Wakil, J.W. Porter and D.M. Gibson, Biochim. Biophys. Acta, 24 (1957) 453.
17 J.W. Porter, S.J. Wakil, A. Tietz, M.I. Jacob and D.M. Gibson, Biochim. Biophys. Acta, 25 (1957) 35.
18 J.W. Porter and A. Tietz, Biochim. Biophys. Acta, 25 (1957) 41.
19 R.O. Brady and S. Gurin, J. Biol. Chem., 199 (1952) 421.
20 P. Hele and G. Popjàk, Biochim. Biophys. Acta, 18 (1955) 294.
21 A. Tietz, Biochim. Biophys. Acta, 25 (1957) 303.
22 D.M. Gibson, E. Titchener and S.J. Wakil, J. Am. Chem. Soc., 80 (1958) 2908.
23 R.O. Brady and S. Gurin, J. Biol. Chem., 189 (1951) 371.
24 G. Popjàk and A. Tietz, Biochem. J., 60 (1955) 147.
25 S.J. Wakil, J. Am. Chem. Soc., 80 (1958) 6465.
26 R.O. Brady, Proc. Natl. Acad. Sci. USA, 44 (1958) 993.
27 K.P. Strickland, in P. Bernfeld (Ed.), Biogenesis of Natural Compounds, Oxford, London, New York, Paris, 1963, p. 83.
28 S.J. Wakil and J. Ganguly, Fed. Proc., 18 (1959) 346.
29 S.J. Wakil and J. Ganguly, J. Am. Chem. Soc., 81 (1959) 2597.
30 J. Ganguly, Biochim. Biophys. Acta, 40 (1960) 110.
31 R.O. Brady, R.M. Bradley and E.G. Trams, J. Biol. Chem., 235 (1960) 3093.
32 F. Lynen, J. Cell. Comp. Physiol., 54, suppl. (1939) 33.
33 M. Waite and S.J. Wakil, J. Biol. Chem., 237 (1962) 2750.
34 D.M. Gibson, E.B. Titchener and S.J. Wakil, J. Am. Chem. Soc., 80 (1958) 2908.
35 R. Bressler and S.J. Wakil, J. Biol. Chem., 236 (1961) 1643.

36 F. Lynen, Fed. Proc., 20 (1961) 941.
37 R.Y. Hsu, G. Wasson and J.W. Porter, J. Biol. Chem., 240 (1965) 3736.
38 J.D. Brodie, G. Wasson and J.W. Porter, J. Biol. Chem., 238 (1963) 1294.
39 A.W. Alberts, P. Goldman and P.R. Vagelos, J. Biol. Chem., 238 (1963) 557.
40 W.J. Lennarz, R.J. Light and K. Bloch, Proc. Natl. Acad. Sci. USA, 48 (1962) 840.
41 S.J. Wakil, E.L. Pugh and F. Sauer, Proc. Natl. Acad. Sci. USA, 52 (1964) 106.
42 P.W. Majerus, A.W. Alberts and P.R. Vagelos, Proc. Natl. Acad. Sci. USA, 51 (1964) 1231.
43 A.W. Alberts, P.W. Majerus, B. Talamo and P.R. Vagelos, Biochemistry, 3 (1964) 1563.
44 P. Goldman, A.W. Alberts and P.R. Vagelos, J. Biol. Chem., 238 (1963) 3579.
45 T.C. Vanaman, S.J. Wakil and R.L. Hill, J. Biol. Chem., 243 (1968) 6409, 6420.
46 I.P. Williamson and S.J. Wakil, J. Biol. Chem., 241 (1966) 2326.
47 R.E. Toomey and S.J. Wakil, J. Biol. Chem., 241 (1966) 1159.
48 A.W. Alberts, P.W. Majerus and P.R. Vagelos, Biochemistry, 4 (1965) 2265.
49 R.E. Toomey and S.J. Wakil, Biochim. Biophys. Acta, 116 (1966) 189.
50 P.W. Majerus, A.W. Alberts and P.R. Vagelos, J. Biol. Chem., 240 (1965) 618.
51 W. Seubert, F. Freull and F. Lynen, Biochim. Biophys. Acta, 11 (1953) 600.
52 S.J. Wakil, J. Lipid Res., 2 (1961) 1.
53 W.R. Harlan and S.J. Wakil, Biochem. Biophys. Res. Commun., 8 (1962) 131.
54 S.J. Wakil, Am. J. Chem. Nutr., 8 (1960) 630.
55 W.R. Harlan and S.J. Wakil, J. Biol. Chem., 238 (1963) 3216.
56 P.W. Holloway and S.J. Wakil, J. Biol. Chem., 239 (1964) 2489.
57 W. Stoffel and K.L. Ach, Z. Physiol. Chem., 337 (1964) 123.
58 D.H. Nutgeren, Biochim. Biophys. Acta, 106 (1965) 280.
59 H. Mohrhauer, K. Christiansen, M.V. Gan, M. Deubig and R.T. Holman, J. Biol. Chem., 242 (1967) 4507.
60 D.K. Bloomfield and K. Bloch, J. Biol. Chem., 235 (1960) 337.
61 K. Bernard, J. von Bülow-Köster and H. Wagner, Helv. Chim. Acta, 42 (1959) 152.
62 S.B. Weiss and E.P. Kennedy, J. Am. Chem. Soc., 78 (1956) 3550.
63 G.C. Buell and R. Reiser, J. Biol. Chem., 234 (1959) 217.
64 S.B. Weiss, E.P. Kennedy and J.Y. Kiyasu, J. Biol. Chem., 235 (1960) 40.
65 T. Baranowski, J. Biol. Chem., 180 (1949) 535.
66 C. Bublitz and E.P. Kennedy, J. Biol. Chem., 211 (1954) 951.
67 O. Wieland and M. Suyter, Biochem. Z., 329 (1957) 320.
68 W.E.M. Lands and P. Hart, J. Lipid Res., 5 (1964) 81.

69   W.E. Lands, Annu. Rev. Biochem., 34 (1965) 313.
70   S. Dagley and D.E. Nicholson, An Introduction to Metabolic Pathways, Oxford and Edinburgh, 1970.
71   S.W. Smith, S.B. Weiss and E.P. Kennedy, J. Biol. Chem., 228 (1957) 915.
72   E.P. Kennedy and S.B. Weiss, J. Biol. Chem., 222 (1956) 193.
73   P. Goldman and P.R. Vagelos, J. Biol. Chem., 236 (1961) 2620.
74   B. Clark and G. Hübscher, Nature (London) 185 (1960) 35.
75   B. Clark and G. Hübscher, Biochim. Biophys. Acta, 46 (1961) 479.
76   J.R. Senior and K.J. Isselbacher, J. Biol. Chem., 237 (1962) 1454.
77   J.M. Johnston and J.L. Brown, Biochim. Biophys. Acta, 59 (1962) 500.

Chapter 60

Biosynthesis of tetrapyrroles and of corrinoids

1. Introduction

It has long been known that there are, in living matter, two pre-dominant pigments. One is the green pigment chlorophyll, the other the red pigment now called haem. In 1880, Hoppe-Seyler recognized that these two pigments were chemically related.

During the 1920s and 1930s the chemistry of chlorophyll and of haem was actively studied by a number of researchers, among whom were Nencki and Zaleski, Piloty, Küster, Willstätter, etc. (see Lieben pp. 288—295).

That haem and chlorophyll were both tetrapyrroles (porphyrins) was recognized by Willstätter and his colleagues. Küster, in 1913, proposed that the four pyrrole rings of a porphyrin are linked by four carbon atoms to form a 16 carbon ring. This suggestion was opposed by Willstätter and by Hans Fischer, and Küster aban-doned it in 1923, but the brilliant work of Fischer in the field of the chemical synthesis of porphyrins proved in 1929 that Küster had been right:

Pyrrole   Porphine

Plate 202. Richard Willstätter.

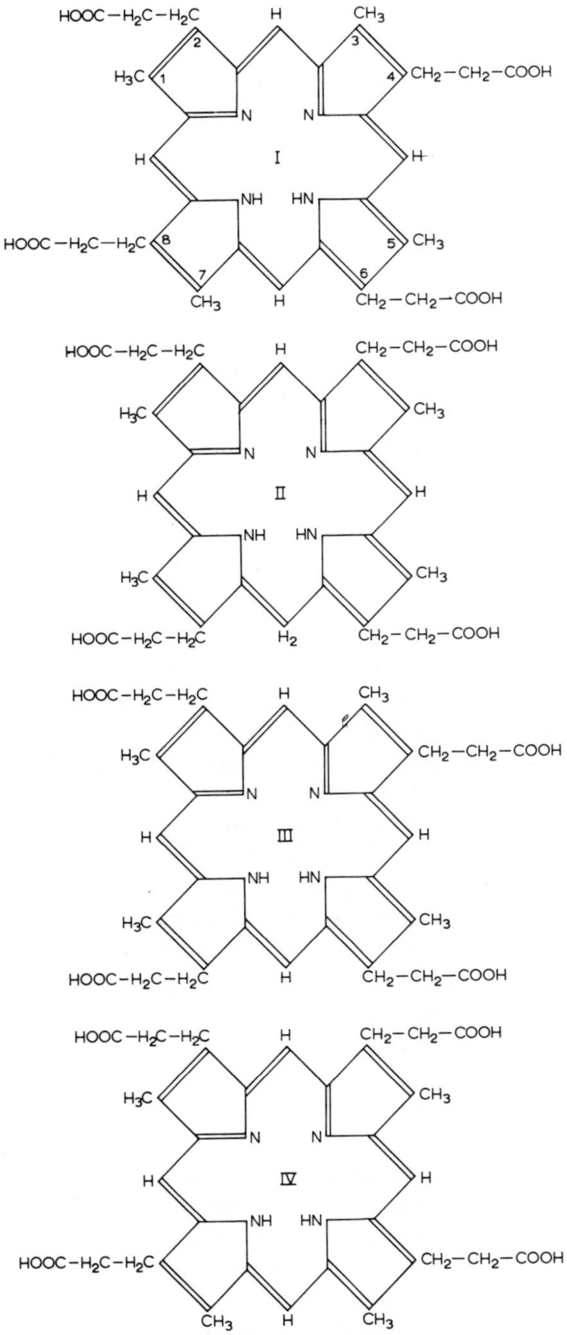

Fig. 13. Coproporphyrins.

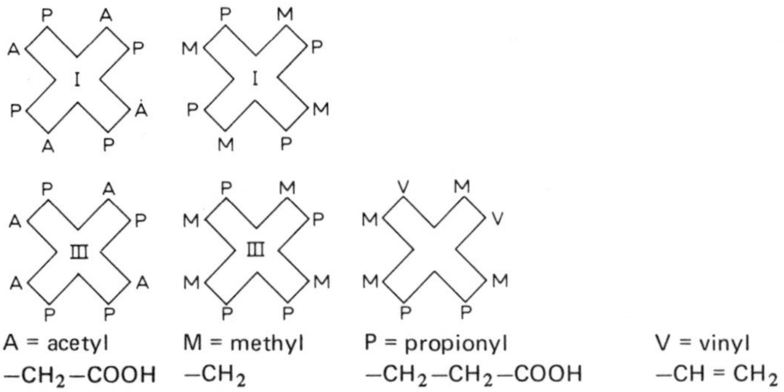

A = acetyl   M = methyl   P = propionyl   V = vinyl
−CH₂−COOH   −CH₂   −CH₂−CH₂−COOH   −CH = CH₂

Fig. 14. Uroporphyrins I and III, coproporphyrins I and III, and protoporphyrin IX (from left to right).

The structure of porphyrins can be schematized by the addition
of diverse radicals to a hypothetical tetrapyrrolic nucleus,
porphine in which the four pyrrole rings are bound through
methene groups (=CH−) designated by the Greek letters α, β, γ
and δ. Relative positions of substituting groups lead to the existence of isomers. For instance, coproporphyrins, a name recalling
the finding of such compounds in faeces, are tetramethyl-tetra-
propionyl-porphines, of which four isomers can exist with different positions of the four methyl-groups:
I: 1, 3, 5, 7; II: 1, 4, 5, 8; III: 1, 3, 5, 8; IV: 1, 4, 6, 7. The
isomers I and III of coproporphyrins have been isolated from
faeces. Uroporphyrins or tetraacetyl-tetrapropionyl-porphyrins are
present in organisms in the form of isomers I and III. On the
chemistry, biochemistry and physico-chemistry of porphyrins, the
reader is referred to Chapters I and II of Vol. 9 of this Treatise.

2. Metabolic sources of the atoms of porphyrins

We have recalled in Chapter 50 that it has long been known that
animals could biosynthesize porphyrins. Speculative schemes
based on comparative chemical data had led to the suggestion that
glutamic acid, proline, or tryptophan might be possible precursors.
    The first direct approach to protoporphyrin biosynthesis was
due to Bloch and Rittenberg [1] who showed, in the Department

of Biochemistry of Columbia University, that the administration of deuterio-acetate to rats results in the formation of deuterio-haemin. This finding showed that carbon atoms of the side chain of haem are derived from acetate for none of the carbon atoms of the pyrrole rings are bonded to hydrogen. During the same year, Shemin and Rittenberg [2,3] observed in a fundamental paper that the precursor of at least some of the nitrogen atoms of proto-porphyrin, in man, is the nitrogen atom of ^{13}N-labelled glycine. As a further study on man was expensive, Shemin and Rittenberg [6] used the rat. They studied the utilization of isotopic glycine, glutamic acid, proline, leucine and ammonia. It had been suggested by Abderhalden [7], on the basis of the similarity of their structure to that of the pyrroles, that proline and pyrrolidonecarboxylic acid might be the precursors of protoporphyrin. Leucine was selected as a representative α-amino acid whose intact carbon chain is unlikely to be used. Ammonia was used to test the non-specific utilization of nitrogen liberated by the deamination of amino acids. The results led to the conclusion that the nitrogen of glycine is used directly while the nitrogen of the other compounds is utilized indirectly by way of glycine. Since the utilization of the nitrogen atom by transamination was exceedingly unlikely, it appeared that the α-carbon atom is also utilized and probably the carboxyl group of glycine as well.

Other nitrogen-labelled compounds such as ammonia, leucine, proline, glutamic acid, histidine (Tesar and Rittenberg [4]), aspartic acid (Wu and Rittenberg [5]), were shown to be ineffective in experiments on rats or on rabbits.

The authors also considered, since glycine contains but two carbon atoms, that other compounds are involved in the biosynthesis of pyrrole. They suggested that it was possible that the biosynthesis involves the condensation of glycine with a β-ketoaldehyde as shown in Fig. 15. According to this formulation the α-carbon of the pyrrole ring and the carbon atoms of the methene bridges would be derived from glycine. The theory was based on an analogy with a reaction described by Fischer and Fink [8], in which a pyrrole-like undefined substance was formed as a result of the condensation of glycine and a β-aldehyde.

That the carboxyl carbon of glycine is not incorporated in haemin was demonstrated by Grinstein, Kamen and Moore [9] who used glycine labelled with ^{14}C on the carboxyl carbon.

References p. 235

Plate 203. David Shemin.

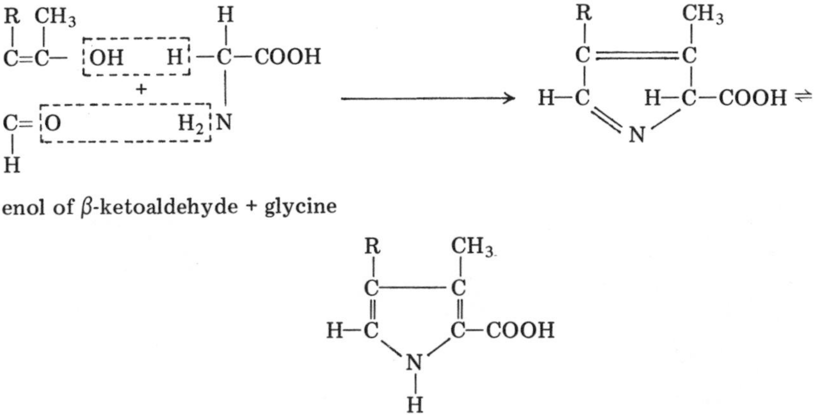

R CH₃ ... H
│ │
C=C— ⌐OH   H⌐—C—COOH   →
       └────┘ │
    +   R   CH₃
C=⌐O   H₂⌐N   │   │
│ └─────────┘   C ══════ C
H   H—C   H—C—COOH ⇌
                              ╲   ╱
                               N

enol of β-ketoaldehyde + glycine

        R   CH₃
        │   │
        C────────C
        ‖   ‖
    H—C   C—COOH
        ╲   ╱
          N
          │
          H

Fig. 15.

The incorporation of α-carbon was confirmed by Altman, Casarett, Masters, Noonan and Salomon [10]. It remained possible that the immediate precursor of part of the haemin molecule is not glycine but a derivative of glycine, one possible pathway appearing in the conversion of glycine to serine and the decarboxylation of this compound to ethanolamine.

About this time, isotopic nitrogen became available to Albert Neuberger at the National Institute of Medical Research, which was then at Hampstead, London and he decided, together with Dr. Helen Muir to test the hypothesis of the participation of ethanolamine, feeding rats with ethanolamine labelled with ¹⁵N, and concluding that the nitrogen of ethanolamine is not used to any marked extent in the biosynthesis of protoporphyrin. Were all the nitrogen atoms specifically derived from glycine? This was shown to be the case and reported in a paper published in 1949 (Muir and Neuberger [11]). Are the carbon atoms of glycine also involved? Muir and Neuberger proceeded to carbon labelling and confirmed in the first place that the carboxyl carbon atom of glycine was not used, but that for each nitrogen atom of glycine two α-carbon atoms were used. One of these gave rise to the meso-carbon atom linking the pyrrolic rings. The results also indicated that all four mesocarbon atoms were derived from the meso-carbon atom of glycine. This was originally reported in a short

Plate 204. Albert Neuberger.

communication of 1949. In their 1950 paper [14], the authors also reported that the methyl carbon of acetate was used in the vinyl group and they speculated that this would occur via α-oxoglutaric acid. Other speculations in this paper of 1950 turned out to be wrong, apart from the suggestion that two α-oxoglutaric groups were incorporated into each pyrrole ring.

In the same year 1950, Shemin's group reported similar results (Shemin [12a], Wittenberg and Shemin [12b], Radin, Rittenberg and Shemin [13]).

Glycine is specifically utilized for porphyrin synthesis. Are the carbon atoms also involved?

Muir and Neuberger, in their paper on the origin of the methene carbon atoms, degraded mesoporphyrin labelled with [2-^{14}C]-glycine and obtained by oxidation three products: carbon dioxide which was mainly derived from the methene carbons, methylethylmaleic acid imide derived from rings I and II and haematinic acid derived from rings III and IV.

Wittenberg and Shemin incubated duck erythrocytes with glycine labelled in the α-carbon atom with ^{14}C. The labelled haemin isolated was degraded to methylethylmaleimide, derived from pyrrole rings A and B and to haematinic acid, derived from pyrrole rings C and D. The activities of both compounds were equal. They degraded these derivatives and determined the carbon atom derived from [2-^{14}C]glycine. In methylethylmaleimide, only carbons A_2 and B_2 were active, and in haematinic acid only C_2 and D_2.

As stated by the authors in their conclusions (p. 115):

"The total activity of these carbon atoms (A_2, B_2, C_2 and D_2) accounts for only 50% of the total activity of the porphyrin. The remaining 50% of the activity must reside in the carbon atoms of the methene bridges".

In the discussion (pp. 112—113), the authors write as follows on the same subject:

"Since the total activity of the porphyrin was 5236, there are 2620 (5236—2616) total counts accounted for. This activity must reside in the carbon atoms of the methene bridges, since these 4 carbon atoms are the only ones which were not examined . . . From the radioactivity of these carbon atoms the number of the methene carbon atoms derived from the α-carbon atom of glycine may be deduced.

"The average activity of the methene carbon atoms is 655 c.p.m. (2620/4), a figure equal to that of the 4 active carbon atoms of the pyrrole units . . . The activity of any of the 4 methene carbon atoms could be equal to, less than, or greater than the average activity. However, since the radioactivity of the carbon atoms in position 2 of the pyrroles is the same as that of the average of the methene carbon atoms, it appears reasonably certain that these 8 carbons all arise from the α-carbon atom of glycine and that the radioactivity of the 4 methene carbon atoms and of the carbon atoms in position 2 of the pyrroles is identical. The location of these carbon atoms in the protoporphyrin molecule is, therefore in positions A_2, B_2, C_2 and D_2 and in the four carbon atoms of the methene bridges as shown in Fig. 16. That indeed 8 carbon atoms of the protoporphyrin are derived from the α-carbon of glycine was indicated by a previous independent approach" [13].

Muir and Neuberger published in 1949 [15] and 1950 [14] the results of the experiments they had performed to find the distribution, within the protoporphyrin molecule, of labelled carbons

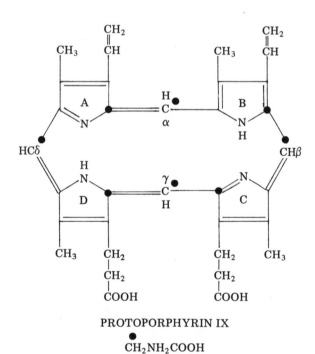

PROTOPORPHYRIN IX
●
CH_2NH_2COOH

Fig. 16. Position in protoporphyrin of the carbons derived from the α-carbon atom of glycine.

derived from likely precursors, such as $^{14}CH_3 \cdot COONa$, $CH_3{}^{14}C_2$ COONa, $NH_3 \cdot {}^{14}CH_2 \cdot COOH$, $^{15}NH_2 \cdot {}^{14}CH_2COOH$, $^{14}CH_2OH \cdot CH(NH_2) \cdot COOH$ and $H^{14}COONa$. They started from the knowledge that the nitrogen atom (Shemin and Rittenberg [3,6]) and the methylene carbon atom (Altman et al. [10]) but not the carboxyl carbon atom (Grinstein et al. [9]) are used in porphyrin biosynthesis. That glycine is the source of at least some of the methene carbons is indicated by the fact that neither of the two acetates, nor formate, nor serine produce significant activity in the methene carbons. That formate is not utilized for the biosynthesis of haem by fowl erythrocytes was shown by Bufton, Bentley and Rimington [16].

At this time it had been shown by Muir and Neuberger and by Shemin and Rittenberg (see above), that glycine must provide at least two of the four nitrogen atoms of protoporphyrin. To quote Muir and Neuberger [14] (p. 101):

"However, as no other nitrogen precursor has yet been found, amongst numerous compounds tested, it is likely that all four nitrogen atoms of the porphyrin are derived from glycine. This is supported by the observation that the newly formed porphyrins of a congenital porphyric have a higher ^{15}N content, even than the glycine excreted as hippuric acid (Gray and Neuberger [17]).

"By using $^{15}NH_2 \cdot {}^{14}CH_2COOH$, it was found that at least twice as many labelled carbon as nitrogen atoms appear in protoporphyrin".

The authors conclude that the activity of the carboxyl carbon of acetate was found chiefly in the haematinic acid, while one or both of the vinyl carbon atoms are derived from the methyl carbon of acetate.

Muir and Neuberger conclude that the methene carbons are mainly, if not exclusively, derived from the methylene carbon of glycine which provides probably eight methylene carbon atoms and four nitrogen atoms for the synthesis. The authors postulate that two molecules of glycine condense with a loss of one nitrogen atom, to a 4-carbon compound. This compound must subsequently lose its two carboxyl groups, as they are not utilized.

To quote Muir and Neuberger (p. 102):

"It is reasonable to suppose that the pyrrole rings are built up by the condensation of the four-carbon fragment with one of the intermediates of the

Krebs cycle. It has been suggested by Lemberg and Legge [18] and Lemberg (private communication) that α-ketoglutarate might be the compound involved. On this basis, the utilization of acetate must occur by its entry into the tricarboxylic acid cycle. Slightly modifying Lemberg's suggestion it is proposed that the four-carbon compound derived from glycine condenses with two molecules of α-ketoglutarate to form a substituted pyrrole-2-carboxylic acid".

Radin et al. [13,19a] demonstrated that 8 carbon atoms of proto-porphyrin are derived from the α-carbon of glycine. They consider that, as the porphyrin molecule contains 34 carbons, other compounds are likely to participate. As it has been observed before by Bloch and Rittenberg that deuteriohaemin is formed when deuterioacetate is fed to rats, they turned to a closer investigation of the role of acetate in the biosynthesis of porphyrins. At this time, ^{14}C had become available and the authors took advantage of the previous finding [19b] that nucleated erythrocytes of birds (ducks) can synthesize haem from glycine. They employed carboxyl and methyl ^{14}C-labelled acetic acid, pyruvate labelled with ^{14}C in the α-position, acetone labelled with ^{14}C in the methyl group, and $^{14}CO_2$. While neither acetone, nor CO_2 were utilized for the biosynthesis, both carbons of acetate as well as the α-carbon of pyruvate were utilized. The authors ascertained by degradation some of the positions in the porphyrin, which are derived from the carbons of the acetate. They concluded that both the carboxyl and the methyl groups of acetate are used in porphyrin biosynthesis, that the carboxyl group of acetic acid is a source of the two carboxyl groups of haem and also contributes to at least four of the carbon atoms of the porphyrin, and that the methyl group of acetate is more widely used for porphyrin synthesis. They concluded that probably all the carbon atoms of haem are derived from acetate and glycine.

After 8 carbon atoms had been, as stated above, identified as originating from the α-carbon of glycine, the origin of the other 26 carbons of the protoporphyrin molecule remained to be determined. When Bloch and Rittenberg [1] had given deuterioacetic acid (CD_3COOH) to a rat, they isolated haemin which showed that at least some of the side-chain carbon atoms were derived from the methyl group of acetate, since these are the only carbon atoms bound to hydrogen.

Shemin and Wittenberg [20] incubated duck erythrocytes

separately with [^{14}C]methyl-labelled acetate and with [^{14}C]carboxyl-labelled acetate. They degraded the resulting ^{14}C-labelled haemin by a method which enabled them to isolate each carbon atom from a particular position. They found that the 26 atoms were all derived from acetate and concluded that a four carbon asymmetric compound arising from citric acid, a succinyl derivative, is the source of the remaining 26 carbon atoms. This supported a previous suggestion of Wittenberg and Shemin [21] that a common precursor pyrrole constitutes the source of all four pyrrole rings. The authors found that ^{15}N-labelled glycine was equally well utilized for the biosynthesis of the different pyrroles of protoporphyrin (methyl and vinyl-bearing or methyl- and propionic acid-bearing). This principle received more confirmation later on. It was shown in the experiments of Wittenberg and Shemin [12b] that the α-carbon atoms of glycine found in the pyrrole ring are in the α-position in the vinyl and propionic acid side chain. As stated by Shemin [22] (p. 262):

"This finding supported the suggestion of a common precursor pyrrole being formed and led to the postulation that the vinyl side chain arose from propionic acid side chains by decarboxylation and dehydrogenation".

An important paper of Shemin and Wittenberg [20] appeared in 1951. When Shemin and Wittenberg started this work, it was known by the work of Muir and Neuberger and of Wittenberg and Shemin, as stated above, that the nitrogen atoms of glycine are utilized for both types of pyrrole rings (rings A and B, rings C and D in Fig. 17). It has been shown that four α-carbons of glycine occupy comparable positions in both types of pyrrole rings (carbon atoms numbered 2 in Fig. 17) and that 4 more carbon atoms from glycine are utilized in methene bridges (Muir and Neuberger, Wittenberg and Shemin). It was also known that the methyl groups (carbon 6 in Fig. 17) and the β carbons of the pyrrole rings to which the methyl groups are attached (carbon 4) are derived from the methyl group of acetic acid, whereas the carboxyl groups of the porphyrin were derived from the carboxyl group of acetate (Radin et al. [13]).
incubated duck erythrocytes with ^{14}CH$_3$COOH and CH$_3$^{14}COOH and degraded the haem completely. With ^{14}CH$_3$COOH they found C-3, C-5; C-4, C-8; C-6, C-9 had the same activity in all four

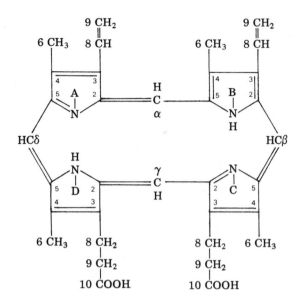

PROTOPORPHYRIN IX

Fig. 17. Numbering of atoms in porphyrin IX according to Shemin and Wittenberg [20].

pyrrole units. With $CH_3{}^{14}COOH$, only the C-10 was highly labelled. C-3 and C-5 had only one-tenth of the activity of C-10; and no labelling was found in C-2, C-4, C-8, C-6 and C-9. To quote the authors [20] (p. 327):

"Examination of the ^{14}C activities of the different carbon atoms in the porphyrin reveals a pattern and relationship among these different carbon atoms. In the porphyrin made from methyl-labelled acetate, not only do the methyl group carbon atoms (A6, B6, C6, D6) of each pair of pyrroles have similar activities, but their activity is also equal to that of the terminal carbon atoms of the vinyl groups (A9, B9) and to the corresponding carbon atoms of the propionic side chains (C9, D9). The methyl-bearing carbon atoms in all the pyrrole rings (A4, B4, C4, D4) have the same activity as the proximal carbon atoms of the vinyl side chains of rings A and B (A8, B8) and their counterparts in the propionic acid side chains of rings C and D (C8, D8). Also the carbon atoms numbered 5 in the pyrrole rings (A5, B5, C5, D5) have the same activity as all the ring carbon atoms to which the longer side chains are attached (A3, B3, C3, D3). Similarly, in the experiment with carboxyl-labelled acetate all carbon atoms numbered 5 and 3 have the same activity. These data strongly suggest that not only are the two types of pyrrole unit in protoporphyrin made from the same precursors but also *that in each pyrrole ring the same compound is utilized for the methyl side of the structure and for the vinyl and propionic acid sides of the structure*. This conclusion is supported by the finding, as pointed out earlier, that the pyruvic acid and α-keto-butyric acid fragments of the pyrrole units have the same activities in the experiments with methyl-labelled acetate and that with carboxyl-labelled acetate''.

By 1951, after the work with isotopes reviewed above, the meta-bolic origin of the atoms of porphyrin IX was established.

3. Relationship of the tricarboxylic acid cycle and porphyrin formation

Shemin and Wittenberg [20] conclude from their experimental data that if each side of each pyrrole utilizes the same compound, the precursor which condenses with glycine can be either a 3- or a 4-carbon compound. Noting the quantitative distribution of ^{14}C among carbon atoms in the experiments, it can be seen that a 3-carbon compound would satisfy the data as precursor of the methyl side of the pyrrole units (C-6, C-4, C-5) as well as precursor of the vinyl side (C-9, C-8, C-3), excluding C-2 known to derive from the α-C of glycine. But for the propionic acid sides (C-10, C-9, C-8, C-3) a 4-carbon compound is required. The distribution of activity in the porphyrin in the experiments eliminates the dicarboxylic acids (succinate, fumarate, malate, oxaloacetate) and pyruvic acid as direct precursors:

"Carboxyl-labelled acetate would label equally the carboxyl group of these dicarboxylic acids, and, if any of these acids were utilized directly, the car-boxyl groups of protoporphyrin (C10, D10) and the carbon atoms numbered 3 and 5 would have had equal activities. If pyruvate formed in the methyl-labelled acetate experiment was utilized for porphyrin formation, carbon atoms 6, 4, 9 and 8 would be equally labelled, whereas carbon atoms 5 and 3 would be lower than carbon atoms 4 and 8" [20] (p. 330).

As stated above, Lemberg and Legge have suggested that 2 moles of α-ketoglutaric acid could combine with glycine with the elimination of the α-carboxyl group of the keto acid, to form a pyrrole bearing acetic acid and propionic acid side chains. We have also mentioned the modified version of this suggestion proposed by Muir and Neuberger who suggested that the keto acid condenses with hydroxy-aspartic acid. This view, as concluded by Shemin and Wittenberg, is not compatible with the distribution of δ-carbon of glycine in the porphyrin as shown in their 1950 paper [12].

Shemin and Wittenberg [20] suggest that a 4-carbon compound arising from both α-ketoglutaric and succinic acid would explain their result. They consider that this compound may be a succinyl-coenzyme complex which may be formed in a manner analogous to the well known formation of acetyl-CoA from pyruvate and acetate.

Experimentally, Shemin and Wittenberg obtained with methyl-labelled acetate evidence of the formation of an unsymmetrical 4-carbon compound from α-ketoglutarate and succinate. The methyl-labelled acetate is converted to α-ketoglutarate with ^{14}C activity in all its carbon atoms except the γ-carboxyl.

"However, the carboxyl groups of protoporphyrin made from methyl-labelled acetate have ^{14}C activity equivalent to that found in positions 3 and 5 in protoporphyrin made from carboxyl-labelled acetate" (p. 332).

While the same mechanism would account for this as well as for the findings related above, the authors consider the possibility of a conversion of pyruvate to acetate, which they discard however on the basis that the reaction does not take place appreciably in the biological system considered (Radin et al. [19a]).

Considering the formation of protoporphyrin IX, Shemin and Wittenberg [20] suggested at this time that the precursor utilized for pyrrole formation is an unsymmetrical 4-carbon compound resulting from α-ketoglutaric and succinic acids (possibly succinyl-CoA). Two molecules of the succinyl derivative would condense with glycine to yield α-carboxy-β-carboxymethyl-β'-carboxyethyl-pyrrole. Four of these pyrroles are condensed with the loss of the α-carboxyl of the pyrrole and with the addition of a compound originating from the α-carbon of glycine. Uroporphyrin III would be formed, which by decarboxylation of the carboxymethyl side

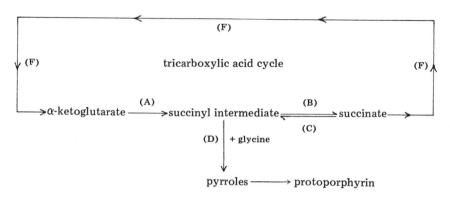

Fig. 18. (Shemin and Kumin [26]). The relationship of the tricarboxylic acid cycle and protoporphyrin formation.

chain would give coproporphyrin III. Protoporphyrin IX would result from decarboxylation and dehydrogenation of the propionic acid side chains of pyrroles A and B of this compound.

In the paper just mentioned, Shemin and Wittenberg suggested that acetate was utilized in pyrrole biosynthesis by being converted into a succinyl intermediate, possibly a succinyl-coenzyme complex, and an intermediate of the α-ketoglutarate-succinate reaction. It happened that this reaction was studied at the time by other authors. Kaufman [23] found that the oxidation of α-ketoglutarate is coenzyme A-dependent. Sanadi and Littlefield [24,25] showed the formation of an "active" succinate from α-ketoglutarate in the presence of CoA, and the ability of this "active" compound to succinylate sulphanilamide. Since, from the knowledge acquired in 1952, it appeared that the "active" succinate involved in pyrrole biosynthesis arises not only from α-ketoglutarate but also from succinate (Shemin and Wittenberg [20]), Shemin and Kumin [26] represented the relationship of the citric acid cycle formation by the diagram of Fig. 18. Shemin and Kumin experimented with both carboxyl- and methylene-labelled succinate, with both intact and haemolysed duck erythrocytes. The position of the problem is stated as starting from the diagram of Fig. 18 suggesting the existence of a reaction (reaction C) not yet described.

As stated by the authors [26] (pp. 827—829):

"In this communication direct evidence is presented for the above scheme:
that a 4-carbon atom compound is the precursor of 2-carbon atoms of protoporphyrin and that this succinyl intermediate can arise from succinate via
both Reactions C and F (Fig. 18) in the tricarboxylic acid cycle.

"It may be well to point out that, although succinic acid should give rise
to the succinyl intermediate by the two pathways (Fig. 18), the labelling of
protoporphyrin from radioactive succinate will depend on the position of
^{14}C in the succinate. It was possible to demonstrate both Reactions C and F
and to differentiate between them by investigating the synthesis of protoporphyrin from ^{14}C-carboxyl-labelled and ^{14}C-methylene-labelled succinate
in the presence and absence of malonate. By determining the ^{14}C distribution
in protoporphyrin synthesized from ^{14}C-carboxyl-labelled succinate with and
without malonate more evidence for Reaction C was obtained.

"Washed duck red blood cells or hemolyzed duck red blood cells were
incubated with equal concentrations of ^{14}C-carboxyl-labelled and ^{14}C-
methylene-labelled succinate in the presence and absence of malonate.
Theoretically ^{14}C-carboxyl-labelled succinate cannot give rise to ^{14}C-labelled
protoporphyrin via Reaction F, but only via Reaction C. On entering the
tricarboxylic acid cycle, i.e. in the oxidative direction of the cycle (Reaction
F, Fig. 18), carboxyl-labelled succinate would give rise to α-carboxyl-labelled
α-ketoglutarate. The resulting succinyl intermediate arising by oxidative
decarboxylation and utilized for porphyrin formation would therefore contain no ^{14}C. However, if Reaction C occurs, carboxyl-labelled succinate
would produce labelled protoporphyrin. If this postulation is correct, then
carboxyl-labelled succinate should produce labelled protoporphyrin only by
Reaction C, and therefore malonate, which blocks Reaction F, should have
little or no influence on the ^{14}C activity of protoporphyrin.

"Methylene-labelled succinate, in contrast to carboxyl-labelled succinate,
should produce labelled protoporphyrin via two pathways: (1) Reaction C
and (2) the oxidative direction of the tricarboxylic acid cycle (Reaction F).
Methylene-labelled succinate should produce α-ketoglutaric acid labelled in
its α- and β-carbon atoms by Reaction F. As a result the succinyl intermediate
formed by this pathway, and utilized for porphyrin formation, should have 2
of its carbon atoms labelled with ^{14}C. Malonate should, in this case, have a
marked inhibitory effect on the utilization of methylene-labelled succinate
for porphyrin formation, and any ^{14}C activity in protoporphyrin made from
methylene-labelled succinate in the presence of malonate will be the result of
Reaction C."

The authors degraded the labelled haemin obtained from carboxyl-
labelled succinate and obtained the distribution pattern shown in
Fig. 19. Haemin was biosynthesized from carboxyl-labelled
succinate in the presence of malonate. When the labelled haemin
was degraded it was found that the ^{14}C distribution pattern was

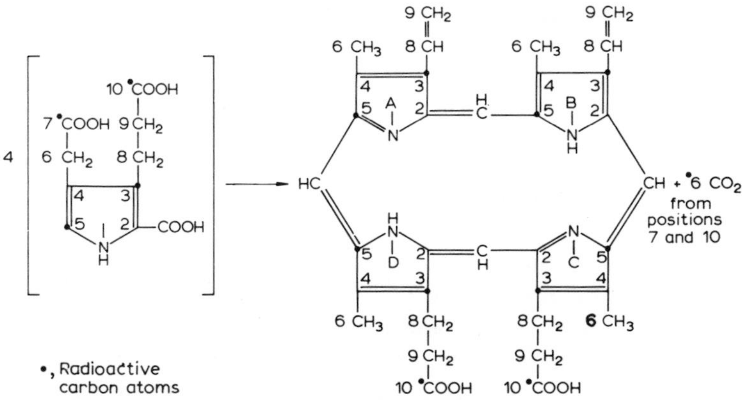

Fig. 19. (Shemin and Kumin [26]). The distribution pattern of ^{14}C in proto-porphyrin synthesized from [^{14}C]carboxyl-labelled succinate.

the same as in the absence of malonate. On the other hand, malonate inhibited only the labelling of the haemin synthesized from methylene-labelled succinate. The distribution of ^{14}C in protoporphyrin synthesized from [^{14}C]carboxyl-labelled succinate showed that succinate was utilized, as a unit, for 26 of the porphyrin carbons. The authors concluded that their data confirmed the participation of an "active" succinate (still not defined), condensing with glycine to form pyrrole and that this "active" succinate arises from both succinate and α-ketoglutarate in the tricarboxylic acid cycle.

That coenzyme A was involved in the biosynthesis of γ-amino-laevulinic acid received indirect support from a study of haem and porphyrin formation in pantothenate-deficient ducks (Schulman and Richert [27]) and *Tetrahymena vorax* (Lascelles [28]).

In the system studied by Laver, Neuberger and Udenfriend [29] (particles from lysates of chicken erythrocytes, synthesizing aminolaevulinic acid from glycine and succinate) an addition of coenzyme A produced a stimulation of activity, but when synthetic succinyl-CoA was substituted for succinate, no δ-aminolaevulinic acid was formed. Gibson, Laver and Neuberger [30] found that if the particles prepared as just stated were freeze-dried, they lost most of their capacity to form δ-aminolaevulinic acid from glycine and succinate. However, they can form δ-aminolaevulinic acid from synthetic or enzymatically generated succinyl-CoA and

"it appears that succinyl CoA is indeed the reactive derivative of the C_4 compound" (p. 71).

In order to further document the relationship of porphyrin biosynthesis with the tricarboxylic acid cycle, Wriston, Lack and Shemin [31] studied the utilization of [5-^{14}C]α-ketoglutarate, [1,2-^{14}C]α-ketoglutarate and [1,5-^{14}C]citrate for porphyrin formation in the presence of haemolysates of duck erythrocytes and confirmed their utilization. In 1955 the same authors turned to the degradation of the products along the hypothesis of the participation of the tricarboxylic acid cycle. In this case, [1,5-^{14}C]-citrate should produce [1,5-^{14}C]α-ketoglutarate exclusively and consequently

"both substrates studied (citrate-1,5-^{14}C and α-ketoglutarate-5-^{14}C) should be equivalent in their biological utilization pattern for porphyrin formation, since the α-carboxyl group of α-ketoglutarate is given off in its conversion to "active" succinate. Further, since carboxyl-labelled acetate produces α-ketoglutarate-1,5-^{14}C, the pattern of labelling in the protoporphyrin synthesized from citrate-1,5-^{14}C and α-ketoglutarate-5-^{14}C should be the same as was previously found in protoporphyrin synthesized from carboxy-labelled acetate" (p. 604).

Wriston et al. [31], from their study of the utilization of [1,5-^{14}C]-citric acid, [5-^{14}C]α-ketoglutaric acid and [1,2-^{14}C]α-ketoglutaric acid in haemolysed preparations of duck erythrocytes, observed that the ^{14}C distribution patterns found in the porphyrin were in complete agreement with their previous formulations regarding porphyrin formation from "active" succinate and glycine. The reaction postulated by Shemin, between glycine and two molecules of a succinyl compound considered to derive from the tricarboxylic acid cycle, was analogous to Knorr's reaction for chemical synthesis of pyrroles. Fruton [32] (p. 477) has also recalled that Hans Fischer, in 1914, drew attention, as possible routes of biosynthesis, to the ready condensation of glycine with formylacetone.

4. Recognition of the role of δ-aminolaevulinic acid and of porphobilinogen

Acute porphyria is a condition accompanied by periodic attacks of abdominal pain and by mental symptoms. In the urine of the patients, Sachs [33], in 1931, identified a compound turning red with Ehrlich's reagent (p-dimethylaminobenzaldehyde in acid solution). Being insoluble in $CHCl_3$, this compound was differentiated from urobilinogen by Waldenström [34].

Porphobilinogen

Porphobilinogen is, when acidified and heated, transformed into porphyrins I and III, hence its name. It has been isolated from the urine of patients with acute porphyria, and crystallized by Westall [35] in 1952. Cookson, Rimington and Kennard [36] and Cookson [37] recognized porphobilinogen as β, β'-dialkyl-α-aminomethylpyrrole. The structural formula was confirmed by X-ray crystallography (Kennard [38]).

In 1950 Neuberger, Muir and Gray [39] suggested, on the basis of the isotope data acquired at the time, a mechanism of porphyrin biosynthesis in which the first pyrrolic compound formed was a dicarboxylic pyrrole with two free α-positions.

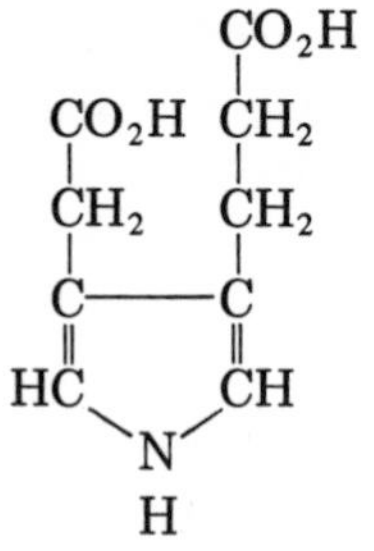

$$
\begin{array}{cc}
 & \text{CO}_2\text{H} \\
 & | \\
\text{CO}_2\text{H} & \text{CH}_2 \\
| & | \\
\text{CH}_2 & \text{CH}_2 \\
| & | \\
\text{C}\!\!-\!\!-\!\!-\!\!\text{C} \\
\| \quad\quad \| \\
\text{HC} \quad\quad \text{CH} \\
\diagdown\;\text{N}\;\diagup \\
\text{H}
\end{array}
$$

The authors assumed that the methene carbon atoms, derived from glycine as stated above, were added later. As the authors state (p. 1093):

"in this way the predominance in Nature of porphyrin related to aetioporphyrin III and the occurrence of smaller quantities of series I porphyrins could be explained".

This view turned out to be wrong.

As early as 1951, Shemin and Wittenberg [20] had suggested the involvement of an unsymmetrically substituted succinate such as succinyl-CoA. In his search for the compound resulting from the condensation of this unsymmetrical succinate and glycine, Shemin looked for a mechanism which would explain the loss of the carboxyl group of glycine. The condensation of succinate onto the δ-carbon atom of glycine would produce a β-keto acid, δ-amino-β-keto adipic acid, which would readily decarboxylate to δ-aminolaevulinic acid. This led to the synthesis of δ-aminolaevulinic acid by Shemin and Russell [40] who showed by the use of [^{15}N]- and [δ-^{14}C]labelled aminolaevulinic acid that this compound was incorporated into haemin in the duck red cell system. While in the midst of these experiments, as stated above, porphobilinogen was isolated by Westall and its structure was elucidated and confirmed by X-ray crystallography. In their paper, Shemin and Russell [40] suggested that their precursor pyrrole from 2 molecules of δ-aminolaevulinic acid was the same as porphobilinogen.

It was known at the time, by the observations of Wittenberg

and Shemin and of Muir and Neuberger, that protoporphyrin is, in the nucleated blood cells of birds, synthesized by the condensation of 8 glycines with 8 molecules of a compound derived from α-oxoglutarate. Succinic acid may be this compound. In this perspective, Neuberger and Scott [41] set out, in 1953, to synthesize compounds of glycine with succinic acid, which might be intermediates and which after labelling might be more highly incorporated into haemin than glycine or succinate.

To quote Neuberger and Scott (p. 1093):

$$
\begin{array}{cc}
\text{CO}_2\text{H} & \text{CO}_2\text{H} \\
| & | \\
\text{CH}_2 & \text{CH}_2 \\
| & | \\
\text{CH}_2 & \text{CH}_2 \\
| & | \\
\text{CO}\diagdown\underset{\text{N}}{}\diagup\text{CH}_2\cdot\text{CO}_2\text{H} & \text{CO} \\
\text{H} & \underset{\text{H}_2\text{N}}{\diagup}\text{CH}\cdot\text{CO}_2\text{H} \\
\text{I} & \text{II}
\end{array}
$$

$$
\begin{array}{cc}
\text{CO}_2\text{H} & \text{CO}_2\text{H} \\
| & | \\
\text{CH}_2 & \text{CO}_2\text{H}\quad\text{CH}_2 \\
| & |\qquad\ | \\
\text{CH}_2 & \text{CH}_2\quad\text{CH}_2 \\
| & |\qquad\ | \\
\text{CO} & \text{CH}_2\quad\text{CO} \\
\underset{\text{H}_2\text{N}}{\diagup}\text{CH}_2 & \text{CO}\diagdown\underset{\text{N}}{}\diagup\text{CH} \\
& \text{H} \\
\text{III} & \text{IV}
\end{array}
$$

$$
\begin{array}{cc}
\text{CO}_2\text{H} & \text{CO}_2\text{H} \\
| & | \\
\text{CO}_2\text{H}\quad\text{CH}_2 & \text{CO}_2\text{H}\quad\text{CH}_2 \\
|\qquad\ | & |\qquad\ | \\
\text{CH}_2\quad\text{CH}_2 & \text{CH}_2\quad\text{CH}_2 \\
|\qquad\ | & |\qquad\ | \\
\text{C}\text{---}\text{C} & \text{C}\text{---}\text{C} \\
\| \qquad \| & \| \qquad \| \\
\text{HC}\diagdown\underset{\text{N}}{}\diagup\text{CH} & \text{H}_2\text{N}\cdot\text{CH}_2\cdot\text{C}\diagdown\underset{\text{N}}{}\diagup\text{CH} \\
\text{H} & \text{H} \\
\text{V} & \text{VI}
\end{array}
$$

References p. 235

"A carboxyl group of succinic acid may be condensed with glycine either by N-succinylation or by an α-C-succinylation, analogous to the Dakin-West reaction. In the first case succinamido-acetic acid (I) and in the second case α-amino-β-oxoadipic acid (II) would be formed. The latter compound would be expected to decarboxylate readily to form δ-amino-laevulinic acid (III). By addition of a second molecule of succinic acid, succinamido-acetic acid or δ-amino-laevulinic acid could be converted to δ-succinamido-laevulinic acid (IV) by a similar process. This compound, by an oxidative self-condensation, might then form a pyrrole of the type shown below (V). A third possibility was that two molecules of δ-aminolaevulinic acid could undergo a Knorr-type condensation to give a pyrrole, bearing in addition an amino-methyl α-substituent (VI).

Neuberger and Scott, in order to test the possibilities just mentioned, prepared a series of labelled compounds: [α-^{14}C]succinamido-acetic acid, δ-succinamido-laevulinic acid and δ-aminolaevulinic acid hydrochloride labelled in each of the five carbons.

Before Neuberger and Scott had completed the biological work with these labelled compounds, Falk, Dresel and Rimington [42] found that porphobilinogen was converted in 60% yield to a mixture of porphyrins when incubated with lysed bird erythrocytes. The biological observations of Neuberger and Scott excluded the possibilities involving either succinamido-acetic or succinamido-laevulinic acid. The labelled δ-aminolaevulinic acid prepared by Neuberger and Scott was tested by Falk and Dresel in the lysed red cell system and they showed that aminolaevulinic acid is equivalent to porphobilinogen in producing a net increase in the levels of uro-, copro-, and protoporphyrin, an observation which confirmed the observations of Shemin and Russell [40] who had demonstrated several months before the involvement of aminolaevulinic acid and the condensation of two molecules to give porphobilinogen. It must also be noted that Shemin's evidence was more detailed and complete. The data of Neuberger and Scott suggested that the first pyrrolic compound formed is in fact compound VI, as the authors announced it in their note in *Nature* of December 12, 1953. The structure advanced by Cookson et al. [36] in May of the same year, and by Cookson [37] in September, was identical with VI.

Falk et al. [42] had shown that porphobilinogen was a highly specific precursor for porphyrins, and Shemin and Russell [40] had showed that δ-aminolaevulinic acid is a specific precursor for haem (confirmed with a substrate synthesized by Neuberger and

Scott [41] to the extent that, on incubation by Dresel and Falk [43] with a chicken erythrocyte haemolysate the yield in uro-, copro- and protoporphyrins from δ-aminolaevulinic acid was identical to that of a corresponding amount of porphobilinogen).

The structure of porphobilinogen arrived at by Cookson and Rimington was in harmony with its formation from two molecules of δ-aminolaevulinic acid (see Shemin and Russell [40], confirmed by Neuberger and Scott [41]).

When the incubated δ-aminolaevulinic acid with chicken erythrocyte haemolysate, Dresel and Falk [43] obtained a formation of porphobilinogen.

5. Biosynthesis of δ-aminolaevulinic acid

By the end of 1953, data had been obtained in favour of a formation of porphyrins through δ-aminolaevulinic acid and porphobilinogen. But the mechanism by which the "active succinate" condensed with glycine to form δ-aminolaevulinic acid, precursor of the pyrrole unit of the porphyrin remained unknown and became a subject of interest.

Neuberger and Scott had, in their papers of 1953 [41] and 1954 [41a] excluded the N-succinylation of glycine, in favour of the α-C-succinylation, analogous to the Dakin—West reaction. In this step, α-amino-β-oxoadipic acid would be formed from which δ-aminolaevulinic acid would derive. As we have stated above, the inclusion of this compound in haemin was demonstrated by Shemin and Russell [40] (confirmed by Neuberger).

Brown [44] approached the subject through studies performed on particles. By gentle haemolysis of chicken erythrocytes and washing, he washed away the soluble enzymes acting on δ-amino-laevulinic acid and was able to observe a formation of δ-amino-laevulinic acid when he added to his particulate preparation glycine, succinic acid, ATP and CoA (the formation of which was known at the time to require a citric acid cycle).

With the purpose of increasing the number of reticulocytes and consequently the enzymatic activity in the system, Laver et al. [29] treated chickens with phenylhydrazine and obtained washed particulate material from their nucleated erythrocytes. This material could synthesize δ-aminolaevulinic acid from an aerated mixture of CoA, $MgCl_2$, pyridoxal-P, glycine, and α-ketoglutaric acid. The freeze-dried particles contained a pyridoxal-P enzyme called δ-aminolaevulinic acid synthetase by the authors, which is required for the synthesis of δ-aminolaevulinic acid from succinyl-CoA and glycine.

Gibson et al. [30] demonstrated the activity of the δ-aminolaevulinic acid synthetase in the form of dry-frozen material. That the enzymatic synthesis of δ-aminolaevulinic acid by the condensation of glycine and "active succinate" could be accomplished by particle-free extracts of the photosynthetic bacteria *Rhodopseudomonas spheroides* and *Rhodospirillum rubrum* grown anaerobically in the light was demonstrated by Shemin, Kikuchi and Bachmann [45] and by Kikuchi, Shemin and Bachmann [46]. Sawyer and Smith [47] used extracts of anaerobically grown *R. spheroides*, and Gibson [48] has observed the biosynthesis in extracts of aerobically grown *R. spheroides*.

Kikuchi, Kumar, Talmage and Shemin [49] have demonstrated that succinate is converted to succinyl-CoA and that this activated form of succinate condenses with a pyridoxal phosphate derivative of glycine to form δ-aminolaevulinic acid. The condensation of succinyl-CoA on the α-carbon of glycine yields α-amino-β-ketoadipic acid. On spontaneous decarboxylation, this acid yields δ-aminolaevulinic acid, in which the δ-carbon is originally the α-carbon of glycine (Shemin, Abramsky and Russell [50]).

The enzyme catalysing the condensation of succinyl-CoA and glycine activated by pyridoxal now commonly called δ-aminolaevulinic synthase, has received the systematic name [Succinyl-CoA : glycine *C*-succinyltransferase (decarboxylating); 2.3.1.37].

Shemin and Russell [40] had postulated that δ-aminolaevulinic acid is the source of all the atoms of porphyrins. From the mode of condensation of succinate and glycine to form α-amino-β-ketoadipic acid, yielding δ-aminolaevulinic acid on decarboxylation, the origin of the carbon atoms can be deduced. Carbons 1 to 4 should arise from succinate and carbon 5 (bearing the amino group) should arise from the α-carbon of glycine.

Shemin et al. [50] and Shemin, Russell and Abramsky [51] have demonstrated that $[5\text{-}^{14}C]\delta$-aminolaevulinic acid is converted to radioactive protoporphyrin. The labelling pattern is the same as was previously found in the porphyrin biosynthesized from $[2\text{-}^{14}C]$glycine. If we accept the suggestion formulated above,

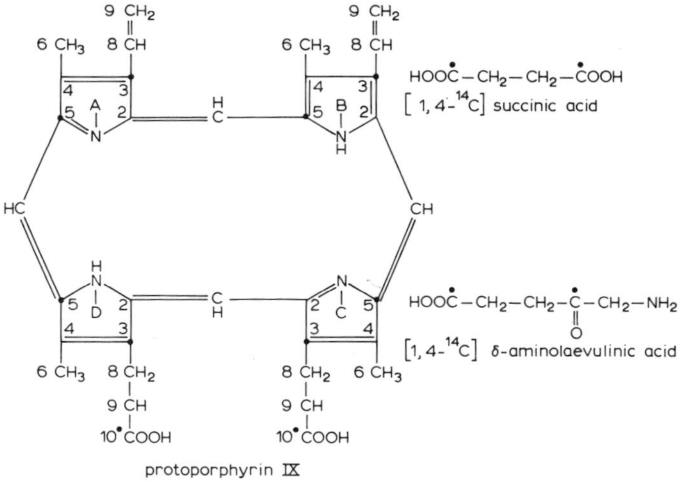

Fig. 20. The carbon atoms (●) of protoporphyrin which arise from the carboxyl groups of succinic acid and from carbon atoms 1 and 4 of δ-aminolaevulinic acid.

according to which the δ-aminolaevulinic acid is the source of all carbons of the porphyrin, we must recognize that it should be experimentally confirmed that the succinyl moiety of δ-aminolaevulinic acid, an amino ketone, has entered the porphyrin molecule in the same positions as those known to arise from succinate. If this hypothesis is correct, [1,4-14C]δ-aminolaevulinic acid should produce precisely the same labelling pattern which was found for carboxylated succinate. Ten atoms of carbon should, in protoporphyrin, occupy the position shown in Fig. 20, as demonstrated by Shemin et al. [51] and by Shemin and Kumin [26].

Schiffmann and Shemin (1957) (in a report part of the Ph.D. dissertation of Schiffmann, the results being also reported by Shemin at the Ciba symposium) showed that indeed the protoporphyrin synthesized from [1,4-14C]δ-aminolaevulinic acid contained the same labelling pattern as that previously found in the porphyrin biosynthesized from [1,4-14C]succinate. It is thus confirmed that δ-aminolaevulinic acid is indeed the source of all the carbon atoms of the porphyrin.

References p. 235

6. From δ-aminolaevulinic acid to porphobilinogen

An enzyme was soon found to exist in nature which converts δ-aminolaevulinic acid to porphobilinogen. First detected by Dresel and Falk [43], the enzyme was named aminolaevulinate

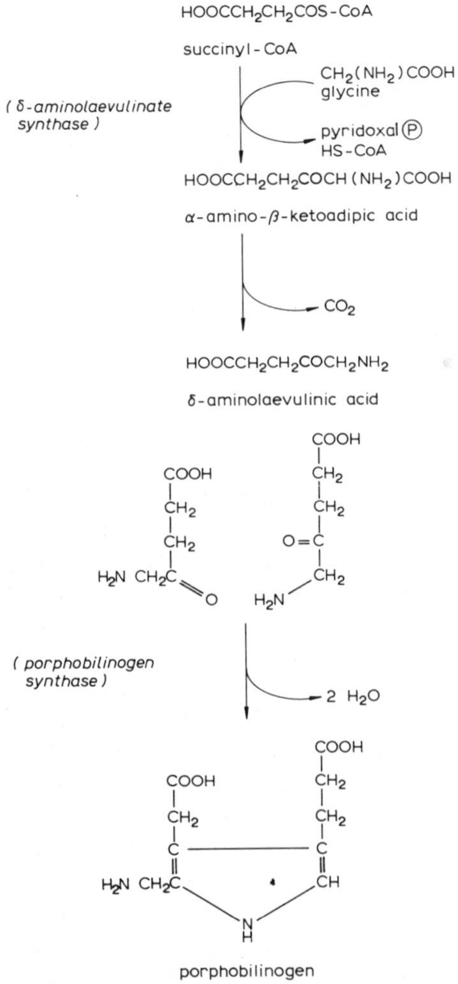

HOOCCH$_2$CH$_2$COS-CoA

succinyl-CoA

(δ-aminolaevulinate synthase)

CH$_2$(NH$_2$)COOH
glycine

pyridoxal Ⓟ
HS-CoA

HOOCCH$_2$CH$_2$COCH(NH$_2$)COOH

α-amino-β-ketoadipic acid

CO$_2$

HOOCCH$_2$CH$_2$COCH$_2$NH$_2$

δ-aminolaevulinic acid

(porphobilinogen synthase)

2 H$_2$O

porphobilinogen

Fig. 21. From succinyl-CoA and glycine to porphobilinogen (modified from Dagley and Nicholson [56], p. 209).

dehydratase by Gibson et al. [52]. Its wide distribution in animals, plants and bacteria was recognized [53—55]. The common name now given to this enzyme is phosphobilinogen synthase, and its systematic name is [5-Aminolaevulinate hydro-lyase (adding 5-aminolaevulinate and cyclizing); 4.2.1.24].

7. From porphobilinogen to tetrapyrroles

Indirect evidence obtained with isotopically labelled materials (Shemin and Wittenberg [20]) led to the suggestion of two alternatives for this pathway. In the first, four identical monopyrrole precursors condense, forming uroporphyrin III. This compound, by successive decarboxylations and dehydrogenations, leads via coproporphyrin III to protoporphyrin IX and haem. The alternative pathway would involve modifications of the side chains of the monopyrrole precursor before condensation. In this case, some of the porphyrins with a high number of carboxyl groups would be bypassed.

The first pathway has received support from a number of observations. For instance, porphyrins with eight to two carboxyl groups readily accumulate in systems in vitro, as was observed by Dresel and Falk [57] and by Bogorad and Granick [58].

On the other hand, such accumulations have also been reported in vivo in normal and pathological conditions, by Rimington [59] and by Dresel and Falk [60], in Rimington's laboratory. They studied the role of some of the natural free porphyrins in the biosynthesis of haem by a chicken-erythrocyte haemolysate and concluded that the free porphyrins of the III isomer series are not utilized. Although they are biosynthesized de novo from glycine, the capacity for this biosynthesis is weak, compared with the biosynthesis of porphobilinogen and haem. The authors thus conclude that they must be side products. They suggest the following scheme (Fig. 22):

Plate 205. Sam Granick.

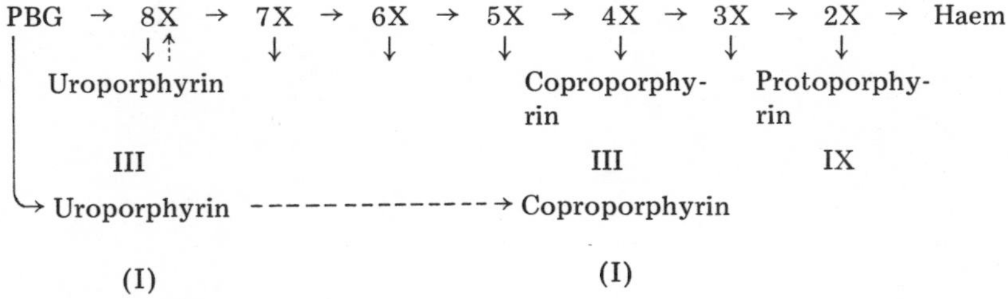

PBG → 8X → 7X → 6X → 5X → 4X → 3X → 2X → Haem
  ↓↑̣ ↓ ↓ ↓ ↓ ↓ ↓
 Uroporphyrin   Coproporphy- Protoporphy-
             rin   rin
   III   III   IX
 → Uroporphyrin − − − − − − − − − − − − → Coproporphyrin

   (I)   (I)

Fig. 22. Suggested scheme for mechanism of haem biosynthesis from PBG (Dresel and Falk [60]).

This scheme is commented on as follows by Dresel and Falk:

"The number in each intermediate indicates the number of side chains possessing a carboxyl group, and X denotes a porphyrin-like structure common to all the intermediates which are not, however, true free porphyrins. A slight reversibility of the side reaction leading to uroporphyrin III is indicated. Further, a direct pathway for handling the I series porphyrins is suggested in view of our findings with the supernatant preparation.

"It is of interest that Bogorad [61], using extracts of spinach leaf and *Chlorella*, has recently come to the conclusion that a colourless precursor of uroporphyrin, rather than true uroporphyrin, is the intermediate between PBG and porphyrins with seven to three carboxyl groups. Thus there is independent evidence for a non-porphyrin intermediate corresponding to 8X in Fig. 22. The origin of these porphyrins with seven to three carboxyl groups was not studied, but it is conceivable that they also arose as the result of side reactions as in Fig. 22."

It is as porphyrinogens, rather than as their oxidized derivatives, porphyrins, that uroporphyrinogen and coproporphyrinogen (and presumably those intermediates with seven, six, five and three carboxyl groups per molecule, respectively) can be used as substrates for subsequent steps in the biosynthesis of protoporphyrin and haem. The porphyrinogens can be readily oxidized to porphyrins and it is in the porphyrin form that they are detected and estimated.

Porphyrinogens, the colourless precursors, fully reduced derivatives of porphyrins (Fischer, Bartholomäus and Röse [62]; Fischer and Zerweck [63]; Fischer and Stern [64]; Mauzerall and Granick [65]), are intermediates in the biosynthesis of protoporphyrin from porphobilinogen by enzyme systems isolated from plants,

Plate 206. Lawrence Bogorad.

animals and bacteria (Bogorad [61,66,67]; Granick [68]; Granick and Mauzerall [69]; Hoare and Heath [70]).

In the biosynthetic pathway of protoporphyrin IX the direct precursor was shown to be the hexahydro form of coproporphyrin isomer III (coproporphyrinogen III) in a frozen and thawed *Euglena* preparation prewashed with buffer (Granick and Mauzerall [71]) and in haemolysed chicken erythrocytes (Granick and Mauzerall [72]). Bogorad and Granick [58] found that porphobilinogen is converted into porphyrins by suspensions of *Chlorella vulgaris* which were subjected to freezing and thawing. This formation has also been studied with spinach preparation by Bogorad [61] and with chicken-erythrocyte haemolysates by Dresel and Falk [57], and by Lockwood and Rimington [73]. In Rimington's laboratory, Heath and Hoare [74] established that preparations from the photosynthetic bacterium *Rhodopseudomonas spheroides* biosynthesize coproporphyrin III from porphobilinogen.

It was also shown by these authors that under certain conditions, porphobilinogen was converted into uroporphyrin, and that uroporphyrinogens I and III were decarboxylated to yield, ultimately, coproporphyrins I and III, respectively. That at least three enzymes are involved in the conversion of porphobilinogen into coproporphyrin III was proven by the experiments of Bogorad [67] with enzyme preparations from spinach, wheat germ and *Chlorella* and those of Granick and Mauzerall [69] and Mauzerall and Granick [65] with enzyme preparations from chick and rabbit erythrocytes.

The conversion of porphobilinogen into uroporphyrin I is catalysed by the enzyme purified by Bogorad [67] from spinach extracts, called by him porphobilinogen deaminase, and which now bears the name uroporphyrinogen I synthase. Its systematic name is [Porphobilinogen ammonia-lyase (polymerizing); 4.3.1.8], catalysing the reaction

4 porphobilinogen = uroporphyrinogen I + 4 NH_3

In the presence of a second enzyme (uroporphyrinogen III co-synthase) uroporphyrinogen III is formed:

Therefore, the former porphobilinogenase of Lockwood and Rimington [73] (1957) is now composed of two enzymes: uroporphyrinogen I synthase and uroporphyrinogen III co-synthase (see Levin and Coleman [75]). Uroporphyrinogen isomerase was iso-

Plate 206a. C. Rimington.

lated from extracts of wheat germ by Bogorad [76]. It catalyses the formation of uroporphyrinogen III from porphobilinogen in the presence of uroporphyrin I synthase and uroporphyrinogen III co-synthase. The formation of coproporphyrin from uroporphyrinogen was found to be catalysed by crude preparations of *Chlorella* (Bogorad [77]). A partially purified uroporphyrinogen decarboxylase was obtained by Mauzerall and Granick [78] from chick and rabbit reticulocytes. Hoare and Heath [70] have separated uroporphyrinogen decarboxylase from uroporphyrinogen I synthase in preparations derived from *R. spheroides*. This enzyme, acting on a number of porphyrinogens, has received the systematic name [Uroporphyrinogen-III carboxy-lyase; 4.1.1.37].

Bogorad and Granick [58] first obtained evidence that mitochondria contained what they first called coproporphyrinogenase (the enzyme that converts coproporphyrinogen III to protoporphyrin IX) using *Chlorella* preparations which could synthesise coproporphyrin isomer III but not protoporphyrin from δ-aminolaevulinic acid, but which could form protoporphyrin IX if supplemented with rat liver mitochondria (the rat mitochondria being inactive with λ-aminolaevulinic acid).

Rat liver mitochondria were found by Rimington and Booij [79] to catalyse the conversion of δ-aminolaevulinic acid to protoporphyrin IX by a haemolysate of human red cells. The same enhancing effect was detected by Sano and his collaborators [80, 81] on the same conversion using chicken erythrocyte preparations. Sano and Granick [82] partially purified the coproporphyrinogen oxidase of beef liver. This enzyme has received the systematic name [Coproporphyrinogen : oxygen oxidoreductase (decarboxylating); 1.3.3.3] catalysing the reaction

$$\text{coproporphyrinogen III} + O_2 = \text{protoporphyrinogen} + 4\,CO_2$$

8. Metalloporphyrins

(a) The iron branch

The presence of an enzyme catalysing the incorporation of iron in porphyrins under conditions similar to those taking place in organisms was first detected by Labbe and Hubbard [83]. It has

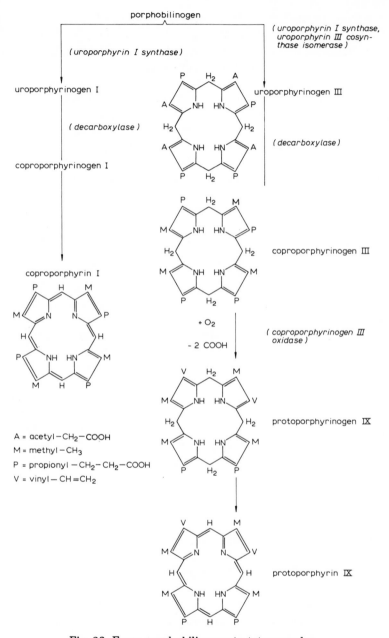

Fig. 23. From porphobilinogen to tetrapyrroles.

received the systematic name of ferrochelatase [Protohaem ferro-lyase, 4.99.1.1].

It is widely distributed in nature and some of its properties have been studied (literature in Burnham [84]).

(b) The magnesium branch

After it had been indicated by Granick [85] that the pyrrole skeleton of chlorophyll may be formed by biosynthetic reactions similar to those discovered for protoporphyrin IX, Della Rosa, Altman and Salomon [86], at Rochester, confirmed the existence of such similarities.

Granick [85,87,88] has isolated different mutants of *Chlorella* unable to biosynthesize chlorophyll. These mutants accumulate magnesium protoporphyrin, magnesium vinyl phaeoporphyrin a_5 and the monomethyl ester of magnesium protoporphyrin. Granick [88] has proposed a scheme for the biosynthesis of magnesium protoporphyrin monomethyl ester.

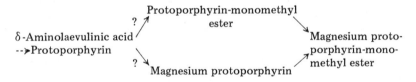

As well as being considered by Granick as intermediates in chlorophyll biosynthesis, they have also been considered by Granick and Mauzerall [89] as intermediates in bacteriochlorophyll biosynthesis. Tait and Gibson (1961), in Neuberger's laboratory, have extracted from *Rhodopseudomonas spheroides* a system which enzymatically methylates magnesium protoporphyrin. Later this enzyme was commonly called magnesium-protoporphyrin-methyltransferase. It has received the systematic name [*S*-Adenosyl-L-methionine : magnesium-protoporphyrin-*O*-methyltransferase; 2.1.1.11]. It catalyses the reaction

S-adenosyl-L-methionine + magnesium protoporphyrin

= *S*-adenosyl-L-homocysteine + magnesium protoporphyrin

monomethyl esters

It methylates magnesium protoporphyrin but not protoporphyrin [90,90a]. Jones [91] has isolated magnesium protoporphyrin monomethyl ester from the medium of wild type R. *spheroides* and the same author [92] found, by adding 8-hydroxyquinoline to the medium of the same microorganism, a decrease in the intracellular concentration of bacteriochlorophyll (but not of haem pigments or carotenoids). He observed at the same time an excretion of tetrapyrrole pigments in the medium.

On the other hand, Gibson, Neuberger and Tait [93] mentioned briefly that when ethionine was added to illuminated suspensions of R. *spheroides*, large amounts of coproporphyrins accumulated in the medium, although growth and bacteriochlorophyll biosynthesis were inhibited.

The same authors [94] studied the effect in more detail, since it was surprising that this effect was not, as in the case of other inhibitors, an inhibition of growth as well as of the formation of porphyrin and bacteriochlorophyll.

The authors tried a number of other compounds and only threonine was found to have the same effect as ethionine. Now, in the presence of ethionine or threonine, the synthesis of bacteriochlorophyll can be restored and porphyrin excretion reduced by methionine or homocysteine. This suggested that ethionine and threonine exert their effects by interfering with the utilization or the synthesis of methionine *. Gibson et al. [94] did illuminate the microorganisms in the presence of [*Me*-^{14}C]methionine. Bacteriochlorophyll was isolated and it was degraded to yield methanol, phytol and bacteriochlorin. High radioactivity was found in the bacteriochlorophyll. The radioactivity was almost exclusively found in the methyl ester group. Gibson et al. [94] demonstrated that the concentration of ethionine required for inhibition of bacteriochlorophyll formation is well below the concentration required for inhibition of protein biosynthesis. They [90] showed that inhibition of enzyme methylation of magnesium protoporphyrin by S-adenosylethionine is competitive with S-adenosylmethionine, which is in favour of a blocking action on the esterifi-

---

* The effect of threonine on porphyrin and bacteriochlorophyll synthesis in R. *spheroides* was shown to be due to its inhibiting homoserine, and hence methionine, biosynthesis (Gibson, Neuberger and Tait [94a]).

cation of Mg-protoporphyrin. We have related above the experiments accomplished on R. spheroides, in Australia, by Jones [95]. Among the tetrapyrroles accumulated in the medium in the presence of 8-hydroxyquinoline, Jones identified 2,4-divinylphaeoporphyrin a_5 monomethyl ester (magnesium being spontaneously lost from those unstable molecules in the preparations). Pathways to these pigments may be accomplished through a series of sequences proposed on a chemical basis (see Burnham [84], p. 452).

2-Vinylphaeoporphyrin a_5 monomethyl ester (protochlorophyllide) was found to be formed from δ-aminolaevulinic acid by etiolated leaves by Granick [88]. It has also been found by the same author [85] to accumulate in a Chlorella mutant unable to form chlorophyll in the dark.

The conversion of 2,4-divinylphaeoporphyrin a_5 monomethyl ester to 2-vinylphaeoporphyrin a_5 monomethyl ester is the result of a simple reduction of the vinyl group at position 4.

The conversion of protochlorophyllide to chlorophyllide a is, in many plants and algae a light-catalysed reaction, by a reduction at positions 7 and 8.

Krasnovsky and Kosobutskaya [96] have observed the conversion in homogenates from etiolated leaves in the dark. The enzymology of this process remains to be studied.

Most evidence acquired so far and stated above leads to the construction of the following pathway for chlorophyll a (and probably for bacteriochlorophyll, or at least very similar ways).

→ Protoporphyrin IX $\xrightarrow{Mg}$ Mg-protoporphyrin IX →

→ Mg-protoporphyrin monomethyl ester → Mg-2,4-

divinylphaeoporphyrin a_5 monomethyl ester →

→ Mg-2-vinylphaeoporphyrin a_5 monomethyl ester (protochloro-

phyllide a) $\xrightarrow{+2}$ chlorophyllide a $\xrightarrow[\text{alcohol}]{+\text{phytol}}$ chlorophyll a.

9. Biosynthesis of corrinoids

Not long after the isolation of vitamin B_{12} chemical studies led to the demonstration (1948) that it was a tetrapyrrole. (On the

References p. 235

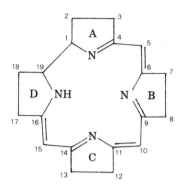

Fig. 24. Corrin.

chemistry of vitamin B_{12}, see this Treatise, Vol. 11, Chapter VIII, by A.F. Wagner and K. Folkers.) Cyanocobalamin (vitamin B_{12}) is not, *stricto sensu*, the product of a biosynthesis, since it is a derivative of the natural compound.

The isotopic method has been relied upon by supplying to a microorganism a labelled suspected intermediate and determining the extent and position of the labelling in the vitamin B_{12} obtained. Corcoran and Shemin [97] and Shemin, Corcoran, Rosenblum and Miller [98] observed that an actinomycete used ^{14}C-labelled δ-aminolaevulinic acid to make vitamin B_{12}. Bray and Shemin [98a,b] further demonstrated that δ-aminolaevulinic acid is the precursor and, furthermore, showed that the "extra" methyl groups are derived by C-methylation by methionine. Schwartz, Ikeda, Miller and Watson [99] demonstrated that in an actinomycete, porphobilinogen is a precursor of vitamin B_{12} and its derivatives. The nucleus is called corrin. It contains six double bonds (Fig. 24).

Besides the tracer studies reported above, endeavours have been made with the purpose of isolating and identifying intermediates on the path to corrinoids. A number of authors consider that the first stage is complete with cobyrinic acid (Fig. 25) (Bernhauer, Muller and Wagner [100]; Wagner [101]; Bernhauer, Wagner, Beisbarth, Rietz and Vogelmann [102]; Friedrich and Sandek [103]; Friedrich [104]; Di Marco and Spalla [105]). The many data so far obtained and pertaining to the complex pathway leading to corrinoids have been collected by Burnham [84].

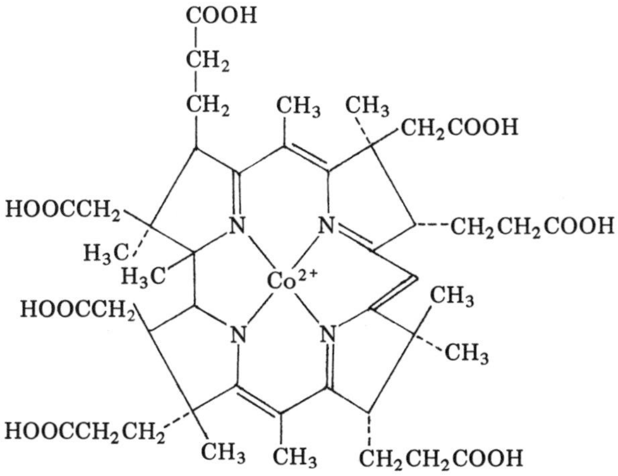

Fig. 25. Cobyrinic acid.

Wagner [101] has proposed a tentative pathway between coby-rinic acid and vitamin B_{12}. Friedrich [104] has indicated that *Propionibacterium stermanii* can amidate cobyrinic acid triamide, tetraamide and pentaamide to form cobyric acid, from which cobinamide and vitamin B_{12} can be derived.

10. Prodigiosin

In this chapter on the biosynthesis of tetrapyrroles and of corrin-oids, we shall also consider the tripyrrolic red pigment, prodigio-sin, produced by the bacterium *Serratia marcescens*. The forma-tion of this tripyrrolic pigment involves neither δ-aminolaevulinic acid, nor porphobilinogen as precursors (Marks and Bogorad [106]). That the final step in the formation of prodigiosin was a conden-sation of a dipyrrolic segment with methylamylpyrrole was dem-onstrated by Santer and Vogel [107], Wasserman, McKeon and Santer [108] and Rizki [109]. Prodigiosin is thus a pyrrolic com-pound which is not derived from δ-aminolaevulinic acid.

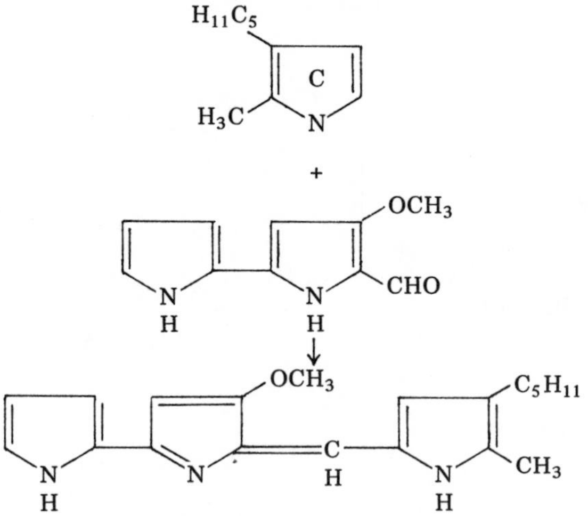

Fig. 26. Biosynthesis of prodigiosin from methylaminopyrrole and a dipyrrolic segment.

REFERENCES

1   K. Bloch and D. Rittenberg, J. Biol. Chem., 159 (1945) 45.
2   D. Shemin and D. Rittenberg, J. Biol. Chem., 159 (1945) 567.
3   D. Shemin and D. Rittenberg, J. Biol. Chem., 166 (1946) 627.
4   C. Tesar and D. Rittenberg, J. Biol. Chem., 170 (1947) 35.
5   H. Wu and D. Rittenberg, J. Biol. Chem., 179 (1949) 847.
6   D. Shemin and D. Rittenberg, J. Biol. Chem., 166 (1946) 621.
7   E. Abderhalden, Lehrbuch der physiologischen Chemie, Berlin und Wien, 1909, p. 744.
8   H. Fischer and E. Fink, Z. Physiol. Chem., 280 (1944) 123.
9   M. Grinstein, M.D. Kamen and C.V. Moore, J. Biol. Chem., 174 (1948) 767.
10 K.I. Altman, G.W. Casarett, R.E. Masters, T.R. Noonan and J. Salomon, J. Biol. Chem., 176 (1948) 319.
11 H.M. Muir and A. Neuberger, Biochem. J., 45 (1949) 163.
12a D. Shemin, Cold Spring Harbor Symp. Quant. Biol., 14 (1949) 161.
12b J. Wittenberg and D. Shemin, J. Biol. Chem., 185 (1950) 103.
13 N.S. Radin, D. Rittenberg and D. Shemin, J. Biol. Chem., 184 (1950) 745.
14 H.M. Muir and A. Neuberger, Biochem. J., 47 (1950) 97.
15 H.M. Muir and A. Neuberger, Biochem. J., 45 (1949) XXXIV.
16 A.W.P. Bufton, R. Bentley and C. Rimington, Biochem. J., 44 (1949) XLIX.
17 C.H. Gray and A. Neuberger, Biochem. J., 47 (1950) 81.
18 R. Lemberg and J.W. Legge, Haematin Compounds and Bile Pigments, New York, 1949, p. 637.
19a N.S. Radin, D. Rittenberg and D. Shemin, Fed. Proc., 8 (1949) 240 (Meeting of the Am. Soc. of Biological Chemists at Detroit, April 1949).
19b D. Shemin, I.M. London and D. Rittenberg, J. Biol. Chem., 173 (1948) 799.
20 D. Shemin and J. Wittenberg, J. Biol. Chem., 192 (1951) 315.
21 J. Wittenberg and D. Shemin, J. Biol. Chem., 178 (1949) 47.
22 D. Shemin, Harvey Lect., 1954—1955, Series L, 258.
23 S. Kaufman, in W.D. McElroy and B. Glass (Eds.), Phosphorus Metabolism, Baltimore, 1951, Vol. I, p. 370.
24 D.R. Sanadi and J.W. Littlefield, J. Biol. Chem., 193 (1951) 683.
25 D.R. Sanadi and J.W. Littlefield, Fed. Proc., 11 (1952) 280.
26 D. Shemin and S. Kumin, J. Biol. Chem., 198 (1952) 827.
27 M.P. Schulman and D.A. Richert, J. Biol. Chem., 226 (1957) 181.
28 J. Lascelles, Biochem. J., 66 (1957) 65.
29 W.G. Laver, A. Neuberger and S. Udenfriend, Biochem. J., 70 (1958) 4.
30 K.D. Gibson, W.G. Laver and A. Neuberger, Biochem. J., 70 (1958) 71.
31 J.C. Wriston, L. Lack and D. Shemin, Fed. Proc., 12 (1953) 294.
32 J. Fruton, Molecules and Life, New York, 1972.
33 P. Sachs, Klin. Wochenschr., 10 (1931) 1123.
34 J. Waldenström, Acta Med. Scand., 83 (1934) 281.

35 R.G. Westall, Nature (London), 170 (1952) 614.
36 G.H. Cookson, C. Rimington and O. Kennard, Nature (London), 171 (1953) 875.
37 G.H. Cookson, Nature (London), 172 (1953) 457.
38 O. Kennard, Nature (London), 171 (1953) 876.
39 A. Neuberger, I.H.M. Muir and C.H. Gray, Biochem. J., 47 (1950) 542.
40 D. Shemin and C.S. Russell, J. Am. Chem. Soc., 75 (1953) 4873.
41 A. Neuberger and J.J. Scott, Nature (London), 172 (1953) 1093.
41a A. Neuberger and J. Scott, J. Chem. Soc., (1954) 1820.
42 J.E. Falk, E.I.B. Dresel and C. Rimington, Nature (London), 172 (1953) 292.
43 E.I.B. Dresel and J.E. Falk, Nature (London), 172 (1953) 1185.
44 E.G. Brown, Biochem. J., 70 (1958) 313.
45 D. Shemin, G. Kikuchi and B.J. Bachmann, Fed. Proc., 17 (1958) 310.
46 G. Kikuchi, D. Shemin and B.J. Bachmann, Biochim. Biophys. Acta, 28 (1958) 219.
47 E. Sawyer and R.A. Smith, in Bacteriological Proceedings, Soc. Am. Biochem., Baltimore, 1958, p. 111.
48 K.D. Gibson, Biochim. Biophys. Acta, 28 (1958) 451.
49 G. Kikuchi, A. Kumar, P. Talmage and D. Shemin, J. Biol. Chem., 233 (1958) 1214.
50 D. Shemin, T. Abramsky and C.S. Russell, J. Am. Chem. Soc., 76 (1954) 1204.
51 D. Shemin, C.S. Russell and T. Abramsky, J. Biol. Chem., 215 (1955) 613.
52 K.D. Gibson, A. Neuberger and J.J. Scott, Biochem. J., 58 (1954) XLI.
53 S. Granick, Science, 120 (1954) 1105.
54 R. Schmidt and D. Shemin, J. Am. Chem. Soc., 77 (1955) 506.
55 M.P. Schulman, Fed. Proc., 14 (1955) 277.
56 S. Dagley and D.E. Nicholson, An Introduction to Metabolic Pathways, Oxford and Edinburgh, 1970.
57 E.I.B. Dresel and J.E. Falk, Biochem. J., 63 (1956) 80.
58 L. Bogorad and S. Granick, Proc. Natl. Acad. Sci. USA, 39 (1953) 1176.
59 C. Rimington, Rep. Progr. Chem., 51 (1954) 311.
60 E.I.B. Dresel and J.E. Falk, Biochem. J., 63 (1956) 388.
61 L. Bogorad, Science, 121 (1955) 878.
62 H. Fischer, E. Bartholomäus and H. Röse, Z. Physiol. Chem., 84 (1913) 262.
63 H. Fischer and W. Zerweck, Z. Physiol. Chem., 137 (1924) 242.
64 H. Fischer and A. Stern, Die Chemie des Pyrrols, Band II, Hälfte 2, p. 420, Leipzig, 1940.
65 D. Mauzerall and S. Granick, J. Biol. Chem., 232 (1958) 1141.
66 L. Bogorad, Fed. Proc., 14 (1955) 184.
67 L. Bogorad, J. Biol. Chem., 233 (1958) 501.
68 S. Granick, Abstr. Pap. Am. Chem. Soc., 128 (1955) 69.
69 S. Granick and D. Mauzerall, J. Biol. Chem., 232 (1958) 1119.
70 D.S. Hoare and H. Heath, Biochem. J., 73 (1959) 679.

71 S. Granick and D. Mauzerall, Fed. Proc., 17 (1958) 233.
72 S. Granick and D. Mauzerall, Ann. N.Y. Acad. Sci., 75 (1958) 115.
73 W.H. Lockwood and C. Rimington, Biochem. J., 67 (1957) 8 P.
74 H. Heath and D.S. Hoare, Biochem. J., 72 (1959) 14.
75 E.Y. Levin and D.L. Coleman, J. Biol. Chem., 242 (1967) 4248.
76 L. Bogorad, J. Biol. Chem., 233 (1958) 510.
77 L. Bogorad, J. Biol. Chem., 233 (1958) 516.
78 D. Mauzerall and S. Granick, J. Biol. Chem., 232 (1958) 1141.
79 C. Rimington and H.L. Booij, Biochem. J., 65 (1957) 3p.
80 S. Sano, S. Inone, Y. Tanabe, C. Sumiya and S. Koike, Science, 129 (1959) 275.
81 S. Sano, Acta Haematol. Jap., 21 (suppl.) (1958) 337.
82 S. Sano and S. Granick, J. Biol. Chem., 236 (1961) 1173.
83 R.F. Labbe and N. Hubbard, Biochim. Biophys. Acta, 41 (1960) 185.
84 B.F. Burnham, in D.M. Greenberg (Ed.), Metabolic Pathways, 3rd ed., Vol. 3, 1969, p. 403.
85 S. Granick, J. Biol. Chem., 183 (1950) 713.
86 R.J. Della Rosa, K.L. Altman and K. Salomon, J. Biol. Chem., 202 (1953) 771.
87 S. Granick, J. Biol. Chem., 172 (1948) 717.
88 S. Granick, J. Biol. Chem., 236 (1961) 1168.
89 S. Granick and D. Mauzerall, in D.M. Greenberg (Ed.), Metabolic Pathways (2nd ed. of Chemical Pathways of Metabolism), Vol. 2, 1961, p. 525.
90 K.D. Gibson, A. Neuberger and G.H. Tait, Biochem. J., 88 (1963) 325.
90a A. Neuberger and G.H. Tait, Biochem. J., 90 (1964) 607.
91 O.T.G. Jones, Biochem. J., 86 (1963) 429.
92 O.T.G. Jones, Biochem. J., 88 (1963) 335.
93 K.D. Gibson, A. Neuberger and G.H. Tait, Biochem. J., 83 (1962) 539.
94 K.D. Gibson, A. Neuberger and G.H. Tait, Biochem. J., 83 (1962) 550.
94a K.D. Gibson, A. Neuberger and G.H. Tait, Biochem. J., 84 (1962) 483.
95 O.T.G. Jones, Biochem. J., 89 (1963) 182.
96 A.A. Krasnovsky and L.M. Kosobutskaya, Dokl. Akad. Nauk SSSR, 82 (1952) 761.
97 J.W. Corcoran and D. Shemin, Biochim. Biophys. Acta, 25 (1957) 661.
98 D. Shemin, J.W. Corcoran, C. Rosenblum and I.M. Miller, Science, 124 (1956) 272.
98a R.C. Bray and D. Shemin, Biochim. Biophys. Acta, 30 (1958) 647.
98b R.C. Bray and D. Shemin, J. Biol. Chem., 238 (1963) 1501.
99 S. Schwartz, K. Ikeda, I.M. Miller and C.J. Watson, Science, 129 (1959) 40.
100 K. Bernhauer, O. Muller and F. Wagner, Angew. Chem. Int. Ed. Engl., 3 (1964) 200.
101 F. Wagner, Annu. Rev. Biochem., 35 (1966) 405.
102 K. Bernhauer, F. Wagner, H. Beisbarth, P. Rietz and H. Vogelman, Biochem. Z., 344 (1966) 289.
103 W. Friedrich and W. Sandeck, Biochem. Z., 340 (1964) 465.
104 W. Friedrich, Biochem. Z., 342 (1965) 143.

105   A. Di Marco and C. Spalla, Gionr. Microbiol., 9 (1961) 237.
106   G.S. Marks and L. Bogorad, Proc. Natl. Acad. Sci. USA, 46 (1960) 25.
107   U.V. Santer and H. Vogel, Biochim. Biophys. Acta, 19 (1956) 578.
108   H.H. Wasserman, J.F. McKeon and U.V. Santer, Biochem. Biophys. Res. Commun., 3 (1960) 146.
109   M.T. Rizki, Proc. Natl. Acad. Sci. USA, 40 (1954) 1057.

Chapter 61

The isoprenoid pathway (terpenes, carotenoids, side chain of tocopherols, ubiquinones, dolichols, rubber, sterols and steroids, steroid hormones, bile acids and bile alcohols, arthropod hormones, insect pheromones)

1. Introduction

Isoprenoid compounds have been considered from the point of view of the organic chemist in several parts of this Treatise: Chapter V of Vol. 9, the whole of Vol. 10 and Chapter I of Vol. 20. These compounds are derived by the repeated condensation of 5-carbon branched-chain units related to isoprene

$$CH_2=\overset{\overset{\displaystyle CH_3}{|}}{C}-CH=CH_2$$

which is in fact a product of the pyrolysis of many such compounds. The class includes the components of "essential oils" of many plants; their presence in the oil of turpentine led to their being baptized "terpenes".

The category also includes the carotenoids, the side-chains of tocopherols *, ubiquinones, phytol in chlorophylls, the "dolichols" **, rubber, sterols and steroids, etc., biosynthesized along the isoprenoid pathway.

Wallach [1] having classified the known terpenes on the basis of C_5H_8 units suggested theoretically that the monocyclic monoterpenes could be built up by the union of two properly coiled isoprene units, but many a repeating unit has been suggested (see Nicholas [2], pp. 645—647).

---

* See Vol. 9, Chapter VII.
** See Vol. 20, p. 31.

The "direct approach" (as Nicholas calls it) was used by Fujita [3] who injected in trees compounds considered as possible precursors of terpenes. After several months, the trees were cut and fractions were obtained from them by steam distillation. The interpretation of the results proved difficult.

In Chapter 46, we have referred to the approach to terpene biosynthesis based on chemical kinship. Ruzicka [54], in 1953, published what he called the "biogenetic isoprene rule", as was stated in Chapter 46. To quote Nicholas [2] (p. 654):

"According to this rule, certain key acyclic terpenes, or hypothetical substances closely related to them, can condense by ionic or radical mechanisms to form all of the known mono-, sesqui- and diterpenes, and, in the case of squalene as the acyclic precursor, can give rise to all of the known sterols and triterpenes of plant and animal tissues".

We have retraced, in Chapter 50, the inquiries which led to the recognition of squalene, a triterpene, as precursor of cholesterol in vivo, and the suggestion by Robinson, of a hypothetical folding mechanism for the conversion of linear squalene to the tetracyclic sterol nucleus.

2. "Acetate" as starting point of the isoprenoid pathway

As early as 1936, Schoenheimer and Rittenberg [4] enriched the body fluids of rats with heavy water. When the rats were given a cholesterol free diet, about one-half of the hydrogen atoms of their tissue cholesterol became deuterium labelled. Not only did this finding confirm that cholesterol is biosynthesized in the body of the animals as earlier studies had shown (Chapter 50) but it indicated that it started from small molecules.

The first use of deuteroacetate in studies on sterol biosynthesis was made by Sonderhoff and Thomas [5]. These authors reported in 1937 that approx. 31 atom per cent deuterium was introduced into the ergosterol of yeast when CD_3COONa was fed. Since this figure was approximately twice as high as the degree of incorporation into the higher fatty acids (see Chapter 50), and about 20 times as high as the incorporation into the carbohydrates formed, the conclusion was reached that neither fatty acids nor carbohydrates were important intermediates in the conversion of acetate to sterol. The data of Rittenberg and Schoenheimer [6] led to discarding the possibility of sterol biosynthesis by cyclization of fatty acids.

Bloch and Rittenberg [7] demonstrated that the deuterium content of rat cholesterol was 3 times as high as the deuterium content of the body fluids when the deuterium was introduced as CD_3COONa. Bloch and Rittenberg [8], in later experiments published the same year, excluded propionic, butyric, succinic, pyruvic and acetoacetic acid as precursors of cholesterol. The same authors, by pyrolysis of the cholesterol obtained after deuteroacetate was fed to the rats, showed that both the side chain and the steroid nucleus contained deuterium. Further work of the same authors [9] proved that both carbons of the acetate participate in sterol biosynthesis.

Tatum, Barratt, Fries and Bonner [10] isolated in 1950 an acetate-requiring mutant of *Neurospora crassa*. Using this mutant and acetate doubly labelled with ^{14}C and ^{13}C ($^{14}CH_3\,^{13}CO_2H$) Ottke, Tatum, Zabin and Bloch [11] found that at least 26 out of the 28 carbons of ergosterol were derived from acetate.

From the methodological viewpoint it must be noted that the recognition of acetate as the starting point of cholesterol biosynthesis in animals has been implemented by experiments on whole organisms. As emphasized by Bloch, Borek and Rittenberg [12].

"It can be estimated that in the liver of the rat half of the cholesterol molecules are replaced by newly formed cholesterol in 5 to 10 days. If cholesterol were synthesized as rapidly in surviving liver as in the living rat, the cholesterol concentration could not change more than a few per cent during the period of incubation".

While the first experiments on the role of acetate in the isoprenoid pathway were accomplished with deuterium, when isotopic forms of carbon became available they were used in experimentation.

In their experiments on surviving rat liver, Bloch et al. [12] not only showed that deuteriocholesterol was formed in the presence of either D_2O or deuterio-acetic acid, but also observed that incubation with acetic acid containing ^{13}C resulted in the formation of cholesterol containing heavy carbon.

Little and Bloch [13] used acetic acid labelled in either the carboxyl or the methyl group, or acetic acid labelled by ^{13}C as well as

Plate 207. Konrad Bloch.

[14]C. In this way, the authors were able to determine the relative utilization of the 2 carbon atoms of acetic acid for the biosynthesis of cholesterol as a whole as well as for the aliphatic and cyclic moieties of the molecule. The authors used for their experiments labelled cholesterol obtained through biosynthesis by surviving rat liver, as had been obtained by Bloch et al. [12]. Little and Bloch [13] concluded from their data that both the side chain and the *cyclo*pentanophenanthrene nucleus are formed from acetate and that of the 27 carbon atoms of cholesterol, 15 originate from the methyl and 12 from the carboxyl carbon of acetate. The degradation of the side chain of cholesterol, biosynthesized from [1-^{14}C]- or [2-^{14}C]acetate, revealed to Wuersch, Huang and Bloch [14] a similar labelling pattern to that recognized in rubber by Bonner and Arreguin [15]. The latter showed that the guayule plant uses acetate to biosynthesize rubber which was already known to consist of a long-chain polymer of isoprene units.

Wuersch et al. [14] called attention to the close relationship between sterols and terpenes.

That acetate is utilized for the synthesis of β-carotene by *Phycomyces blakesleeanus* was demonstrated by Grob, Poretti, von Muralt and Schopfer [16] in 1951.

3. Acetoacetate as an intermediate in the biosynthesis of cholesterol

The data of Brady and Gurin [17] suggested that acetoacetate could be incorporated into cholesterol without prior degradation to acetate, although this has never been rigorously proved even though Zabin and Bloch [18,19] seemed to have offered confirmatory evidence. These data suggested that at a stage in the isoprenoid pathway the condensation of two acetate molecules is involved. It has been demonstrated by Buchanan, Sakami and Gurin [20], by Brady and Gurin [17] and by Chen, Chapman and Chaikoff [21] that the isotope of singly labelled acetoacetate is not randomized by incubation with liver slices.

4. Distribution of acetate carbons in cholesterol

In 1947, J. Cornforth (now Sir John) and G. Popjàk both joined the National Institute for Medical Research. Popjàk [22] has

Plate 208. Sir John Cornforth and Lady R. Cornforth.

recalled the beginning of their collaboration:

"Cornforth was interested in the total chemical synthesis of cholesterol and I in its biosynthesis. It was the most natural thing in the world that we should combine our efforts. There were no biochemical reactions known 20 years ago whereby the complex molecule of cholesterol could be built up from acetate. Therefore we planned experiments to determine whether the two carbon atoms of acetate may not be arranged in some definite pattern in the cholesterol molecule and give us clues as to the type of reactions involved in the biosynthesis. We were encouraged in this approach by the manifestly successful work of Shemin and of Neuberger on haem and porphyrin biosynthesis, which was soon to lead to the identification of δ-aminolaevulinate as the precursor of porphyrins. We began the work on cholesterol in 1949 by which time both carboxy-labelled and methyl-labelled acetates became available to us. My task was relatively simple: I only had to find an efficient system to convert sufficient amounts of the two [^{14}C]-labelled acetates into cholesterol so that it could be diluted with several grams of inactive material needed for the chemical manipulations envisaged. The chemical task facing Cornforth was much more formidable since there were no methods available for the carbon-by-carbon degradation of the ring structure of cholesterol. We decided not to tackle the degradation of the side-chain, for which methods were available, because we learnt from Dr. Konrad Bloch, then in Chicago, that he planned to do that work" (pp. 147—148).

The development of this topic has been concisely narrated by Popjàk [22a] as follows:

"The distribution of acetate carbons in the ring-structure of cholesterol synthesized by rat liver slices either from acetate-1-^{14}C or acetate-2-^{14}C has been the subject of systematic study by Cornforth and the writer and their colleagues. Special methods have been developed for the degradation of rings A and B (Cornforth, Hunter and Popjàk [23,24]) and rings C and D (Cornforth, Popjàk and Gore [25]; Cornforth, Gore and Popjàk [26]), which allowed the isolation (with the exception of C-7) of all the carbon atoms of the sterol rings for isotope assay.

"Bloch [27] and Dauben and Takemura [28] reported on the origin of C-7. These investigations established the origin of all the carbons of cholesterol from either the methyl or carboxyl carbon of acetate.

Plate 209. George Joseph Popjàk.

"It was the invariable experience in these investigations that when acetate-1-^{14}C was the precursor of cholesterol only one group of positions of the sterol became labelled. On the other hand, when acetate-2-^{14}C was added to the liver slices, the cholesterol isolated contained one group of carbon atoms highly labelled and another group of lower isotope content (approximately one-tenth of the first group); the former group contained all the positions that did not derive isotope from acetate-1-^{14}C and the latter those that became labelled from acetate-1-^{14}C. The specific activities of the carbon atoms within each of the two groups were found equal to one another within experimental error. The fact that acetate-2-^{14}C contributed some isotope to positions which acquired label from the carboxyl carbon of acetate could be explained readily by the formation of $CH_3{}^{14}COOH$ from $^{14}CH_3COOH$ through the citric acid cycle. These results showed that every position in cholesterol had its origin in one or the other of the carbons of acetate and that of the 27 carbons of cholesterol 15 were derived from the methyl and 12 from the carboxyl carbon; this conclusion was reached by Little and Bloch [13] some time earlier on the basis of the ratio of $^{13}C/^{14}C$ in cholesterol biosynthesized from $^{14}CH_3{}^{13}COOH$" [22a] (pp. 534—535).

5. Squalene confirmed as cholesterol precursor

Enquiries on biosynthetic pathways have, as a rule, been based on chemical knowledge and the experimentation has generally been in a systematic manner, led by hypotheses derived from information previously acquired by organic chemists. The methodology has in the case of the isoprenoid pathway differed from this pattern. In 1952, rubber was known to be a polyterpene, and squalene was recognized as a dihydrotriterpene. Steroids were classified in another chemical family. When, at this time, Wuersch et al. [14] recognized that the degradation of the side chain of cholesterol biosynthesized from [1-^{14}C]- or [2-^{14}C]acetate revealed a pattern similar to the one identified by Bonner and Arreguin [15] in rubber, Bloch suggested a close chemical relationship between sterols and terpenes. We have, in Chapter 50 retraced the recognition of squalene (synthesized by Karrer and Helfenstein in 1931) as a triterpene with six isoprene units. In 1926, Channon, feeding squalene to rabbits, observed an increase in cholesterol of their tissues. After the general acceptance of the Rosenheim-King formula of cholesterol, Robinson [28a] proposed a hypothetical folding mechanism for the conversion of the linear squalene molecule to the tetracyclic sterol nucleus (see Chapter 50).

But this concept of considering squalene as a precursor of

Plate 210. Harold John Channon.

cholesterol lacked the basis of positive data. Balance studies which were carried out to test the concept gave conflicting results (Channon [29]; Channon and Tristram [30]; Kimizuka [31]). First isolated from shark liver oil by Tsujimoto in 1916, squalene has since been recognized to occur in yeast (Täufel, Thaler and Schreyegg [32]), in palm oil (Fitelson [33]), in human dermoid cysts (Dimter [34]) and in human sebum (McKenna, Wheatley and Wormall [35]; Sobel [36]). Attempts to obtain tagged squalene by organic synthetic methods failed because the available methods led to isomeric hydrocarbons which are biologically inactive.

The biological synthesis of squalene from acetate was described by Langdon and Bloch [37] in 1953. Labelled squalene can therefore be prepared biosynthetically and with the aid of such material a very efficient conversion of the hydrocarbon to cholesterol by mammalian (mice) tissues has been demonstrated (Langdon and Bloch [38]). Thus squalene appeared as a much better precursor of cholesterol than any other compound previously tried. Cornforth and Popjàk [39] prepared [^{14}C]squalene by biosynthesis from [2-^{14}C]acetate by rat liver slices and degraded the labelled squalene carbon by carbon.

They inferred from their data that three of the carbons of the isopentene units came from the methyl and two from the carboxyl of acetic acid:

$$\underset{m}{\overset{m}{\diagdown}}C-m-C-$$

6. Biosynthesis of cholesterol by cell-free preparations of rat liver and by microsomes

In 1953, Nancy Bucher [40] reported from Harvard University the finding that cell-free preparations of rat liver are capable of forming radioactive cholesterol from ^{14}C-labelled acetate. This was confirmed by Frantz Jr. and Bucher [41].

N. Bucher and McGarrahan [42] later showed that the biosynthetic process took place in the microsomal fraction (which, as stated in Chapter 16, is an artifact derived from the endoplasmic reticulum).

The same authors observed that following the injection of

labelled acetate, nearly all of the [^{14}C]cholesterol appeared in the microsomal fraction.

7. Role of coenzyme A

The discovery of the requirement for coenzyme A, a derivative of pantothenate, in biological acetylations, was reported in Chapter 33. That the multiple utilization of acetate by mammals in the biosynthesis of cholesterol requires the participation of coenzyme A was recognized by Boyd [43], at Edinburgh, in 1953. Boyd observed that if rats were maintained on a fat-free diet, the liver ester cholesterol concentration and the plasma ester cholesterol concentration depended on dietary panthothenate and hence on the coenzyme A content of the tissues.

8. Cyclisation of squalene

It is clear that at the time when biochemists were first engaged in the studies on the biosynthesis of sterols from squalene, they were handicapped by a lack of chemical knowledge concerning the compounds involved. The epistemological basis for the formation of hypotheses which could be put to the experimental test was increased by the progress made in several laboratories of organic chemistry engaged in the elucidation of the structure of the cyclic triterpenes. It was demonstrated by Voser, Mijovic, Hensen, Jeger and Ruzicka [44] that lanosterol, a C_{30}-sterol of wool fat had the same steroid nucleus as cholesterol, a fact which pointed to the possibility that a trimethyl C_{30}-sterol may be an intermediate in cholesterol formation.

In Chapter 50 we have recalled the theory put forward by Robinson in order to account for a possible folding of squalene, leading to cholesterol. This belonged to the category of guesses based on chemical knowledge but lacking factual substantiation.

Woodward and Bloch [45], in 1953, postulated, on the basis of a series of data (see Popjàk [22a]) that after folding and cyclisation of squalene a methyl group migrated from either position 8 or 14 to position 13 and the subsequent elimination of three methyl groups led to the formation of cholesterol.

As stated by Tchen:

"This scheme can be experimentally differentiated from that of Robinson in the following manner:
"If methyl-labelled acetate were used in the biosynthesis of cholesterol, the labelling pattern of carbon atoms 7, 8, 12, and 13 of cholesterol would be different depending upon the mechanism of cyclisation of squalene, as illustrated in Fig. 27. The carbon atom 13 would be labelled according to the Woodward-Bloch mechanism and unlabelled by the Robinson mechanism. Degradation of cholesterol biosynthesized from methyl-labelled acetate yielding carbon atoms 13 and 19 as acetate proved that carbon atom 13 in cholesterol is derived from the methyl carbon of acetate, in agreement with the Woodward-Bloch mechanism of squalene cyclization. Independently, the same conclusion was reported by Dauben, Abraham, Hotta, Chaikoff, Bradlow and Soloway [47], based on similar experimental evidence. When the degradation of the steroid nucleus was completed, the labelling pattern of carbons 7 and 12 of cholesterol and ergosterol synthesized from labelled acetates was shown to be also in accord with the prediction of the Woodward-Bloch mechanism [27,28,39,48,49].

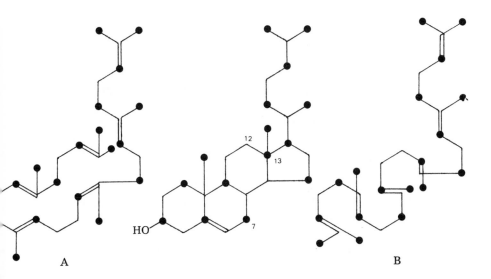

Cholesterol

Fig. 27. (Tchen [46]) Labelling pattern of the carbon atoms of acetate in cholesterol and in squalene. The squalene molecule on the left is folded according to Robinson, and the one on the right according to Woodward and Bloch.

In 1956, the scheme was further confirmed when labelled lanosterol was formed from acetate and was shown to be converted into cholesterol" (Clayton and Bloch [50,51]), (pp. 405—406).

In the first of these papers, Clayton and Bloch [50] introduce the subject in the following terms:

"It has been suggested (Woodward and Bloch [45]) that the cyclization of the aliphatic triterpenoid hydrocarbon to the tetracyclic ring system of the steroids occurs in a manner which would also rationalize the structure of lanosterol, a naturally occurring 4,4', 14-trimethylcholestane derivative (Voser et al. [44]).
   In view of its structural relationship to squalene on the one hand and to cholesterol on the other, it has become of interest to investigate the possible role of lanosterol, or of related structures having 30 carbon atoms, as intermediates in the biosynthesis of cholesterol."

As pure lanosterol was not available, Clayton and Bloch used the so-called "isocholesterol" of wool fat (consisting of lanosterol, dihydrolanosterol, agnosterol and dihydroagnosterol, with lanosterol as main component). From these experiments the authors concluded that lanosterol and agnosterol were synthesized from acetate by rat liver homogenates prepared according to N. Bucher.
   In their second paper, Clayton and Bloch [51] showed that biologically "labelled lanosterol" was converted into cholesterol in rat liver homogenates. These experiments were accomplished with "labelled lanosterol" obtained by diluting a small quantity of biologically labelled "isocholesterol" with non-radioactive lanosterol. The proportion of the total radioactivity which is attributed to lanosterol in the reacting medium as well as the proportion of radioactivity due to cholesterol in the product were obtained indirectly by calculation based on a number of assumptions. It appeared nevertheless that lanosterol was the principal single source of ^{14}C for cholesterol.
   In the last part of the discussion of their results obtained with lanosterol, Clayton and Bloch [51] carefully discuss the state of the theory of squalene cyclization as it was presented in 1956:

"While it is clear from the results of the present investigation that lanosterol can be a precursor of cholesterol, it does not follow that this transformation is a necessary event in the normal course of steroid biogenesis. Experiments with crude tissue preparations of the type used are inherently inconclusive in this respect. Some evidence has been obtained, however, which is consistent with the view that lanosterol is a normal precursor of cholesterol. The intestinal tract tissue of rats killed shortly after injection of acetate-^{14}C was found to contain lanosterol of specific activity several times higher than that of the cholesterol isolated in the same experiments. At longer time intervals after injection, the lanosterol fractions still contained ^{14}C, but their specific activities were now relatively low compared to those of cholesterol. According to Schwenk et al. [52,53], the cholesterol synthesized from acetate in similar experiments of short duration or in liver perfusions is accompanied by "high counting" materials which can serve as precursors of cholesterol. It seems possible from the above results that, in part at least, the activity in these so far unidentified substances may be due to lanosterol.

"Structural reasons make lanosterol particularly attractive as an intermediate in the conversion of squalene to cholesterol. On the one hand, the constitution of lanosterol is suggestive of an origin from a triterpenoid precursor by cyclization; on the other hand, lanosterol shares with the steroids the tetracyclic ring system and can in fact be designated as a 4,4', 14-trimethylcholestadienol. It was on the basis of these relationships that the cyclization scheme given in Fig. 1 was proposed [45]. Moreover, the cyclization of squalene to lanosterol has been rationalized in terms of ionic mechanisms by

squalene, C$_{30}$ lanosterol, C$_{30}$

cholesterol

Plate 211. Leopold Ruzicka.

the suggestion of Ruzicka [54] that, subsequent to an attack by a cationic reagent, the cyclization of the triterpene proceeds synchronously to lanosterol without stabilization of intermediates. The distribution of the carbon atoms of acetic acid in biosynthetic squalene reported by Cornforth and Popjàk [39] and particularly the origin of C_7 and C_{13} [27,45] of cholesterol from methyl carbons of acetate have until now been the most compelling evidence for the postulated mechanism of steroid biogenesis. That squalene itself is a direct precursor of the steroids, though in accord with all the experimental data, has remained a tentative conclusion because squalene is as yet not obtainable by organic synthesis, and because the identity of biologically labeled squalene has not been rigorously established. With the demonstration that lanosterol is synthesized in liver tissue and is efficiently converted to cholesterol, the squalene hypothesis has been considerably strengthened since squalene alone among naturally occurring compounds satisfactorily explains the origin of lanosterol and the role played by a C_{30} sterol in the biosynthesis of cholesterol."

The most remarkable theoretical contribution to the biosynthesis of isoprenoid was made in 1955 by Eschenmoser, Ruzicka, Jeger and Arigoni [55], who proposed a detailed mechanism of the cyclization of squalene to lanosterol. They postulated two hydride shifts (one from C-17 to C-20, another from C-13 to C-17) and two 1 : 2 methyl shifts, one from C-14 to C-13 and another from C-8 to C-14 in addition to the elimination of a proton from C-9 as a final step in stabilizing the cyclization product of squalene and leading to lanosterol (Fig. 28).

The authors expanded the "isoprene rule" into a "biogenetic isoprene rule", predicting the full stereochemical course of the cyclization of squalene to either lanosterol or to the tetracyclic triterpenes. It is the greatness of Ruzicka's "biogenetic isoprene rule" that all its predictions were subsequently demonstrated experimentally.

As stated by Popjàk:

"It is the remarkable feature of this hypothesis that, after certain well-defined assumptions are made, all the known C_{30} cyclic triterpenes and lanosterol may be derived with their full structural and configuration details from an all trans-squalene. It is assumed that all cyclizations of squalene are initiated by the attack of a cation (a hypothetical OH^+) at one end of the squalene molecule folded in a specific way (e.g., chair-boat-chair-boat configuration for lanosterol) and once the reaction commenced it proceeds non-stop by electron shifts in a concerted manner, i.e. without the formation of stable intermediates. The cyclizing molecule of course carries a positive charge leading to the formulation of an unstable carbonium ion which becomes ultimately

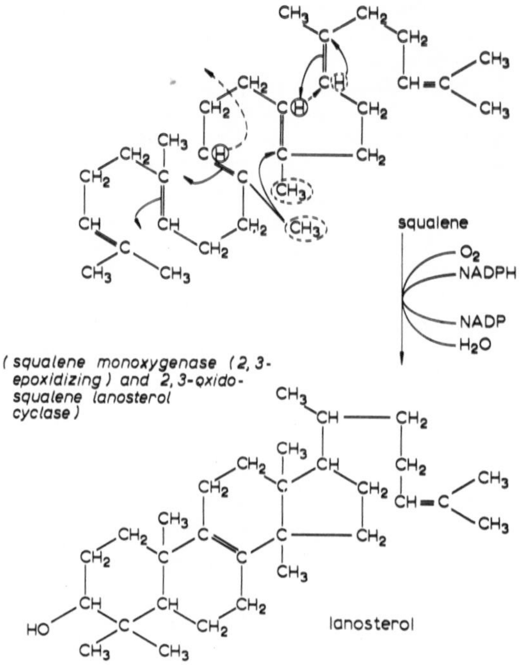

Fig. 28. Conversion of squalene into lanosterol.

stabilized by hydride and methyl shifts and not by hydration or proton elimination".

The cyclisation of squalene to lanosterol marks the beginning of the aerobic phase of sterol biosynthesis. All details cannot be considered as explained (see Danielsson and Tchen [56], pp. 122—124). The enzymes involved are microsomal. As in the case of all oxygenases, many features are still obscure. The relevant enzyme has first been called squalene oxidocyclase or squalene hydroxylase. The reaction has since been recognized as being catalysed by two enzymes: squalene monooxygenase (2,3-epoxidizing) and 2,3-oxidosqualene lanosterol-cyclase. Squalene monooxygenase (2,3-epoxidizing), also called squalene epoxidase, catalyses the reaction

squalene + AH_2 + O_2 = 2,3-oxidosqualene + A + H_2O

(on this enzyme see Corey, Russey and Ortiz de Montellano [57];

Tchen and Bloch [58]; van Tamelen, Willett, Clayton and Lord [59]; Yamamoto and Bloch [60]). It has received the systematic name [Squalene, hydrogen-donor : oxygen oxidoreductase (2,3-epoxidizing); 1.14.99.7].

The second component of what was previously called oxido-cyclase is 2,3-oxidosqualene lanosterol cyclase. It catalyses the reaction:

2,3-oxidosqualene = lanosterol

(on this enzyme see Dean, Ortiz de Montellano, Bloch and Corey [61]). This enzyme has received the systematic name [2,3-Oxido-squalene mutase (cyclizing, lanosterol-forming); 5.4.99.7].

9. Metabolic precursors of the isoprene units of squalene

In 1956 the knowledge of the isoprenoid pathway remained limited to the existence of a derivation of squalene from acetyl-CoA by an unknown mechanism, of a cyclisation of squalene into lanosterol according to the mechanism formulated by Ruzicka, and to a conversion of lanosterol into cholesterol by mechanisms unknown.

The situation was radically changed by the recognition of mevalonic acid as an intermediate in the isoprenoid pathway. This discovery came from an entirely different field: the study of nutritional requirements of lactic bacteria pursued by Karl Folkers at the Merck, Sharpe and Dohme laboratories. Wolf, Hoffman, Aldrich, Skeggs, Wright and Folkers [62] isolated from "distillers' solubles" a new compound which replaced acetate as a growth factor for *Lactobacillus acidophilus*. They determined the structure of the optically active compound, 3-hydroxy-3-methyl-pentano-5-lactone (mevalonic lactone). The parent acid was called mevalonic acid

$$CH_2COOH$$
$$CH_3C(OH)CH_2CH_2OH$$

Optically active mevalonic acid was prepared synthetically by Shunk, Linn, Huff, Gilfillan, Skeggs and Folkers [63].

The acid bears a close chemical relationship to 3-hydroxy-3-

Plate 212. Karl Folkers.

methylglutaric acid

$$CH_2COOH$$
$$CH_3\overset{|}{C}(OH)CH_2COOH$$

This compound was the subject of studies which demonstrated its biosynthesis from acetate and acetoacetate (Rudney [64]; Bachawat, Robinson and Coon [65]). Rabinowitz and Gurin [66], in Philadelphia, described the preparation of a particle-free extract of rat liver capable of incorporating ^{14}C-labelled acetate and 3-hydroxy-3-methyl glutaric acid into cholesterol. With the same enzyme system, Florapearl, Cobey, Warms and Gurin [67] demonstrated that acetate and 3-hydroxy-3-methylglutaric acid could be incorporated into squalene.

This raised the hope to find in 3-hydroxy-3-methylglutaric acid the branched-chain intermediate (isoprenoid unit), with an *iso*pentane skeleton, much sought after as an intermediate between acetate and squalene. But the attempts at labelling it at a specific position with an isotopic carbon failed to give the information looked for, due to the fact that 3-hydroxy-3-methylglutaric acid is in rapid equilibrium, in biological systems, with 3-methylcrotonic acid, 3-hydroxy*iso*valeric acid and 3-methylglutaconic acid as well as with acetyl-CoA and acetoacetyl-CoA (Adamson and Greenberg [68]).

The chemical kinship of mevalonic acid to 3-hydroxy-3-methyl glutaric acid prompted Tavormina, Gibbs and Huff [69] to test it, labelled with ^{14}C, as a possible intermediate in cholesterol biosynthesis. Liver preparations were found to be able to convert the synthetic DL-mevalonic acid labelled in position 2, into cholesterol with an efficiency of 43%. The authors observed that the carboxyl carbon was eliminated as CO_2.

The results obtained by Tavormina et al. [69], and by Tavormina and Gibbs [70] suggested that mevalonic acid may directly donate five of its carbon atoms as isoprenoid units for sterol biosynthesis, a point which was examined in detail by Cornforth, Cornforth, Popjàk and Gore [71]. When these authors degraded [^{14}C]squalene, biosynthesized from [2-^{14}C]mevalonate, they found it to contain only six labelled positions: one in each of the terminal isopropyl groups and four within the chain of the molecule. The branched methyl groups were not labelled. The authors concluded that the results indicated that the C-2 and the β-methyl carbon of mevalonic acid retained their individuality during biosynthesis, and that an asymmetrically labelled isoprenoid unit was derived from mevalonic acid. They concluded therefore that the biosynthesis of squalene from mevalonic acid must proceed by the coupling of C-2 of one molecule to C-5 of another. Using a different method of degradation of squalene, Dituri, Gurin and Rabinowitz [72] reached similar conclusions.

10. From mevalonic acid to isopentenyl pyrophosphate and dimethylallylpyrophosphate

After the discovery of mevalonic acid as a precursor of cholesterol, it was shown that liver preparations (Popjàk, Gosselin, Gore and Gould [73]) as well as yeast extracts (Amdur, Rilling and Bloch [74]; Lynen [75]) required ATP, Mg^{2+} or Mn^{2+} and NADH for the conversion of mevalonic acid into squalene.

To account for the ATP requirement, it was suggested that ATP might be necessary

"to form an active acyl group so that condensation may occur between the adjacent CH_2 group and an aldehyde or an acyl-CoA group of another molecule of the condensing unit" (Tchen [46] p. 400).

This hypothesis was rendered unlikely by the finding that the condensing unit did not have aldehyde or carboxyl groups, and that during the condensation no loss of hydrogen took place from the CH_2 groups derived from the C-2 of mevalonic acid (Amdur et al. [74]; Rilling, Tchen and Bloch [76]; Rilling and Bloch [77]).

Another possible explanation, which proved to be the correct one, was that ATP is required because phosphate esters are involved in the pathway.

The first phosphorylation was demonstrated to take place in yeast autolysate by Tchen [78]. The product was recognized by this author as a phosphate ester of mevalonic acid. By the use of labelled compounds, Chaykin, Law, Phillips, Tchen and Bloch [79] recognized that the phosphate is attached to the 5-hydroxyl group of mevalonic acid. To this indirect demonstration, Henning, Kessel and Lynen [80] added the direct demonstration of chemical synthesis. The formation of 5-phosphomevalonic acid was also identified in the conversion of mevalonic acid in a first phosphorylated derivative by liver enzymes in the presence of ATP and Mg^{2+}, by de Waard and Popjàk [81].

Levy and Popjàk [82] prepared from pig liver the enzyme catalysing the phosphorylation of mevalonic acid into 5-phosphomevalonic acid.

Tchen [83] had already reported the partial purification of the same enzyme from yeast, which he called mevalonic kinase. This enzyme has received the systematic name [ATP : mevalonate 5-phosphotransferase; 2.7.1.36].

In the reaction catalysed by yeast mevalonate kinase, CTP, GTP or UTP has been recognized as able to act also as phosphate donor instead of ATP. Liver mevalonate kinase can function only with ATP or ITP (inosine triphosphate). (On this enzyme see Markley and Smallman [84]).

The formation of a second phosphorylated intermediate: 5-pyrophosphomevalonate (5-diphosphomevalonate) was also demonstrated by the use of yeast extracts (Chaykin, Law, Phillips, Tchen and Bloch [79]; Agranoff, Eggerer, Henning and Lynen [85]). The diphosphomevalonate was also reported by de Waard and Popjàk [81] to be formed in rat-liver preparations. The nature of the product of the phosphorylation of 5-phosphomevalonic acid in the presence of ATP remained uncertain for some time.

Tchen [46] summarizes as follows the evidence from which its structure was deduced:

"1. It contains two moles of P per each mole of mevalonic acid.
  2. It still retains the carboxyl group of mevalonic acid.
  3. One of the two phosphates, the one derived from ATP, is readily hydrolyzed off by acid.
  4. In its subsequent conversion to the pyrophosphate of Δ^3-isopentenol, both phosphate groups are retained, one in an acid-labile linkage.
  5. Its acid hydrolysis gives a product identical to P-mevalonate.
From this evidence, only one reasonable structure can be assigned to this compound, namely 5-pyrophosphomevalonic acid" (pp. 400—401).

Levy and Popjàk [82] proposed for the enzyme involved the name phosphomevalonic kinase. They removed it completely from mevalonic kinase. Partial purification of the enzyme from yeast extracts was described by Henning, Möslein and Lynen [86]; and by Bloch, Chaykin, Phillips and de Waard [87].

Phosphomevalonate kinase has received the systematic name [ATP : 5-phosphomevalonate phosphotransferase; 2.7.4.2].

By the use of an enzymatic system obtained from yeast, Chaykin et al. [79] showed the stepwise accumulation of 5-phosphomevalonic acid, 5-pyrophosphomevalonic acid and isopentenyl pyrophosphate (pyrophospho-3-methyl-but-3-ene-1-ol). Lynen, Eggerer, Henning and Kessel [88] have demonstrated the incorporation of chemically synthesized 5-phosphomevalonic acid and isopentenylpyrophosphate into squalene by a similar yeast enzyme system.

With an enzyme system of rat liver, Witting and Porter [89], at

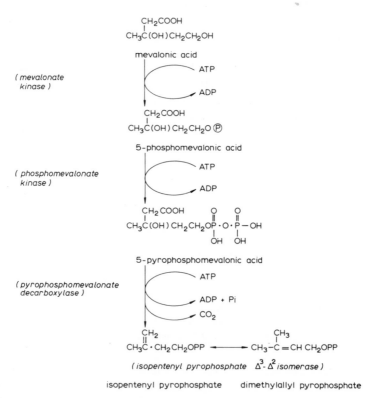

Fig. 29. From mevalonic acid to isopentenyl pyrophosphate and dimethylallyl pyrophosphate.

Madison, have also identified the successive formation of 5-phosphomevalonic acid, 5-pyrophosphomevalonic acid and isopentenyl pyrophosphate.

Tchen [46] summarizes as follows the evidence which has led to identification of isopentenyl pyrophosphate:

"1. It has two moles of P per each mevalonic acid or its equivalent.

2. It no longer contains the carboxyl group of mevalonic acid. Hence it must contain the C-2 of mevalonic acid as a terminal methylene group and cannot have a hydroxyl group (or phosphate) on C-3, . . .

3. Enzymic hydrolysis of the two phosphate groups (by snake venom) gives rise to Δ^3-isopentenol" (pp. 401—402).

Lynen et al. [88] have synthesized isopentenyl pyrophosphate and showed that it is identical to the enzymic product. The enzyme commonly designated as pyrophosphomevalonate decarboxylase, first identified by Bloch et al. [87] has received the systematic name [ATP : 5-diphosphomevalonate carboxy-lyase (dehydrating); 4.1.1.33].

Agranoff, Eggerer, Henning and Lynen [90] identified an enzyme catalysing the isomerisation of isopentenyl pyrophosphate in dimethylallyl diphosphate. This enzyme, isopentenyl pyrophosphate Δ-isomerase has received the systematic name [Isopentenyl-diphosphate Δ^3-Δ^2-isomerase; 5.3.3.2].

11. Identification of the "biological isoprene units" and their conversion to farnesyl pyrophosphate

Lynen et al. [88] have characterized farnesyl pyrophosphate as an intermediate in the conversion of isopentenyl pyrophosphate to squalene in yeast. But the condensation of three molecules of isopentenyl pyrophosphate would lead, not to farnesyl pyrophosphate, but to an isomer of it. In yeast, it has been demonstrated (Bloch et al. [87]; Lynen, Agranoff, Eggerer, Henning and Möslein [91]) that an isomerization of isopentenyl pyrophosphate to dimethylallyl pyrophosphate takes place in presence of the isopentenyl pyrophosphate isomerase mentioned in the previous section. The condensation of isopentenyl pyrophosphate and dimethylallyl pyrophosphate gives, as shown by the same authors, geranyl pyrophosphate. Geranyl pyrophosphate, condensing with isopentenyl pyrophosphate, leads to farnesyl pyrophosphate.

Goodman and Popjàk [92] found that a fraction of soluble enzymes from rat liver homogenates catalyses the formation of farnesylpyrophosphate from mevalonate or from 5-phospho-mevalonate in excellent yield, and that dimethylallyl pyrophosphate and geranyl pyrophosphate are also always formed during the process which indicates that the reactions observed in yeast cells also occur in liver cells. To quote Goodman and Popjàk [92] (p. 298):

"There can be no doubt that farnesyl pyrophosphate is the major (or only) sesquiterpenoid intermediate in squalene biosynthesis, since squalene can be obtained from it in yields as high as 80%. In both the liver and the yeast cells the reaction involved in the formation of farnesyl pyrophosphate (and of the other allyl pyrophosphates) from mevalonate are catalyzed by enzymes found in the soluble fraction of the disrupted cells. The only cofactors needed for the synthesis of the allyl-pyrophosphates from mevalonate or from 5-phos-phomevalonate are ATP and a divalent cation. It was demonstrated in Bloch's laboratory (Chaykin et al. [79]; Bloch et al. [87] that 3 moles of ATP are required for the formation of 1 mole of isopentenyl pyrophosphate from 1 mole of mevalonate, and both Chaykin et al. [79] and Lynen et al. [88] have shown that squalene synthesis can proceed from isopentenyl pyrophos-phate without ATP. It follows, therefore, that in the soluble liver enzyme system the ATP is probably needed only for the formation of isopentenyl pyrophosphate and that the synthesis of geranyl and farnesyl pyrophosphates proceeds by the electrophilic condensation of dimethyl allyl pyrophosphate and of geranylpyrophosphate with isopentenyl pyrophosphate, as proposed by Lynen et al. [88], by Rilling and Bloch [77], and by Cornforth and Popjàk [93], and proved experimentally by Lynen and his colleagues" [91].

"The two 5-carbon alcohol pyrophosphates (isopentenyl pyrophosphate and its isomerized product dimethylallyl pyrophosphate) may be considered as representing the long sought after "biological isoprene units"."

Popjàk and R.H. Cornforth [94] have proposed in 1960 that the terpene alcohols should be called *prenols*, with the prefix mono-, di-, tri (or poly-, etc.) according to the number of isoprenoid units they contain. Consequently they use the term prenyl pyrophos-phate to describe collectively dimethylallyl pyrophosphate, geranyl pyrophosphate and farnesyl pyrophosphate. They also use the term prenyltransferases for dimethylallyltransferase and geranyltransferase. These transferases were first recognized in yeast autolysates and purified about 20-fold by Lynen et al. [91] who named them farnesyl pyrophosphate synthetases. Isopentenyl pyrophosphate isomerase was isolated from pig liver by Shah, Cleland and Porter [95] in 1965.

Benedict, Kett and Porter [96] isolated geranyltransferase from pig liver in 1965.

"Prenyltransferases" were purified 100-fold from pig liver by Holloway and Popjàk [97] in 1967. The same authors confirmed the presence of isopentenyl pyrophosphate isomerase in pig liver and purified it 100-fold.

The enzyme dimethylallyltransferase, also transferring geranyl and farnesyl residues (also called farnesylpyrophosphate synthetase or prenyltransferase) has received the systematic name [Dimethylallyldiphosphate : isopentenyldiphosphate dimethylallyltransferase; 2.5.1.1.].

12. Condensation of farnesyl pyrophosphate.
Presqualene

That, both in yeast and in animals, the enzymic dimerization of two molecules of farnesyl pyrophosphate leads to squalene has been well documented by Lynen et al. [88]. Concerning the mechanism of a condensation of two molecules of farnesyl pyrophosphate to such a symmetrical molecule as squalene, hypothetical schemes have been suggested by Lynen et al. [88] and by Conforth and Popjàk [93] on the theme of the rearrangement of a molecule of farnesyl pyrophosphate to a nucleophilic substance, such as a derivative of nerolidol, followed by a condensation of this substance with farnesyl pyrophosphate. The scheme proposed by Cornforth and Popjàk [93] was supported by experimental data which were obtained by Rilling and Bloch [77]. According to these authors, squalene biosynthesized by a yeast autolysate from mevalonate-5-D_2 contained 10 atoms of deuterium instead of the theoretically possible 12; the two missing deuteriums having been lost from the two central atoms of squalene. Furthermore, it was reported by Rilling and Bloch [77] that when squalene was synthesized from [2-^{14}C]mevalonate in a medium of 99% D_2O, 4 atoms of deuterium were incorporated into the hydrocarbon; 2 of these atoms were attached to the 2 central carbon atoms and the other 2 to the terminal carbon atoms of each end of the squalene molecule.

References p. 316

To quote Popjàk, Goodman, J.W. Cornforth, R.H. Cornforth and Ryhage [98]:

"Because farnesyl pyrophosphate was found to be the immediate precursor of squalene and because C-1 of farnesyl pyrophosphate is derived from C-5 of mevalonate, the data of Rilling and Bloch could be interpreted to indicate either a change in the level of oxidation of C-1 of farnesyl pyrophosphate before condensation or an exchange of one of the hydrogen atoms attached to this carbon atom in each of the two farnesyl pyrophosphate molecules during condensation. According to the hypothesis of Cornforth and Popjàk [93], loss of 2 hydrogen atoms occurred from the two central carbon atoms of an unstable molecule, which was derived from the condensation of farnesyl pyrophosphate with a derivative of nerolidol (nerolidyl pyrophosphate); this led to dehydrosqualene (an analogue of phytoene), the reduction of which with NADPH to squalene was assumed" (p. 1934).

To test the correctness of this hypothesis, Popjàk et al. [98], studied the squalene synthesis from $[2\text{-}^{14}C]$mevalonate-5-D_2 and measured the incorporation of tritium, from either the water of the incubation medium, or from tritium-labelled NADPH into

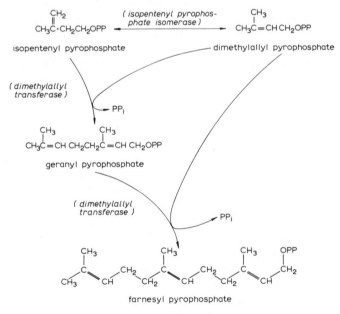

Fig. 30. Biosynthesis of farnesyl pyrophosphate.

squalene during its synthesis from [^{14}C]farnesyl pyrophosphate, or from [2-^{14}C]mevalonate.

The results were unexpected. They showed that the number of D$_2$ atoms in squalene biosynthesized from mevalonate-5-D$_2$ by the liver enzyme system was 11 and not 10, and that the one lost from one of the central carbons of squalene was replaced by hydrogen derived directly from NADPH and not from the water of the incubation medium.

"Thus, squalene biosynthesized from mevalonate-5-D$_2$ is asymmetrically labelled with deuterium on its 2 central carbon atoms (—CHD · CH$_{2-}$) indicating an asymmetrical process in the condensation of 2 farnesyl pyrophosphate molecules to squalene" (p. 1934).

In harmony with these data, Childs and Bloch [99] confirmed that 11, out of a total of 12, atoms of deuterium are retained in the conversion and not 10 as reported earlier from their laboratory. From these data, the possible types of mechanisms have been narrowed down. Popjàk et al. [98] have discussed possible hypotheses.

Danielsson and Tchen [56] (p. 127) refer to the

"beautiful and extensive stereochemical studies by Cornforth and Popjàk and their collaborators on the reactions between mevalonate and squalene (Cornforth, Cornforth, Donninger and Popjàk [163]). These workers demonstrated that (1) the isomerization of isopentenyl pyrophosphate to dimethylallyl pyrophosphate is stereospecific, (2) the condensation between an allylic pyrophosphate and isopentenyl pyrophosphate occurs with inversion, and (3) the condensation of two farnesyl pyrophosphates to squalene involves the stereospecific introduction of one hydrogen atom from NADPH onto the C-1 of one of the C$_{15}$ units, and inversion of the C-1 of the other C$_{15}$ unit".

A detailed historical survey of these beautiful studies is found in the first Ciba Medal Lecture delivered at a meeting of the Biochemical Society on 21 October 1966 by Popjàk and J.W. Cornforth [99a].

Rilling [100] has isolated a new intermediate in the biosynthesis of squalene. It is the product of the condensation of two molecules of farnesyl pyrophosphate into the pyrophosphate ester of a 3-carbon alcohol whose carbon skeleton differs from that of squalene by the presence of a cyclopropane ring. Rilling believes that squalene is formed directly by the reduction and rearrange-

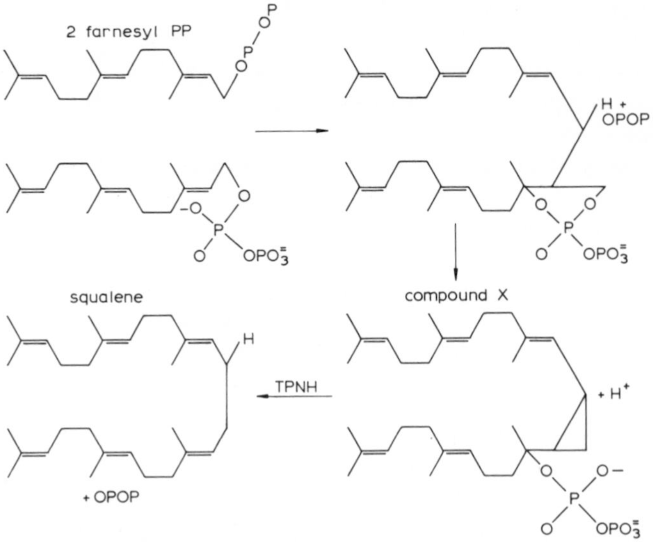

Fig. 31. (Rilling [100]) A mechanism for the synthesis of squalene from farnesyl pyrophosphate showing the formation of Compound X as an intermediate.

ment of this compound. From the knowledge that a reduction is involved in the conversion of farnesyl pyrophosphate to squalene, Rilling concluded that intermediates prior to the reductive step would accumulate in the absence of NADPH. He prepared a yeast enzyme system capable of synthesizing squalene from geranyl pyrophosphate, isopentenyl pyrophosphate and NADPH. In the absence of NADPH this system was recognized as containing a compound different from farnesyl pyrophosphate, which was called compound X. A mechanism suggested by Rilling is shown in Fig. 31.

To quote Rilling [100] (p. 3235):

"In this mechanism the C-1 of the one farnesyl pyrophosphate condenses with the C-2 of a second farnesyl pyrophosphate with the concomitant loss of pyrophosphate yielding a cyclic phosphate intermediate. This compound then loses a proton with the simultaneous formation of a cyclopropane ring and opening of the cyclic phosphate. Finally, a hydride transfer from NADPH cleaves the cyclopropane ring with the elimination of the pyrophosphate-giving squalene".

The cyclic phosphate intermediate between farnesyl pyrophosphate and squalene was called presqualene pyrophosphate by Epstein and Rilling [101]. In their paper on the mechanism of the biosynthesis of squalene from farnesyl pyrophosphate, Popjàk et al. [98] have postulated the possible formation, from farnesyl pyrophosphate, of stable intermediates such as (I) and (II)

$$R-CH_2-\underset{\underset{\displaystyle O}{|}}{\overset{\overset{\displaystyle CH_3}{|}}{C}}-CH=CH-CH_2=\underset{\underset{\displaystyle CH_3}{|}}{C}-CH_2-R \qquad (I)$$

$$R-CH_2-\underset{\underset{\displaystyle O}{|}}{CH}-\overset{\overset{\displaystyle CH_3}{|}}{CH}-CH-CH=\underset{\underset{\displaystyle CH_3}{|}}{C}-CH_2R \qquad (II)$$

Both (I) and (II) are asymmetric and they have both lost one hydrogen atom originally attached to C-1 of one of the two farnesyl pyrophosphate molecules from which they were hypothetically derived. Rilling first ascribed the structure of II to presqualene pyrophosphate. This was one of the structures proposed by Popjàk et al. [98] as the intermediate between farnesyl pyrophosphate and squalene. But it turned out not to be the correct structure.

Edmond, Popjàk, Wong and Williams [101a] showed that presqualene alcohol biosynthesized from [1-D$_2$]farnesyl pyrophosphate contained only three deuterium atoms, two of these being on the carbinol carbon and the third at position 3 of the cyclopropane ring. The authors remark that the elimination of the one pro-S proton from C-1 of one farnesyl phosphate, noted earlier during the overall synthesis of squalene (Popjàk and Cornforth [99a]) occurs during the synthesis of presqualene pyrophosphate. This was first recognized by Rilling [100]. To quote Popjàk [101b] (p. 652A):

"Presqualene pyrophosphate and its hydrolysis product, presqualene alcohol,

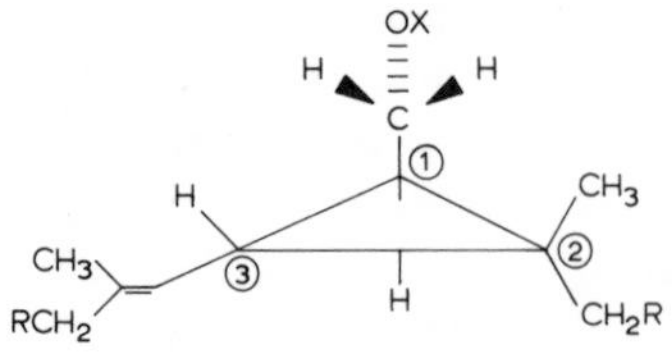

(Ⅲ)

Fig. 32. The stereochemistry of the molecule of presqualene alcohol. ▶, above the plane of the paper; | | | |, below the plane of the paper; ———, in the plane of the paper. The cyclopropane ring itself is perpendicular to the plane of the paper, C-1 of the ring being away from the viewer. Epstein and Rilling [101] were the first to propose the correct structure, but the real proof by NMR, mass spectrometry and chemical degradation and identification of fragments came from Popjàk's laboratory (cf. Edmond, Popjàk, Wong and Williams [101a]) and also by chemical synthesis from several laboratories.

are optically active. Thus the symmetrical molecule of squalene is synthesized through the intermediacy of a truly asymmetric molecule, which is clearly a relative of chrysanthemum monocarboxylic acid that has been known since 1924, as a constituent of pyrethrin I. From a comparison of the NMR Spectra * of presqualene alcohol with those of the *trans*- and *cis*-chrysanthemols (made by LiAlH$_4$ reduction of the methyl esters of the respective acids), we established that the two large substituents on the cyclopropane ring were in an *anti*-position to the carbonyl group whereas the methyl group on the ring was *syn* to it. Further, by an extension of Nakanishi's benzoate chirality rule, we were able to deduce that the absolute configuration of presqualene alcohol at all three asymmetric centers in the cyclopropane was *R* as shown in III . . ." (literature in Popjàk [101b]).

Concerning the mechanism of the biosynthesis of presqualene pyrophosphate, Epstein and Rilling [101] and Edmond et al. [101a] have proposed a similar mechanism. This mechanism was epitomized by Popjàk [101b] as follows (p. 653A):

"It assumes the parallel alignment of two farnesylpyrophosphate molecules on the enzyme and the initiation of the reaction by a nucleophilic group on the enzyme polarizing the allylic double bond of the farnesyl pyrophosphate lying at the top and resulting in the formation of C-C bond between C-2 of

———

* Nuclear Magnetic Resonance spectra, see Vol. 3, Chapter VIII, by C.D. Jardetzky and O. Jardetzky.

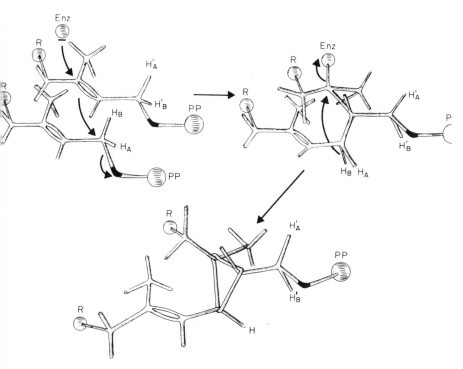

Fig. 33. Hypothetical mechanism of the biosynthesis of presqualene pyrophosphate (Edmond et al. [101a]). The pyrophosphate groups are represented
by the large globes.

one farnesyl pyrophosphate and C-1 of the other, with elimination of the
pyrophosphate anion from the second. Stereospecific elimination of the one
proton that was originally the pro-S hydrogen atom at C-1 of this second substrate molecule, with closure of the cyclopropane ring, completes the
sequence.

"This mechanism (Fig. 33) accounts not only for the deduced absolute
configuration of presqualene alcohol, but also for the observations I already
mentioned that the carbinol carbon of presqualene pyrophosphate is derived
from C-1 of one farnesyl pyrophosphate molecule and that C-1 of the second
farnesyl pyrophosphate is at position 3 of the cyclopropane ring and carries
one of the two hydrogen atoms originally at C-1 of that second farnesyl pyrophosphate. It also follows from this mechanism that the absolute configuration at the carbinol carbon of presqualene pyrophosphate or alcohol must be
the same as at C-1 of farnesyl-PP and that the oxygen in the C-O-P bond of
presqualene pyrophosphate is the one originally in farnesyl pyrophosphate".

References p. 316

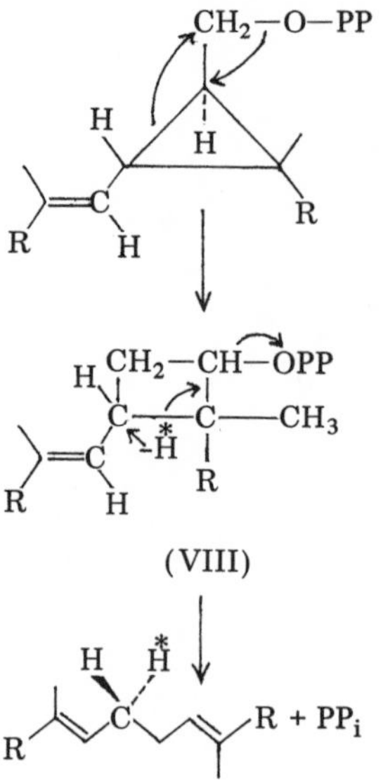

Fig. 34. Hypothetical mechanism of the conversion of presqualene pyrophosphate into squalene (Edmond et al. [101a]).

With farnesyl pyrophosphate labelled stereospecifically with deuterium and also with ^{18}O (Popjàk, Ngan and Agnew [101c]) the suggested mechanism was demonstrated.

A mechanism (Fig. 34) for the conversion of presqualene-PP into squalene through the intermediary of a cyclobutyl pyrophosphate has been suggested by Edmond et al. [101a]. The authors comment on this scheme as follows:

"We suggest a two-step reaction for the transformation the first one being the rearrangement of presqualene-PP, by ring expansion and migration of the pyrophosphate group, to the tautomeric cyclobutyl derivative (structure VII). Reduction by NADPH in the second step leads to ring cleavage, generation of the double bond, and elimination of the pyrophosphate anion. The attack by the hydride ion of NADPH must be directed from behind the cyclobutyl ring in order that it may attain the correct steric position in squalene as well as that the *trans* double bond may be generated" (p. 6270).

Space limitations prevent from reporting here the fascinating studies performed on the stereospecificity of enzymatic reactions involved in the isoprenoid pathway. The reader is referred, for a clear presentation of the way enzymes deal with "symmetrical" substrates in an asymmetrical manner, and for the terminology adopted in studies on chirality *, to Goodwin's essay entitled *Prochirality in Biochemistry* [101d] in which reactions involving the C-4, C-5 and C-2 prochiral centres of mevalonate, respectively, are analysed.

13. Conversion of lanosterol to cholesterol

It has been established that lanosterol, the product of the cyclization of squalene, is converted into cholesterol. This conversion consists in a removal of three methyl groups (positions 4, 4' and 14); a shift of the nuclear double bond from position 8,9 to position 5,6 and a reduction of the side chain double bond 24,25.

Many sterol intermediates, convertible to cholesterol by tissue preparations, or isolated from biological sources, or both, have been recognized [102–113]. But no exact sequence or sequences have so far been determined. From a series of data related to the succession of chemical changes taking place during the conversion of lanosterol to cholesterol, schemes of this derivation have been suggested (see for instance Popjàk and Cornforth [114], or Dagley and Nicholson [115]).

The stereospecific aspects involved in the conversion of lanosterol into cholesterol have been analysed in the section "Desaturation in Sterol Biosynthesis" of Goodwin's essay mentioned above.

14. Biosynthesis of mevalonic acid from acetic acid

It was only in 1959 that an enzymic formation of mevalonate from acetate by a tissue extract was demonstrated by Witting,

---

* "A compound is said to be *chiral* and to possess *chirality* if it cannot be superimposed on its mirror image, either as a result of simple reflection or after rotation and reflection. If superposition can be achieved then the molecule is said to be *archiral*" (Goodwin [101d] p. 106). The term *chirality* is equivalent to Pasteur's *dissymmetry*.

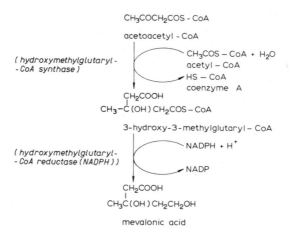

Fig. 35. From acetoacetyl-CoA to mevalonic acid.

Knauss and Porter [116]. Activation of acetate to acetyl-CoA and condensation of this to acetoacetyl-CoA are well known metabolic reactions.

A condensation of acetoacetyl-CoA with acetyl-CoA has formally been studied in the laboratories of Rudney at Cleveland (Rudney [117,118]; Rudney and Ferguson [119—121]), and of Lynen [122]. The enzyme involved was called β-hydroxy-β-methylglutaric acid coenzyme A ester condensing enzyme by Ferguson and Rudney [123], who partially purified it from yeast. Now commonly called hydroxymethylglutaryl-CoA synthase, this enzyme has received the systematic name [3-Hydroxy-3-methyl-glutaryl-CoA acetoacetyl-CoA-lyase (CoA-acetylating); 4.1.3.5].

When, as narrated above, mevalonic acid was recognized as an efficient precursor of cholesterol and related compounds, it became interesting, in view of its chemical similarity with hydroxy-methylglutaric acid to determine whether the latter could be enzymatically reduced to mevalonic acid. Investigations on yeast extracts demonstrated the direct synthesis of mevalonic acid from hydroxymethylglutaryl-CoA (Ferguson, Durr and Rudney [124]; Lynen [125]).

The enzyme was purified from yeast by Knappe, Ringelmann and Lynen [126] who called it β-hydroxy-β-methyl reductase. Details on the reduction of β-hydroxy-β-methylglutaryl-CoA to mevalonic acid were described by Durr and Rudney [127] who called the enzyme hydroxymethylglutaryl-CoA reductase. They showed that NADPH was the reductant and that the reduction occurs on the thiol-esterified carboxyl group of hydroxymethyl-glutaryl-CoA. Now called hydroxymethylglutaryl-CoA reductase (NADPH), the enzyme has received the systematic name [Mevalo-nate : NADP$^+$ oxidoreductase (CoA-acylating); 1.1.1.34].

This enzyme has further been purified from yeast by Kirtley and Rudney [128] and by Knappe et al. [126]. Kawachi and Rudney [129] have succeeded in preparing it from rat liver micro-somal fraction. Relying on chemical knowledge, it was foreseen that the reduction should involve a two stage process with mevaldic acid as an intermediate.

$$CH_2COOH$$
$$CH_3C(OH)CH_2CHO$$

Mevaldic acid

Neither the school of Rudney, nor the school of Lynen ob-served such an intermediate. The direct pathway is therefore believed to be the major pathway of mevalonate biosynthesis. (On data pointing to other possible pathways, see Popjàk and Corn-forth [114], and Danielsson and Tchen [56].)

15. From mevalonic acid to rubber

After the position of mevalonic acid and isopentenyl pyrophos-phate as key intermediates in the biosynthesis of sterols was estab-lished as reviewed above, several investigators showed that mevalonic acid and isopentenylpyrophosphate are incorporated into high molecular weight rubber by freshly tapped *Hevea brasiliensis* latex (Park and Bonner [130]; Kekwick, Archer, Barnard, Higgins, McSweeney and Moore [131]; Lynen and Hen-ning [132]; Archer, Ayrey, Cockbain and McSweeney [133]; Archer, Audley, Cockbain and McSweeney [134]). Park and Bonner [130] showed that the degradation of the biosynthesized rubber yields a pattern of labelling similar to that found for squalene.

16. Biosynthesis of plant terpenes

In the chapter on Isoprenoid Compounds written by Haagen-Smit and C.C. Nimmo in this Treatise (Vol. 9; Chapter V) a systematic survey of terpenoids has been given.

After the discovery of mevalonic acid, the use of ^{14}C permitted the development of a large number of researches which, in the wide field of plant terpenes appeared in agreement with the "biogenetic isoprene rule" (literature in Nicholas [2]). Where labelled mevalonic acid has been applied to plant tissues, the degradation of terpenes has given support to the "biogenetic isoprene rule". Birch, Boulter, Fryer, Thomson and Willis [135] fed [2-^{14}C]-sodium acetate and [2-^{14}C]mevalonic lactone to isolated terminal branches of *Eucalyptus citriodora* and *E. globulus Lab.* By steam distillation, citronellal was isolated from *E. citriodora* and cineole from *E. globulus* (both in the form of complexes with other compounds) and degradation confirmed the pattern of distribution for biogenesis from an isoprenoid unit.

A scheme of derivation of sesquiterpenes from farnesol has been given by Ruzicka [54] as part of his "biogenetic isoprene rule". Of diterpenes, many forms (acyclic, bicyclic, tricyclic, tetracyclic and pentacyclic forms) are known. On the basis of structure considerations, Ruzicka has suggested a geranylgeraniol type of structure as biogenetic precursor of diterpenes.

At the present time, very few examples are described of terpenoid synthesis in cell-free extracts of higher plants. The most interesting experiment was accomplished by Graebe, Dennis, Upper and West [136] with cell-free homogenates of wild cucumber endosperm. These authors showed the conversion of mevalonic acid to geranylgeraniol and a number of diterpenes.

But the introduction of proper methods of obtaining active cell-free extracts of all types of higher plants would greatly stimulate the study of terpenoid synthesis in these plants, a field which remains largely submitted to reasoning by analogy on the basis of chemical structure rather than based on experimentation with relevant biological material.

17. Nature and biosynthesis of the C_{40} carotenoid precursor (phytoene)

The complete synthesis of β-carotene was accomplished in the course of 1950 by Karrer and Eugster [137], by Inhoffer, Bohlman, Bartran, Rummert and Pommer [138], as well as by Milas, Davis, Belic and Fles [139].

It is only in plants that carotenoid synthesis has been shown to take place de novo.

The epistemological background of the elucidation of the biosynthetic pathway of carotenoids is to be found in the "biogenetic isoprene rule" of Ruzicka [140], an extension of his "isoprene rule" [141]. According to the "biogenetic isoprene rule", Ruzicka formulated a guess based on chemical structure which received confirming evidence from biochemical experimentation progressing side by side along with chemical information. Ruzicka [140] formulated a theory stating that terpenoids are formed from aliphatic compounds such as geraniol (C_{10}), farnesol (C_{15}), and geranylgeraniol (C_{20}), by cyclisations and rearrangements.

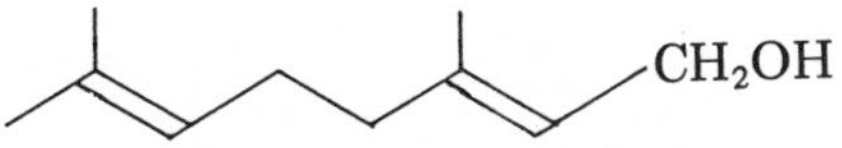

Geraniol (C_{10})

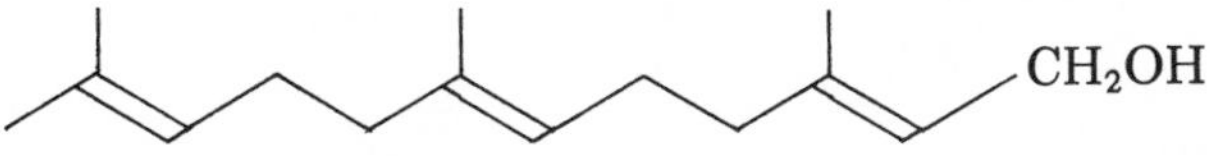

Farnesol (C_{15})

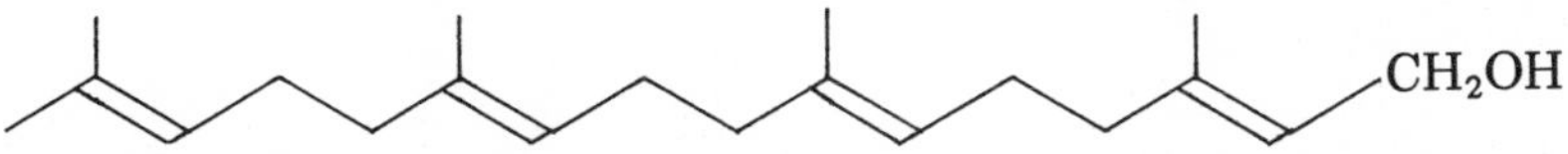

Geranylgeraniol (C_{20})

The theory requires the existence of a universal "biological isoprene unit". As stated in Section 11, this unit is represented by isopentenylpyrophosphate and dimethylallylpyrophosphate, condensing with the formation of geranyl pyrophosphate (C_{10}).

By adding two further isopentenylpyrophosphate molecules, stepwise, to geranylpyrophosphate, farnesylpyrophosphate (C_{15}) is

References p. 316

Plate 213. T.W. Goodwin.

formed, which, by dimerization, leads to squalene (C_{30}) the precursor of triterpenoids, including steroids.

That the labelling pattern of β-carotene after the addition of [^{14}C]acetate to cultures of the fungi *Mucor hiemalis* and *Phycomyces blakesleeanus* is the same as that observed in sterols and other terpenoids was established by Grob [142,143], by Braithwaite [144] and by Goodwin [145].

In experiments with *P. blakesleeanus*, the incorporation of mevalonic acid into β-carotene was reported by MacKinney, Chandler and Lukton [146] at the 4th International Congress of Biochemistry (Vienna, 1958), by Braithwaite and Goodwin [147,148] and by Yokoyama, Chichester and MacKinney [149]. This incorporation has also been demonstrated with a variety of materials, such as other moulds, carrot root preparations, etc. (literature in Goodwin [150], p. 148).

The incorporation of [^{14}C]isopentenyl pyrophosphate into β-carotene by fungal extracts, and into lycopene by tomato extracts, was reported by Varma and Chichester [151]. Before this, [^{14}C]farnesylpyrophosphate had been reported to be incorporated into β-carotene and other polyenes in a *P. blakesleeanus* cell-free system by Yamamoto, Yokoyama, Simpson, Nakayama and Chichester [152]. It was confirmed in a carrot plastid system by Anderson and Porter [153]. As Yamamoto et al. [152] had observed that the incorporation of [^{14}C]farnesyl pyrophosphate in β-carotene was stimulated by the addition of mevalonic acid or of isopentenyl pyrophosphate, it was logical to assume the formation of geranyl-geranyl pyrophosphate as an intermediate in the biosynthesis of carotenes, a hypothesis which was also in harmony with the "biogenetic isoprene rule".

The experimental demonstration of the formation of geranylgeranyl pyrophosphate by carrot and by pig liver was reported by Wells, Schelble and Porter [154].

That the C_{20} terpenyl pyrophosphate, geranylgeranyl phosphate, participated in the biosynthesis of carotenoids was, in 1964, supported by several lines of evidence. The observation, stated above, that the labelling pattern of β-carotene biosynthesized by *Mucor hiemalis* from [^{14}C]acetate conforms with that found in squalene derived from the same precursor, suggested that a C_{40} precursor of the carotenes is formed by the condensation of two C_{20} terpenoid intermediates in a manner analogous to the way

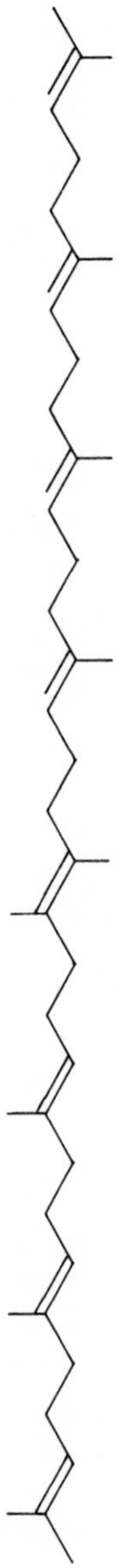

Lycopersene (C_{40})

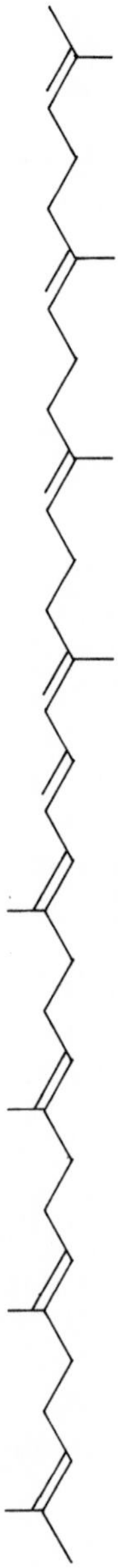

Phytoene (15,15'-dehydrolycopersene (C_{40}))

Fig. 36.

in which squalene is formed by the condensation of two farnesyl residues (see Section 12). The stimulation, reported above, of the incorporation of farnesyl pyrophosphate, by mevalonic acid or by isopentenyl pyrophosphate suggested further lengthening of the C_{15} pyrophosphate by a C_5 unit before its utilization in carotenogenesis. The conversion of farnesyl pyrophosphate to geranyl pyrophosphate is catalysed by an enzyme system of baker's yeast, as stated by Grob, Kirschner and Lynen [155].

Kandutsch, Paulus, Levin and Bloch [156] have reported the isolation from *Micrococcus lysodeikticus*, a yellow bacterium, of a soluble, purified enzyme catalysing the synthesis of geranylgeranyl pyrophosphate from isopentenyl pyrophosphate and either dimethylallyl, geranyl or farnesyl pyrophosphate. They called the enzyme geranylgeranyl pyrophosphate synthetase.

The knowledge of the biosynthetic pathway leading to triterpenes suggested that a similar sequence of reactions with geranylgeranyl pyrophosphate (C_{20}) would yield lycopersene (C_{40}) (Fig. 36).

This was for some time accepted on the basis of an alleged isolation of a particulate enzyme from *Neurospora crassa*, carrying out the dimerization of geranylgeranyl pyrophosphate with a production of lycopersene (Grob et al. [155]). On the other hand Grob and Boschetti [157] reported on the presence of lycopersene in the mycelium of *N. crassa*.

Further researches from Goodwin's laboratory failed to detect lycopersene in carotenogenic systems (Davies, Jones and Goodwin [158]; Mercer and Goodwin [159]). The same negative result was reported by Pennock, Hemming and Morton [160]. Neither did Anderson and Porter [153] observe a formation of lycopersene in a system of carrot plastids synthesizing phytoene from terpenyl pyrophosphate. The recognition of phytoene (not lycopersene) as the first C_{40} unit formed was supported by a number of experimental data. Diphenylamine is known to inhibit the production of the β-carotene group of pigments in microorganisms. This inhibition is accompanied by an accumulation of phytoene, not lycopersene (literature in Chichester and Nakayama [161] and in Goodwin [150]).

A series of "carotenoid-less" mutants of microorganisms or plants accumulate phytoene (for literature, see Chichester and Nakayama [161]). A direct conversion of geranylgeranyl phosphate into phytoene by tomato plastids was reported by Wells et al. [154]. As stated by Goodwin [150]:

"The formation of phytoene and not lycopersene as the first C_{40} compound formed allows the formulation of a reason for the absence of polycyclic tetraterpenoids analogous to the polycyclic triterpenoids (sterols, etc.) in Nature. The presence of the central double bond in phytoene would prevent its folding similarly to squalene and thus extensive cyclization initiated by OH^+ or H^+ would not be possible" (p. 152).

Goodwin and Williams [162] have studied the biosynthesis of phytoene in carrot root preparations, using the two forms of tritium-labelled mevalonic acid synthesized by Cornforth, Cornforth, Donninger and Popjàk [163] and they have found that, as with squalene and rubber, the loss of the C-4 hydrogen in the path from mevalonic acid to phytoene was stereospecific.

That phytoene is the first C_{40} carotenoid formed in cells is now generally accepted.

18. Biochemical sequence in C_{40} conversion

Porter and Lincoln [164] have proposed a scheme for the conversion of phytoene into lycopene via phytofluene, ζ-carotene and neurosporene (Fig. 37). This scheme, based on studies of mutants of tomato was first formulated at a time when the exact structures of the intermediates had not yet been elucidated. The pattern of pigment distribution in *Chlorella* mutants confirmed the theory (Claes [165]). Other facts were in favour of the scheme. For instance, extracts of *Staph. aureus* were obtained, converting phytoene into δ-carotene (Suzue [166]). A conversion of phytoene into phytofluene was also observed in the presence of a preparation of tomato plastids (Beeler and Porter [167]).

With respect to the mechanism of cyclization leading to cyclic carotenoids, such as the monocyclic γ-carotene or the bicyclic β-carotene, the problem has not been entirely solved (see Goodwin [150]; MacKinney [168]; Chichester and Nakayama [161]) and neurosporene, or lycopene are considered as possible precursors in competing theories.

The concept according to which the more highly oxygenated carotenoids such as the xanthophylls result from an insertion of

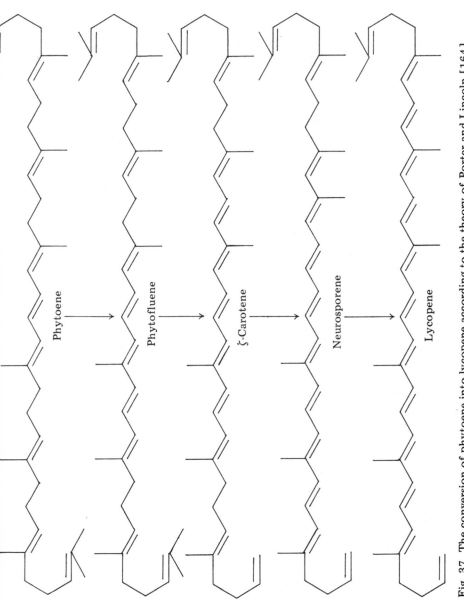

Phytoene → Phytofluene → ζ-Carotene → Neurosporene → Lycopene

Fig. 37. The conversion of phytoene into lycopene according to the theory of Porter and Lincoln [164].

oxygen in the last step of the biosynthetic pathway is generally accepted. A mutant of *Chlorella vulgaris*, when grown in the dark under heterotrophic conditions, synthesizes mainly phytoene. If it is illuminated in the absence of oxygen, it synthesizes coloured carotenes. If the cultures are placed in the dark, in the presence of oxygen, xanthophylls are formed while carotenes disappear (Claes [165]).

That the oxygen of lutein arises in *Chlorella* from molecular oxygen and not from water has been demonstrated by Yamamoto, Chichester and Nakayama [169] with the help of $^{18}O_2$ and of $H_2{}^{18}O$.

19. From mevalonic acid to isoprenoid alkaloids

That a number of steroidal and mono-, sesqui- and diterpenoid alkaloids are of mevalonoid origin was demonstrated by the incorporation of [2-^{14}C]mevalonate, as described in this Treatise (Vol. 20, Chapter VI, by J.D. Spenser), to which we refer the reader. As in the case of phytosterols, the pathway of isoprenoid alkaloids is not completely clarified.

20. From mevalonic acid to isoprenoid quinones *, chromanes and chromers

In Chapter 36 we retraced the discovery of ubiquinones (coenzymes Q). It was observed by Parson and Rudney [170] that *p*-hydroxybenzoic acid is a precursor for ubiquinone in various microorganisms. The same result was obtained in animals by Parson and Rudney [170]; Olson, Bentley, Aiyar, Dialameh, Gold, Ramsey and Springer [171] and Aiyar and Olson [172].

The polyisoprenoid side chain is formed by the C_5-isoprenoid condensation which is common to farnesyl pyrophosphate and squalene biosynthesis.

Friis, Daves and Folkers [173] formulated in 1966 the com-

---

* On isoprenoid quinones (terpenoid quinones and chromanes) see Chapter V, in Vol. 9 of this Treatise.

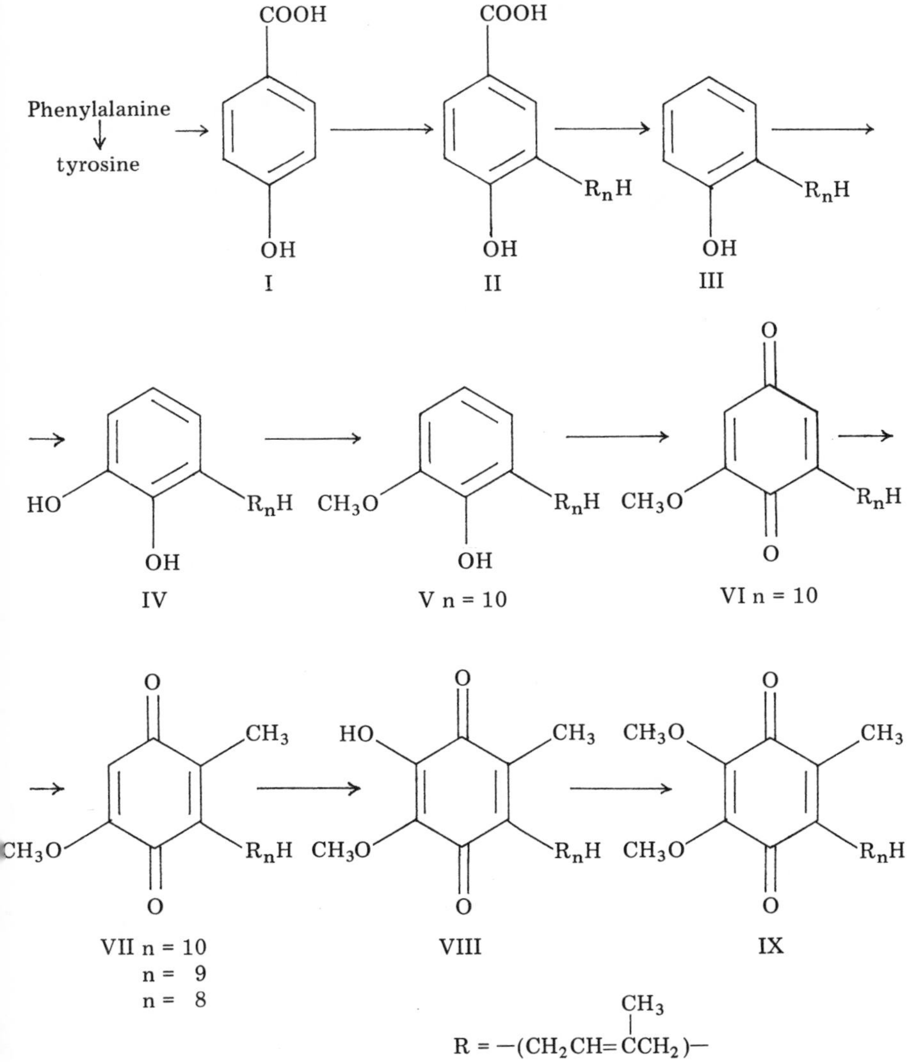

Fig. 38. (Friis, Daves and Folkers [173]) Biosynthesis from p-hydroxybenzoic acid (I) to ubiquinone (IX).

plete biosynthetic sequence reproduced in Fig. 38. They isolated four new quinones by extensive fractionation of a lipid extract from *Rhodospirillum rubrum*. Structural studies showed that these products were 2-decaprenyl-6-methoxy-3-methyl-1,4-benzoquinone (VII, n = 10), 2-nonaprenyl-6-methoxy-3-methyl-1,4-benzoquinone (VII, n = 9), 2-octaprenyl-6-methoxy-3-methyl-1,4-benzoquinone (VII, n = 8), and 2-decaprenyl-6-methoxy-1,4-benzoquinone (VI, n = 10). The compounds VII (n = 10) and VI (n = 10) are apparent precursors of ubiquinone-10. The compounds VII (n = 9) and VII (n = 8) are apparent precursors of ubiquinone-9 and ubiquinone-8, respectively. To quote the authors:

"The finding of 2-decaprenylphenol (III, n = 10) has been reported from *R. rubrum* [174] and III has been established as a precursor of Q_{10} [175]. Isolation, also from *R. rubrum*, of the structurally related 2-decaprenyl-6-methoxyphenol (V, n = 10) [176], another precursor, led to a partial biosynthetic sequence (I—V) [176].

"The isolation of the new 2-decaprenyl-6-methoxy-3-methyl-1,4-benzoquinone (VII, n = 10) and the detection of the new 2-decaprenyl-6-methoxy-1,4-benzoquinone (VI, n = 10) allow this sequence to be completed (Fig. 38). The intermediacy of 2-decaprenyl-5-hydroxy-6-methoxy-3-methyl-1,4-benzoquinone (VIII, n = 10) is obvious. This series of transformations is clearly in accord with available knowledge of biological hydroxylations and methylations."

Ubichromenol-50 has been isolated from human kidney (Laidman, Morton, Paterson and Pennock [177]) and shown to be a cyclic isomer of ubiquinone-50

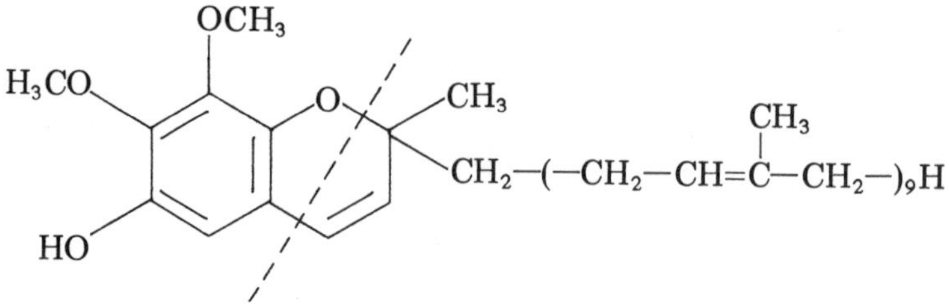

Ubichromenol

Green, Diplock, Bunyan and McHale [178] have obtained, in rats, results which indicated that either ubichromenol arises from ubiquinone or that the two have a common precursor. It may be noted also that the authors obtained a biosynthesis of ubiquinones in vitro by homogenates.

21. From cholesterol to steroid hormones

When ^{14}C-labelled acetyl-CoA is incubated with slices or homo-genates of any tissue producer of steroid hormones, these hormones become radioactive [179—183].

The distribution of the methyl carbon and of the carboxyl carbon is the same as was observed in cholesterol. If ATPase is inhibited, mevalonic acid can be converted to sterols and steroid hormones (Salokangas, Rilling and Samuels [184]). That cholesterol itself can be converted to the steroid hormones has been demonstrated (Zaffaroni, Hechter and Pincus [180]; Werbin,

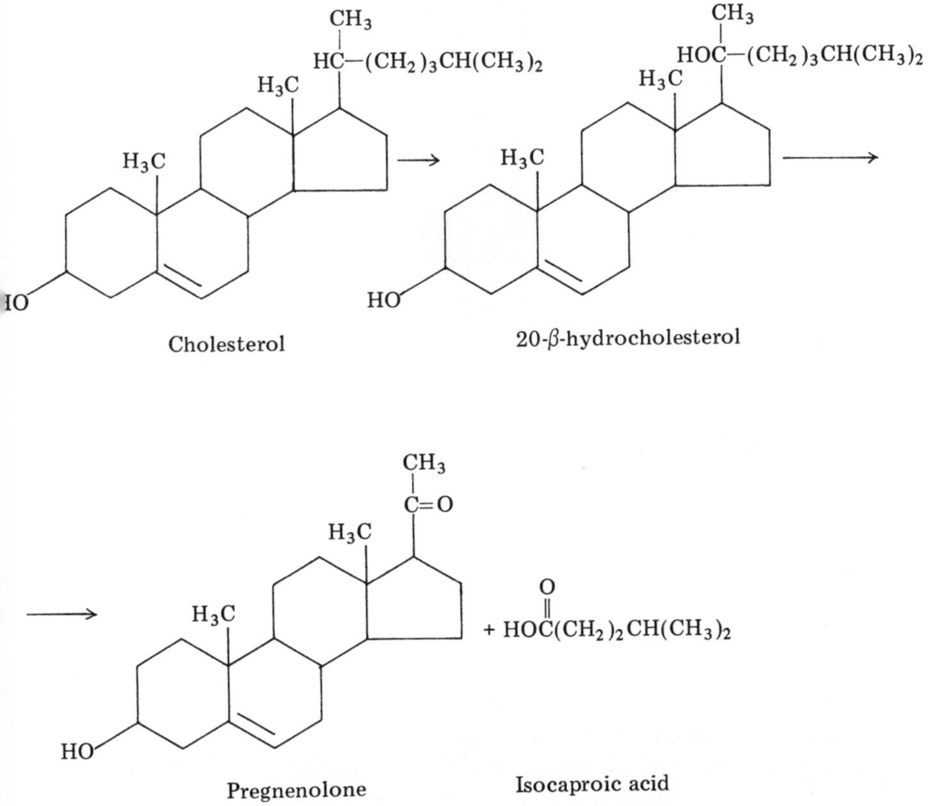

Cholesterol   20-β-hydrocholesterol

Pregnenolone   Isocaproic acid

Fig. 39. From cholesterol to pregnenolone.

Plate 214. Leo T. Samuels.

Plotz, Le Roy and Davis [185]; Hall and Eik-Nes [186], Mason and Savard [187]).

Pregnenolone had been known as a product of the chromic acid oxidative process, used in the pharmaceutical industry to produce derivatives of cholesterol acting as gonadal hormones. It was postulated by Selye [188] that pregnenolone could be an intermediate in the biosynthesis of steroid hormones. That all tissues forming steroid hormones contain an enzyme system able to convert pregnenolone into progesterone was shown in 1951 by Samuels, Helmreich, Lasater and Reich [189]. Formation of pregnenolone was demonstrated in the different tissues producing steroid hormones (Saba, Hechter and Stone [190]; Staple, Lynn and Gurin [191]; Solomon, Vande Wiele and Lieberman [192]; Shimizu, Hayano, Gut and Dorfman [193]).

It was shown that progesterone is readily converted into cortisol and corticosterone by adrenal perfusion (Levy, Jeanloz, Jacobsen, Hechter, Schenker and Pincus [194]); into androstenedione and testosterone by testis tissue (Slaunwhite and Samuels [195]), and to oestrogens in vivo (Davis and Plotz [196]).

These different observations were at the origin of the concept of a general pathway of steroid hormone biosynthesis common in its basic features in the different tissues producing steroid hormones and leading to different kinds of hormones as a result of the relative dominance of a given enzyme system.

22. Biosynthesis from pregnenolone in the adrenal cortex

The earliest views of the hormonal secretion of the adrenal cortex assumed the existence of a single cortical hormone, designated as cortin by Hartman, Brownell, Hartman, Dean and MacArthur [197]. Following the classical endocrinological methods several authors tried to isolate and characterize different defined compounds and from these researches emerged the conclusion that the activity tended to concentrate in the lipid fractions of the extract (Swingle and Pfiffner [198]). This important conclusion was the base of developments in organic chemistry in the laboratories of Kendall and Mason, Wintersteiner and Pfiffner, and Reichstein. This led to the isolation, from extracts of the adrenal cortex, of a wide array of compounds of isoprenoid nature. Ringold and

Bowers, in Chapter III, Section C of Vol. 10 of the present Treatise have dealt with the chemical properties and chemical synthesis of these compounds and their derivatives. The present knowledge of their biosynthesis, as well as that of the other steroid hormones has been dealt with by Gower in Chapter II of Vol. 20. Among those compounds, three important hormones, secreted in the blood stream and all 11-β-hydroxy compounds have been recognized, the ability to hydroxylate in this position appearing as characteristic of the adrenal cortex. Of these three hormones, in Vertebrates, two were recognized as important in the regulation of organic metabolism and inflammation (cortisol, corticosterone) and one as the main regulator of the electrolyte balance (aldosterone). The other steroid compounds of the adrenal cortex have been recognized as intermediates, or derivatives either of the pathways of biosynthesis of the three main hormones, or of accessory or unrelated pathways.

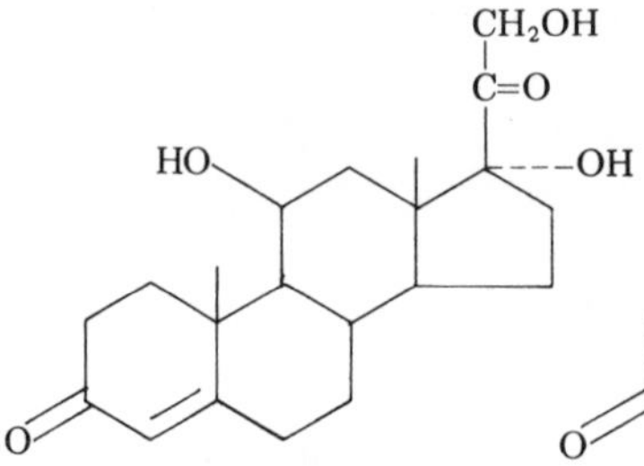

Cortisol

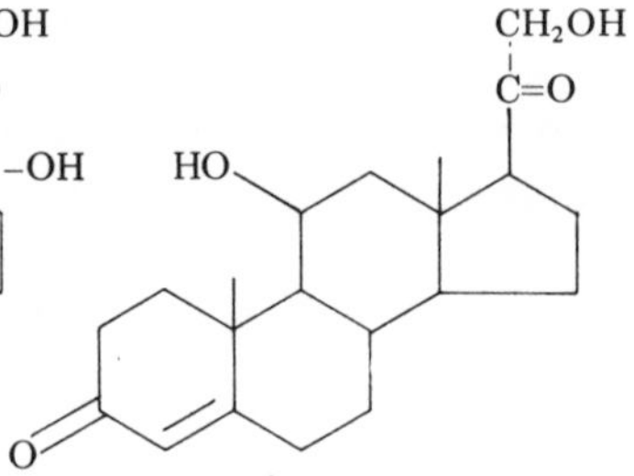

Corticosterone

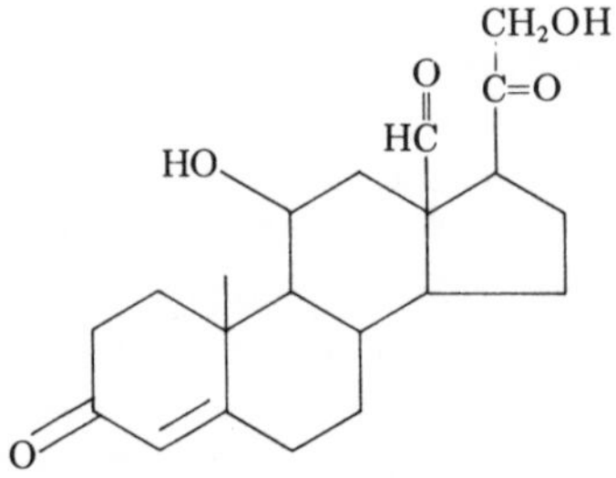

Aldosterone

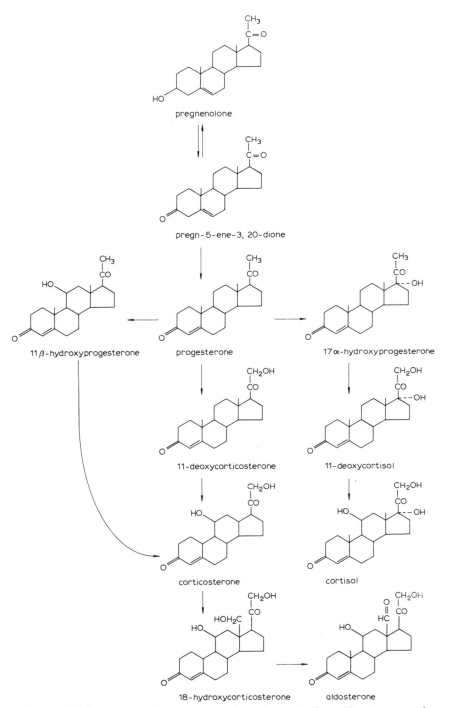

Fig. 40. Major sequence from pregnenolone to cortisol, corticosterone and aldosterone (C_{21} steroids).

The first insight into the biosynthesis of adrenocortical hormones was accomplished by Hechter and Pincus (see Samuels and Eik-Nes [199]) who found that perfusion of 11-deoxy-corticosterone (DOC) through bovine adrenals led to the production of high glucocorticoid activity, which suggested the existence, in the adrenal cortex, of 11-oxygenating capacity.

Levy, Jeanloz, Jacobsen, Hechter, Schenker and Pincus [194], perfused bovine adrenals with 11-deoxycorticosterone and obtained corticosterone.

That adrenal tissue possesses systems for the step-wise introduction of hydroxyl groups at C-17, C-21 and C-11 of the progesterone molecule was established by perfusing bovine adrenals with a number of different steroids (Hechter, Zaffaroni, Jacobsen, Levy, Jeanloz, Schenker and Pincus [200]).

These data, obtained in a biochemical laboratory (the Worcester Foundation for Experimental Biology, Shrewsbury, Massachusetts) led to an interest in the enzymology of the biosynthesis.

This advance came in a context of a deep interest of the pharmaceutical industry for cortisone. This had started in May 1949 with the discovery by Hench, Kendall, Slocumb and Polley [201], at the Mayo Clinic that cortisone exerts striking therapeutic effects in a number of diseases (rheumatoid arthritis, rheumatic fever, etc.).

It obviously was impossible to derive from compounds obtained from slaughtered animals sufficient cortisone to satisfy the need for it.

As stated by Hechter and Pincus [202]:

"It was recognized that one of the key problems in the synthesis of cortisone was the introduction of an oxygen function into the C-11 position of the steroid nucleus. The remaining functional groups presented no special problem thanks to the efforts of the steroid chemists who participated in the U.S. war-time project which attempted to synthesize cortisone and related corticosteroids" (p. 462).

The enzymological approach at Shrewsbury, which led to the identification of the enzymatic activities in homogenates of adrenals (literature in Samuels [203]) led the pharmaceutical industry towards the search for "screening procedures" leading to the isolation of microorganisms able to promote specific oxidations at C-11 and other points.

"Within a few years, a number of laboratories succeeded in this objective, the first being the group of the Upjohn Company led by Peterson and Murray [204]. During this period, the chemical lines to approach the synthesis of cortisone (total synthesis as well as partial synthesis from readily available plant sources) likewise were crowned with success (see Ciba Foundation Colloquia on Endocrinology, Vol. 7, London, 1953). The "cortisone shortage" problem was thus solved by a variety of lines of approach" (Hechter and Pincus [202], p. 463).

While the methods of synthesis in vitro were studied as just stated, the biochemical problem concerned with biosynthesis in the biological context were pursued on the basis of the determination of corticosteroids present in adrenal vein blood. Before the perfusion experiments on bovine adrenals of Hechter and Pincus, referred to above, as soon as 1943, Vogt [205], in experiments on dogs and other animals, had found that the adrenals produce large amounts of corticoid relative to the amount extractable from tissues. The interval of time between this pioneer work and the perfusion experiments of Hechter and Pincus on bovine adrenals is explained by the necessity of developing methods of analysis on the microscale.

Hechter and Pincus have aptly analysed the methodological difficulties inherent in reaching, from indirect or direct approaches, a conclusion concerning the biosynthetic pathway as it occurs in the tissue.

The pathway of Fig. 40 is derived from the knowledge of the rates of formation of the intermediates (literature in Samuels and Eik-Nes [199]).

In the case of aldosterone, the pathway described (and not universally accepted) derives from the data obtained by Raman, Sharma, Dorfman and Gabrilove [206] who concluded that aldosterone is formed by 18-hydroxylation of corticosterone followed by dehydrogenation to the 18-aldehyde. It results from the sum total of information obtained that the adrenal cortex has all the enzymes for the synthesis of any steroid hormone. Aside from those involved in the biosynthesis of corticoids, the enzymes are present at very low activity, or converted in the cells into inactive products. Nevertheless, in pathological cases such as congenital adrenal hyperplasia, the production of C_{19} compounds in the adrenal cortex may be sufficient to produce virilism (literature in Samuels and Eik-Nes [199]).

References p. 316

23. Biosynthesis from pregnenolone in the testis

Fieser and Fieser [207] have retraced the history of the events which led to the recognition of the nature of the hormonal secretion of the testis. The chemistry of androgens and other C_{19} steroids has been treated by Fujimoto and Ledeen in Chapter III, Vol. 10 of the present Treatise, and the nature of their biosynthetic pathway described by Gower in Chapter II of Vol. 20. The two androgens which are liberated in the circulation are Δ^4-androstenedione and testosterone.

The history of unravelling the pathways of biosynthesis of the hormones of the testis mainly resulting from experiments with homogenates, is described as follows by Samuels and Eik-Nes [199]:

"Samuels, Helmreich, Lasater and Reich [189] first demonstrated the conversion of pregnenolone to progesterone by testis tissue. Later Talalay and Wang [208] found that this reaction involved two enzymes, a 3β-hydroxy-steroid dehydrogenase that converted pregnenolone to pregn-5-ene-3,20 dione with NAD as the preferred acceptor, and a Δ^5-ketosteroid isomerase that catalyzed the migration of the double bond to the Δ^4-position without the presence of any known cofactor. Progesterone is converted to 17-hydroxy-progesterone by testis homogenates in the presence of molecular oxygen and NADPH, and the latter steroid in turn splits to Δ^4-androstenedione and acetic acid, oxygen and NADPH again being required (Slaunwhite and Samuels [195]; Lynn and Brown [209]). There is no evidence of a hydrolytic cleavage of 17-hydroxyprogesterone to testosterone and acetic acid. Testosterone can be formed from Δ^4-androstenedione by a 17β(testosterone)-dehydrogenase present in the Leydig cells, NADPH being the preferred hydrogen donor" (p. 187—188).

Data have been published in favour of alternative routes still under discussion (see Samuels and Eik-Nes [199]).

As in the case of other tissues producing steroid hormones, the specificity of the secretion of the testis depends on the relative activities of the synthesizing enzymes rather than on their presence or absence.

As shown by Rabinowitz [210], it is not uncommon to find that in some species the testes are able to form small amounts, generally insignificant, of oestradiol-17β and oestrone (see below). Nevertheless, in the Equidae, due to the presence in the testis of the enzyme catalysing the production of oestradiol-17β and

pregnenolone   17-hydroxypregnenolone   dehydroepiandrosterone

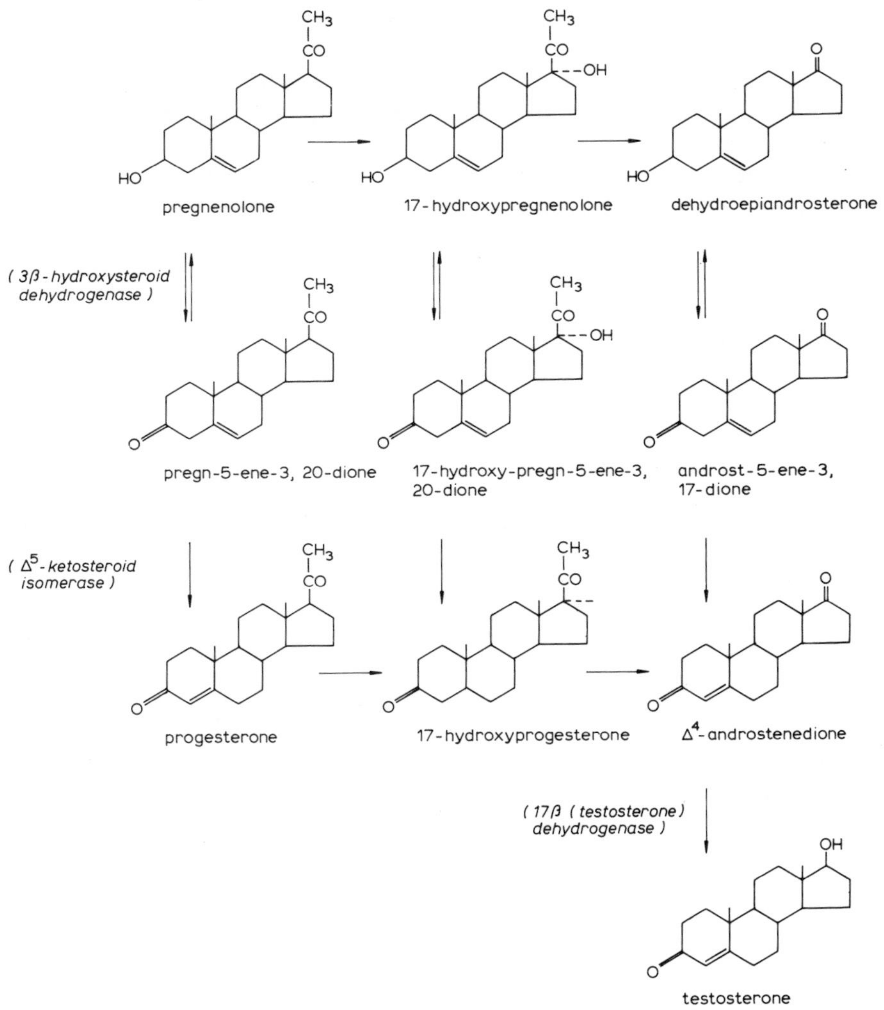

(3β-hydroxysteroid dehydrogenase)

pregn-5-ene-3, 20-dione   17-hydroxy-pregn-5-ene-3, 20-dione   androst-5-ene-3, 17-dione

(Δ^5-ketosteroid isomerase)

progesterone   17-hydroxyprogesterone Δ^4-androstenedione

(17β (testosterone) dehydrogenase)

testosterone

Fig. 41. Biosynthetic pathway from pregnenolone to Δ^4-androstenedione and testosterone (C_{19} steroids).

oestrone from testosterone (Baggett, Engel, Balderas and Lanman [211]), the yield is greater. This explained the presence of large amounts of oestradiol-17β and oestrone in stallion urine which made this urine a potent commercial source of oestrogens before the chemical synthesis of those compounds was accomplished.

References p. 316

24. Biosynthesis from pregnenolone in the ovary

A first approach to the knowledge of ovarian hormones consisted in the recognition of variations in the urinary excretion of those hormones during the ovarian cycle.

As the ovarian follicle reaches maturity a cavity develops which is filled with a fluid containing ovarian steroid hormones. (Litera-

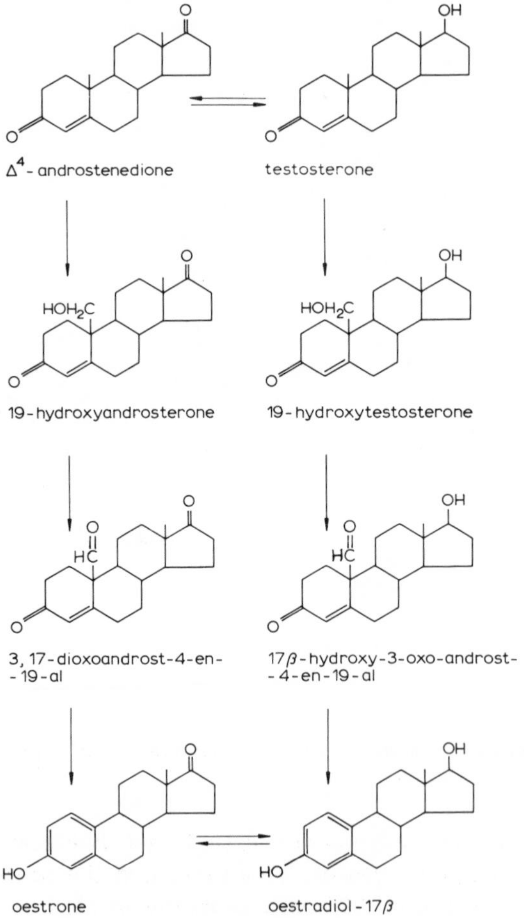

Δ^4 - androstenedione   testosterone

19 - hydroxyandrosterone   19 - hydroxytestosterone

3, 17 - dioxoandrost - 4 - en -   17β - hydroxy - 3 - oxo - androst -
- 19 - al   - 4 - en - 19 - al

oestrone   oestradiol - 17β

Fig. 42. Biosynthetic pathway from Δ^4-androstenedione and testosterone to oestrone and oestradiol 17β (C_{18} steroids).

ture in Fieser and Fieser [207]). As in the equine large follicles are formed, it was possible to demonstrate in the ovarian follicle, the presence of the enzymes for androstenedione and testosterone synthesis by either of the two routes represented in Fig. 42. These compounds undergo aromatization to oestrone and oestradiol-17β. Ryan and Smith [212] incubated [1-^{14}C]acetate with human follicular linings and found cholesterol, androstenedione, testosterone, oestradiol-17β and oestrone marked with the isotope. When equal quantities of [^{3}H]pregnenolone and [^{14}C]progesterone were introduced in the same incubation flasks, the ^{3}H : ^{14}C ratio in the oestrogens indicated that the Δ^4 and Δ^5 routes were approximately of equal importance.

More recent knowledge of the biosynthesis of C_{18} steroid is reviewed by D.B. Gower in Chapter II of Vol. 20 of the present Treatise.

25. Biosynthesis from pregnenolone in the corpus luteum

From the results obtained by several authors who studied the compounds derived from pregnenolone or precursors of it, in the presence of a homogenate of corpus luteum, it can be deduced that this tissue forms large amounts of progesterone and of 20α-hydroxypregn-4-en-3-one. It was also observed that there is a formation of smaller amounts of 17-hydroxyprogesterone, oestrone and oestradiol-17β but no 17-hydroxypregnenolone or dehydroepiandrosterone (Ryan and Smith [212]; Hammerstein, Rice and Savard [213]; Hayano, Lindberg, Wiener, Rosenkrantz and Dorfman [214]; Ryan [215]; Samuels and Eik-Nes [199]).

Ryan determined the isotope ratios in 17-hydroxyprogesterone and oestradiol after he incubated a homogenate of corpus luteum with a mixture of [^{3}H]pregnenolone and [^{14}C]progesterone and confirmed that these compounds were formed entirely from progesterone.

26. Biosynthesis from pregnenolone in the placenta

Pregnancy is unaffected if the ovaries are, in the human species, removed at the end of the third month of pregnancy, and the levels of oestrogens in urine continue to rise. On the other hand if abortion occurs at any time the excretion of those steroids drops. This is in accord with the production of hormones by the foetal-placental unit (see Fieser and Fieser [207]). That the placenta is the major source of the high oestrogen levels in pregnancy is substantiated by observations of Maner, Saffan, Wiggins, Thompson and Preedy [216], showing that the oestrogen content of the uterine vein blood in the pregnant woman is higher than that of the uterine artery, which is not observed in nonpregnant subjects. The blood of the umbilical vein flowing to the foetus is higher in oestrogen content than the blood returning via the umbilical artery.

The ovary of the nonpregnant woman does not secrete oestriol which, in the blood flowing from the pregnant uterus, is the major oestrogen. Oestriol was first isolated from human pregnancy urine by Marrian [217] in 1930 (literature on oestriol in Fieser and Fieser [207], p. 316 and in the chapter on the chemistry of the oestrogens by P.A. Katzman and W.H. Elliot in Vol. 10 of the present Treatise, Chapter III, Section b, p. 64).

Concerning the biosynthetic pathway of oestriol it was observed by Ryan [218] that placental microsomes plus the supernatant fraction converted 16α-hydroxyandrostenedione to oestriol but no oestrone or oestradiol were obtained. On the other hand the same author observed that the same placental preparation converted dehydroepiandrosterone, Δ^4-androstenedione and testosterone to oestrone and oestradiol-17β but in these conditions there was no formation of oestriol.

The placental production of progesterone from acetate, via pregnenolone, has been demonstrated by Levitz, Condon and Dancis [182], and from cholesterol by Solomon, Vande Wiele and Lieberman [192].

The foetus appears to be the major source of Δ^5-3β-hydroxy-steroids precursors of the oestrogens (literature in Samuels and Eik-Nes [199], p. 192).

That in the foetus the adrenal is the primary source of those precursors was confirmed by a number of observations. A direct proof was given by the demonstration, by Colas and Heinrichs [219], that the levels of 16α-hydroxydehydroepiandrosterone are low in the umbilical arterial blood returning from anencephalic foetuses with atrophic adrenals. Another argument was produced

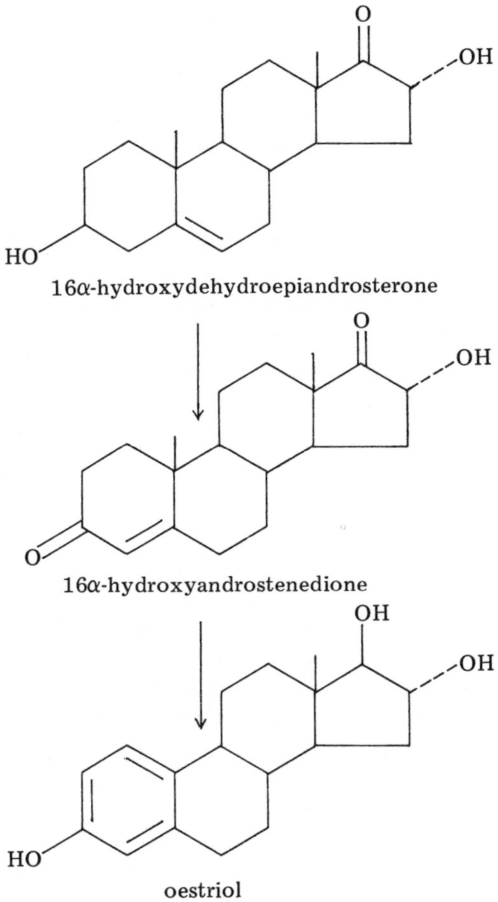

16α-hydroxydehydroepiandrosterone

16α-hydroxyandrostenedione

oestriol

Fig. 43. Pathway from 16α-hydroxydehydroepiandrosterone (of foetal origin) to oestriol.

in favour of the foetal origin of placental oestrogens was provided by the observation that the levels of Δ^5-3β-hydroxycompounds in blood and urine rapidly decrease when the foetal zone of the adrenal disappears (Cathro, Birchall, Mitchell and Forsyth [220]; Reynolds [221]).

The human foetal adrenal cortex is made of two zones, the adult cortex and the foetal zone, comprising about 80% of the total adrenal volume during most of pregnancy (Lanman [222]).

After birth the foetal zone degenerates rapidly, while the adult cortex transiently diminishes in size at birth.

27. Enzymes for the biosynthesis of steroid hormones

D.B. Gower, in Chapter II of Vol. 20 of the present Treatise, has reviewed the history of the knowledge of the enzymes involved in the conversion of cholesterol to pregnenolone and of pregnenolone to progesterone, of the 3α- and 3β-hydroxysteroid dehydrogenases and the Δ^4-5α- and 5β-reductases, of 3-(or 17β)-hydroxysteroid dehydrogenase, of 17β-oestradiol dehydrogenase, of Δ-dehydrogenases, of the aromatizing system, of 11β-hydroxylase, of 17α-hydroxylase, of 21-hydroxylase, and the reader is referred to his excellent chapter.

28. Biosynthesis of bile acids and bile alcohols

In the intestine, the digestion of lipids is accomplished by the enzymes of the digestive juices in the presence of detergents present in bile (bile salts) which facilitate the emulsification of glycerides. The hydrolysis of triglycerides leads to a large proportion of β-monoglycerides mixed with fatty acids. Monoglycerides and fatty acids associate with bile salts, forming "soluble micelles" (Borgström [223]). These attach themselves to the intestinal mucosa into the cells of which the monoglycerides and the fatty acids penetrate, liberating the bile salts, which are reabsorbed at the level of the lower ileum (Senior [224]; Dietschy, Salomon and Siperstein [225]).

The history of the biochemistry of bile has been the subject of an exhaustive presentation in Mani's excellent book on liver [226] (pp. 268–323). Bile salts (which appear to occur only in vertebrates) are the conjugation products of bile acids or bile alcohols. C_{24} bile acids occur in the higher sections of the systematic of vertebrates (most bony fishes, snakes, birds, mammals).

In more primitive vertebrates (cartilaginous fishes, amphibia, some reptiles) bile acids with 27 carbon atoms and/or C_{26} or C_{27} steroid alcohols occur in the bile (on the history of the distribution of bile alcohols and bile acids among vertebrate species and its correlation with phylogeny, see Haslewood [227]).

The most common bile acids are cholic acid, chenodeoxycholic acid and deoxycholic acid. These acids are C_{24} 5β-cholanoic acids carrying α-hydroxyl groups in the C-3, C-7 and/or C-12 positions. Other C_{24} bile acids carry hydroxyl groups in some or all of the positions just mentioned and in the C-6, C-16 or C-23 positions.

While steroid alcohols are present in bile as sulphate esters, the bile acids are conjugated with taurine and, in mammals, also with glycine.

The biosynthesis of bile acids has been approached by experiments with homogenates, microsomes, mitochondria and other liver preparations. Compounds have also been added to the diet and their derivatives isolated from faeces. But the experiments have mostly been accomplished on fistula-bearing rats and mice to which [14]C- and ^{3}H-labelled steroids have been administered.

That cholic acid derives from cholesterol was demonstrated by Bloch, Berg and Rittenberg [228] in 1943. Later on, whatever bile acid has been studied after labelled cholesterol had been given to bile fistula animals, this acid has been labelled.

The conversion of cholesterol to a primary bile acid (i.e. made by the liver from cholesterol; not resulting from bacterial actions in the intestine and subsequent reabsorption. Primary bile acids are found in fistula bile while the other bile acids are not) i.e. cholic acid and chenodeoxycholic acid involves a number of changes:

saturation of the Δ^5 double bond,

transformation of the 3β-hydroxyl group to the 3α group,

introduction of hydroxyl groups in C-7 and C-12 positions,

degradation of the cholesterol side chain, forming a carboxyl group at C-24.

The pathway of cholic acid formation has been studied by considering the changes in vivo (in the bile fistula rat) and in vitro, of cholesterol or other oxygenated C_{24} steroids labelled with ^{14}C or [3]H.

The step from cholesterol to 7α-hydroxycholesterol has been individualized by Danielsson and Einarsson [229]. Before that, Mendelsohn and Staple [230] had obtained from rat liver an enzyme system capable of converting cholesterol to $3\alpha,7\alpha,12\alpha$-trihydroxy-5β-cholestane (VI). This compound has been shown by Danielsson [231] to be oxidized by rat-liver homogenates, and by

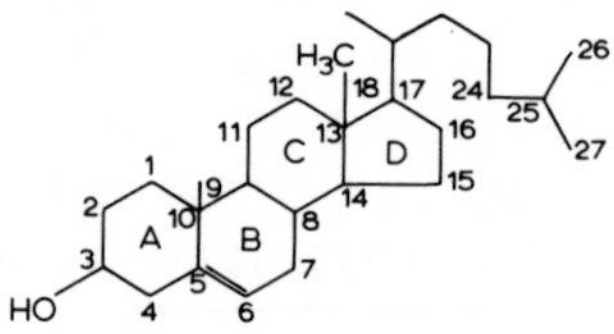

cholesterol

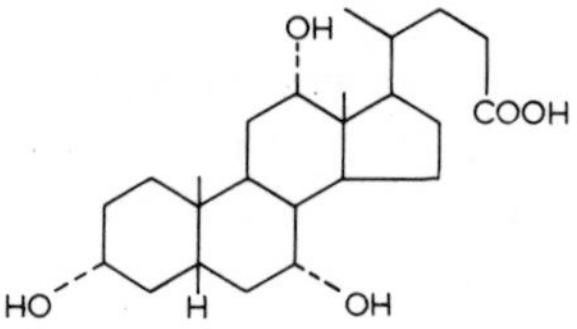

cholic acid

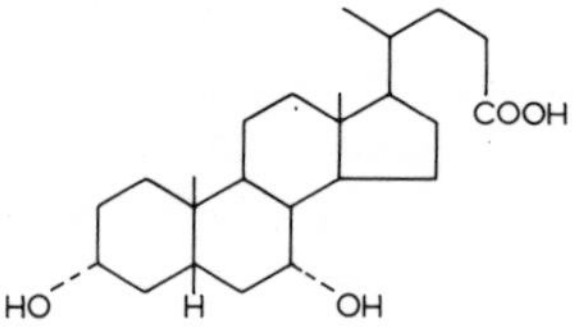

chenodeoxycholic acid

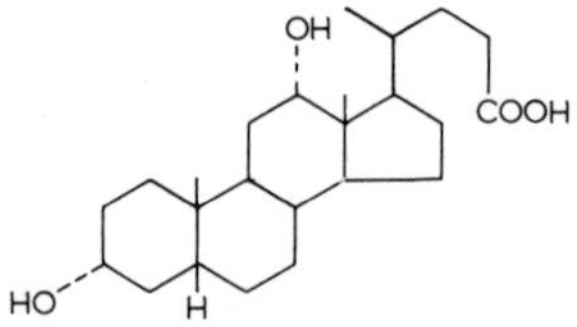

deoxycholic acid

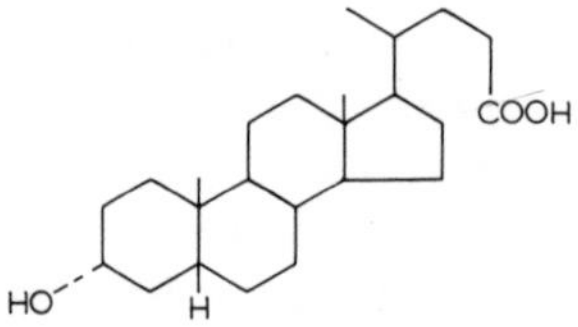

lithocholic acid

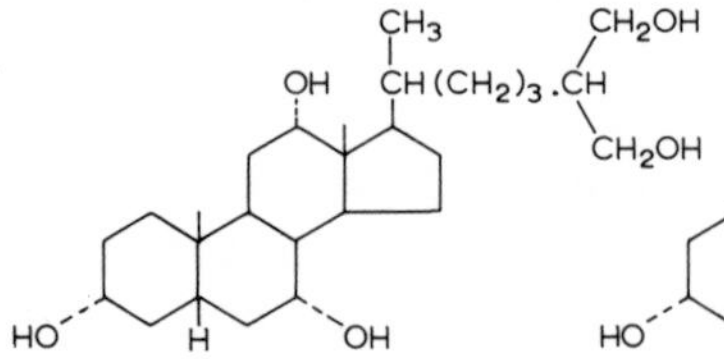

5 β-cyprinol

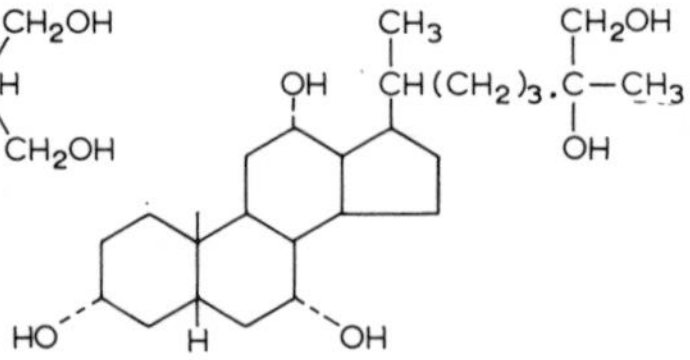

5 β-bufol

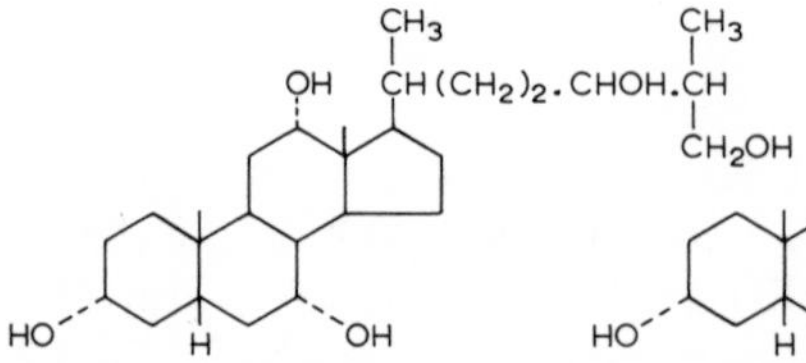

chimaerol

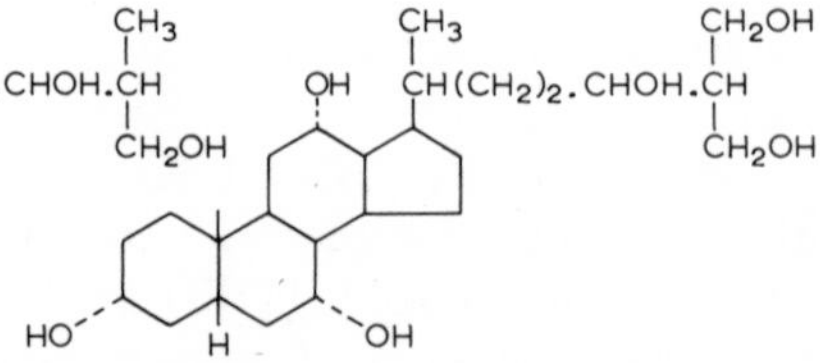

scymnol

Fig. 44. Bile acids and bile alcohols.

washed rat-liver mitochondria to 3α, 7α, 12α, 26-tetrahydroxy-5β-cholestane (VII), i.e. with side-chain terminal oxidation. Compound IX has been obtained from compound VII, in the presence of rat mitochondrial preparations by Danielsson [231] and by Suld, Staple and Gurin [232]. That, in this conversion, compound VIII is an intermediate, was shown by Okuda and Danielsson [233], who showed that it is oxidized to compound IX by the mitochondrial, microsomal and soluble fractions of rat-liver.

That the final step by which compound IX loses three carbons, resulting in cholic acid occurs as a β-oxidation was shown in rats by Bergström, Danielsson and Samuelsson [234] and in humans by Staple and Rabinowitz [235] and by Carey [236]. The three carbons are removed as propionyl-CoA, as shown by Suld et al. [232].

Concerning the changes taking place in the nucleus during the path from cholesterol to cholic acid, the conversion of cholesterol to compound III has been demonstrated in mouse-liver homogenates by Danielsson [237].

For the sequence II—V in rat liver, evidence has been brought about by Danielsson [238]; Berséus [239], and Danielsson and Einarsson [240]. If the 12α hydroxylation of the step III → IV does not take place, the compounds XI and XII (Fig. 44) will be formed. By 26 hydroxylation compound XIII will result and side-chain degradation will lead to chenodesoxycholic acid (XIV). The steps in this conversion have been demonstrated in experiments with mitochondria [237,238,241,242].

Other bile acids have been recognized as biosynthesized from cholesterol, but also from other sterols such as cholestanol or coprostanol (see Chapter II, Vol. 20, of this Treatise).

In 1954, Bergström, Pääbo and Rumpf [243] have shown that coprostane-3α,7α,12α-triol is rapidly converted into cholic acid in bile fistula rats. Tritium-labelled samples of the same compound were added to mouse and rat liver homogenates by Danielsson [231] who obtained coprostane-3α,7α-12α,26-tetrol as the main product. He also detected smaller amounts of 3α,7α,12α-trihydroxycoprostanic acid. The fact that this acid is converted to cholic acid in vitro (Bergström, Bridgwater and Gloor [244]) and in vivo (Bridgwater and Lindstedt [245]) completes the basis for the pathway represented in Fig. 45.

In the scheme of Fig. 45 the steps VII-VIII-IX accomplish an

Plate 215. Henry Danielsson.

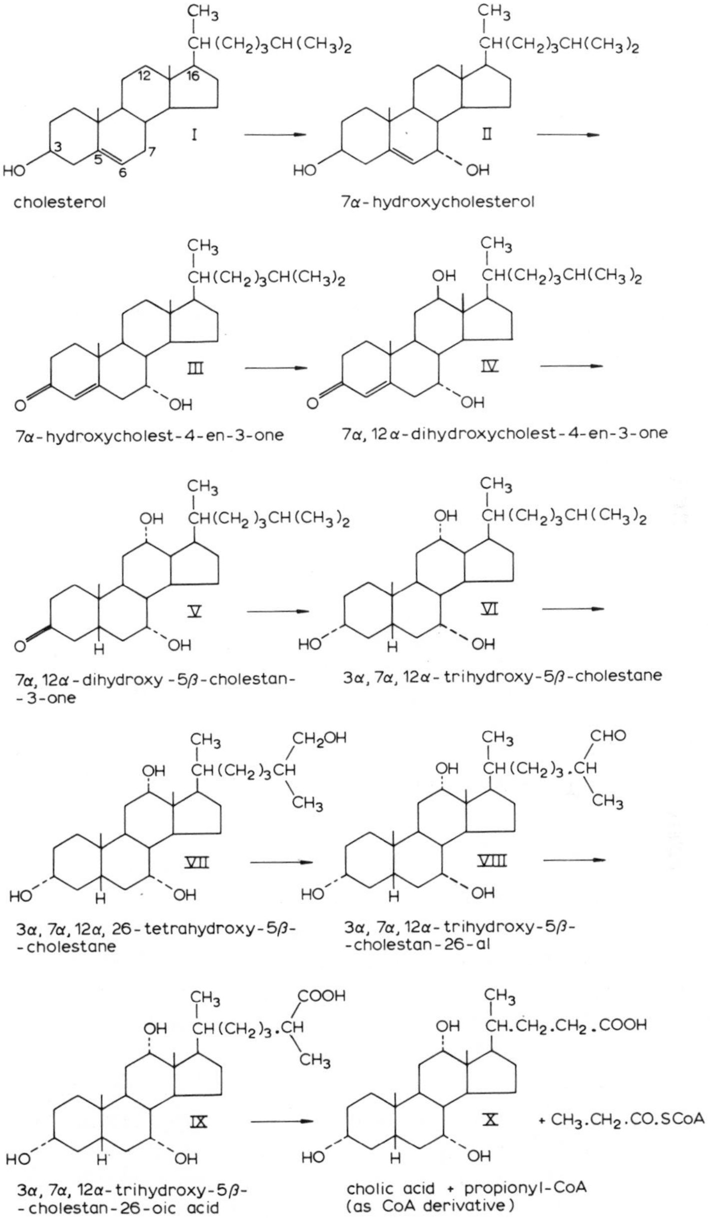

Fig. 45. From cholesterol to cholic acid.

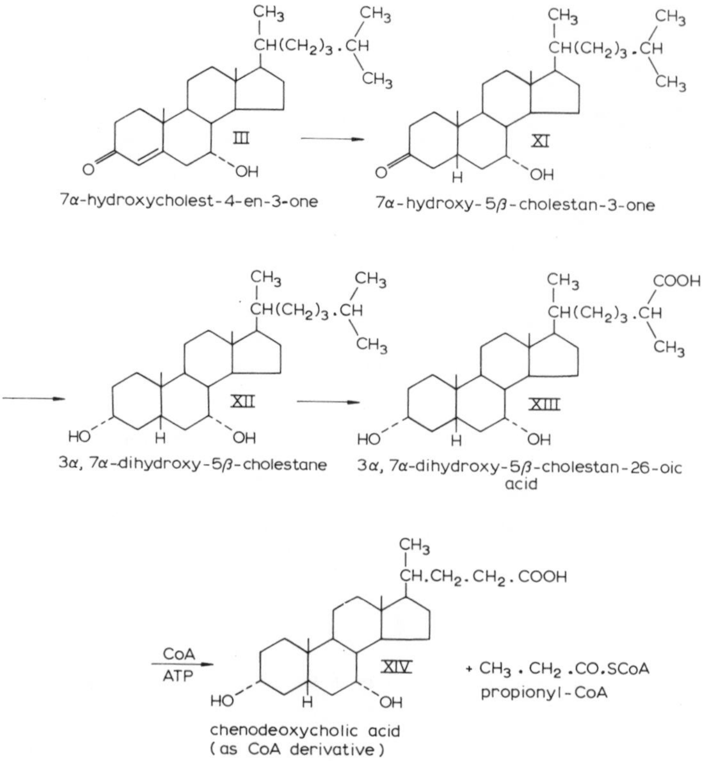

Fig. 46. Biosynthesis of chenodeoxycholic acid.

oxidation at C-26. Instead of this oxidation a further hydroxylation at C-27 of compound VII would lead to 5β-cyprinol, at C-25
to 5β-bufol, at C-24 and C-27 to scymnol. That such transformations are possible has been demonstrated by a number of indirect
studies (literature in Chapter II of Vol. 20, p. 117). While a number of biosynthetic studies on bile acids and bile alcohols have
been carried out by the classical methods, with a background of
organic chemistry, Haslewood has in this field introduced a new
epistemological dimension by obtaining hypotheses related to biosynthesis by a consideration of the colinearity of the evolution of
the organisms containing them and the molecular evolution of bile
acids and bile alcohols (Haslewood [246—248]). It results from
such studies that the general molecular evolution has the following

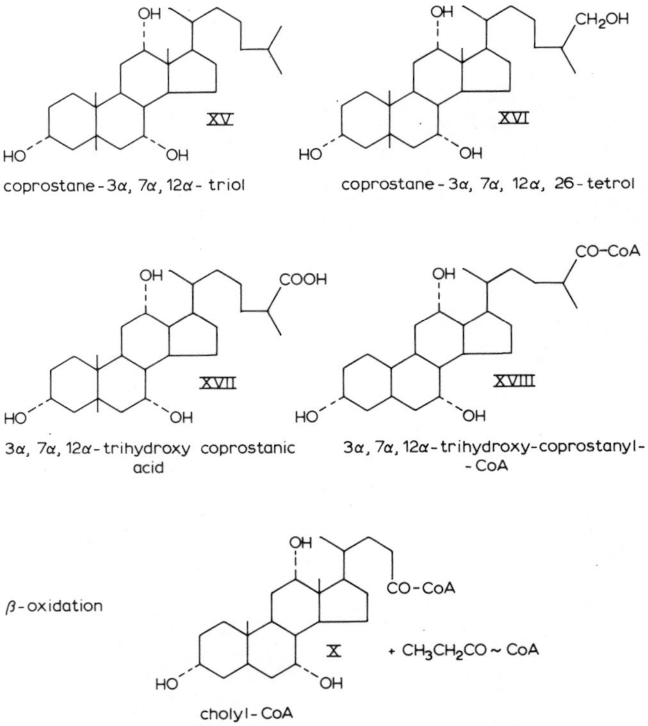

coprostane-3α, 7α, 12α- triol

coprostane-3α, 7α, 12α, 26-tetrol

3α, 7α, 12α-trihydroxy coprostanic acid

3α, 7α, 12α-trihydroxy-coprostanyl-CoA

β-oxidation

cholyl-CoA

Fig. 47. From coprostane to cholic acid.

course: C_{27} (and C_{26}) alcohols (sulphates) → C_{27} acids (taurine conjugates) → C_{24} acids (taurine conjugates) → C_{24} acids (glycine conjugates).

Danielsson and Kazuno [249] have shown, from studies on bile-fistula rats that 5β-ranol can be a precursor of cholic acid.

As stated above, coprostanol can be a precursor of bile acids. One of the bile alcohols, found in *Chimaera* by Haslewood, chimaerol (see Fig. 44) is a coprostane-3α,7α,12α,24,26-pentol. To quote Haslewood [246] (p. 195)

"There is plenty of evidence to suggest that the pathway in advanced vertebrates for cholic acid biosynthesis involves intermediates that serve as principal bile acids in more primitive animals or can be converted by a single biochemical step into the chief bile-alcohols of such types. Such a situation

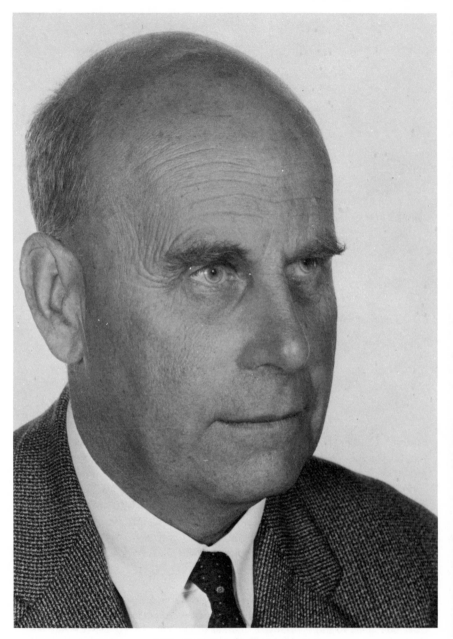

Plate 216. G.A.D. Haslewood.

strongly suggests that the evolution of the pathway of bile salt biosynthesis has taken place."

This pathway seems to branch at the level of a common intermediate, 7α,12α-dihydroxycholest-4-en-3-one (IV) which by hydrogenation can lead to the α or the β series of alcohols and also to cholic acid.

The most primitive vertebrate, the hagfish, has myxinol disulphate as chief bile acid. It is a C_{27} steroid still having the cholesterol skeleton. It also has special features such as a 16α-OH group. The lampreys, constituting the second cyclostome group have a bile alcohol sulphate, petromyzonol sulphate, a C_{24} compound

myxinol disulphate   petromyzonol sulphate

Primitive teleosts have C_{27} bile alcohols while more advanced teleosts have C_{24} bile acids.

In Amphibia and fishes the bile acids are more primitive than in mammals. The acids are conjugated with taurine while the C_{24} acids of mammals are conjugated with taurine and glycine. Haslewood has formulated the concept of a derivation of bile alcohols such as 5β-bufol, 5β-chimaerol or 5β-cyprinol from an intermediate on the way from cholesterol to cholic acid (Fig. 48).

The more evolved bile salts are therefore considered as an extension, leading to C_{27} and further on to C_{24} acids of the pathway leading to the bile alcohols.

The changes undergone by bile salts when modified in the intestinal tract, eventually reabsorbed in the ileum, and modified in the liver before being excreted as secondary bile salts, have been the

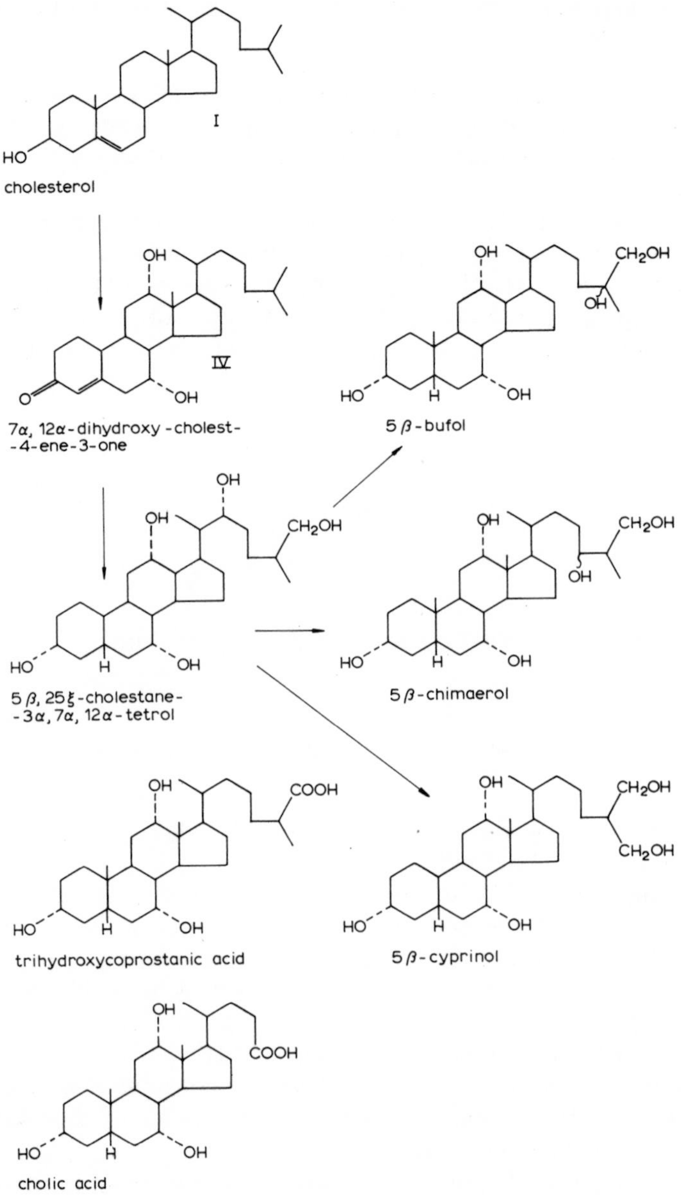

Fig. 48. Derivation of bile alcohols from intermediates on the way from cholesterol to cholic acid.

subject of a review of Danielsson and Tchen [56] in which some historical data will be found.

29. Conjugation of bile acids and alcohols

With rat-liver slices or homogenates, Bergström and Gloor [250], using ^{14}C-labelled bile acid and [^{35}S]taurine showed the conversion of deoxycholic acid to taurodeoxycholic acid. That the conjugation of cholic acid with taurine required ATP and CoA was shown with rat-liver microsomes, by Bremer [251]. Elliott [252] and Siperstein and Murray [253] showed that the bile acid is first activated in the presence of CoA and ATP and subsequently condensed with glycine:

$$R \cdot COOH + CoA \cdot SH + ATP \rightarrow RCO \cdot SCoA + AMP + PP$$

bile acid

$$RCO \cdot SCoA + NH_2 \cdot CH_2 \cdot COOH \rightarrow R \cdot CO \cdot NH \cdot CH_2 \cdot COO^-$$

$$+ CoA \cdot SH \qquad glycine$$

With respect to bile alcohols, Bridgwater and Ryan [254] have studied the conjugation of ranol in the presence of frog-liver homogenates, using $Na_2^{35}SO_4$ and found that in the presence of ATP, the alcohol sulphate was obtained

$$R \cdot OH + HO \cdot SO_3^- \rightarrow R \cdot O \cdot SO_3^- + H_2O.$$

30. Insect and crustacean ecdysones

Molting and adult development are, in arthropods, regulated by the hormones isolated by Butenandt and Karlson [255] from silkworm pupae in 1954 and called ecdysone and 20-hydroxyecdysone (Fig. 49). These hormones are secreted by prothoracic glands in insects and by the Y-organ in crustaceans. In 1956, Karlson [256, 257] obtained in crystalline form an impure preparation of 20-hydroxyecdysone to which he gave the name β-ecdysone. During the same year, 20-hydroxyecdysone was also isolated by Hocks and Wiechert [258] from *Bombyx* extracts, and by Hampshire and

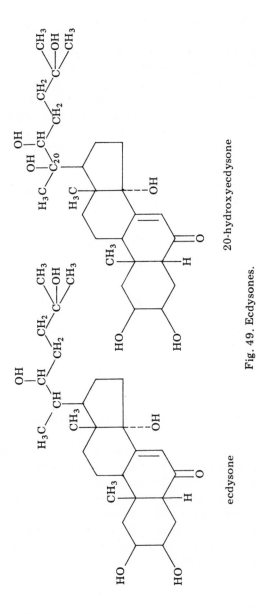

Fig. 49. Ecdysones.

Horn [259] from the crayfish *Jasus lalandei.* The double bond is essential for the molting and growth regulation (Hora, Lábler, Kasal, Černy, Sörm and Slàma [260]).

The biosynthesis of ecdysone from cholesterol was demonstrated by Karlson and Hoffmeister [261] working with *Calliphora* larvae. The last step in this biosynthetic pathway was shown by King and Siddall [262] and by Thomson, Siddall, Galbraith, Horn and Middleton [263] to be the hydroxylation at C-20.

31. Isoprenoids in insects

While molting and adult development is regulated in arthropods in general, the juvenile hormone is among the constituents which belong to the biochemical definition of insects. It is secreted by the corpora allata and when it is secreted along with ecdysone, the larval characters will prevail after the next mould.

Roeller, Dahm, Sweeley and Tros [264] have isolated juvenile hormone and determined its structure. It is a homologue of a terpenoid, with an epoxide ring and two double bonds. Recently, at the Hamburg Congress of Biochemistry, 1976, Roeller and Dahm [265] have stated that

"Three compounds with juvenile hormone activity have been isolated from extracts of insects or from cultures of their corpora allata: JH-I ($R_1 = R_2 = C_2H_5$), JH-II ($R_1 = CH_3$; $R_2 = C_2H_5$) and JH-III ($R_1 = R_2 = -CH_3$). JH-I and JH-II are unique insofar as they possess ethyl side chains in place of the usual methyl branches of terpenoid compounds.

In vivo, the ethyl branches are derived by incorporation of the intact C_3-units of propionate into JH-I and JH-II, a result compatible with the presumption of a normal sesquiterpene biosynthesis pathway with 3′-homomevalonate as one of the intermediates. Homomevalonate was not incorporated into JH under in vivo conditions, but it is an efficient precursor of JH-I and JH-II when applied to in vitro cultures of corpora allata from larvae and adults of some lepidopterous insects".

The isoprenoids of essential oils in plants contribute to the

Plate 217. Peter Karlson.

formation of the mist which plays a role in governing the flight of insects. Many isoprenoids considered to be derived from mevalonic acid are used as pheromones by insects. Investigations with labelled precursors have confirmed that these compounds are derived from a lateral extension of the terpenoid pathway, branching at the level of mevalonic acid. The incorporation of mevalonic acid into alarm substances such as citronellal and citral (Happ and Meinwald [266]) or dolichodial (Meinwald, Happ, Labows and Eisner [267]) has been for example reported.

REFERENCES

1   O. Wallach, Terpene and Camphor, 2nd ed., Leipzig, 1914.
2   H.J. Nicholas in P. Bernfeld (Ed.), Biogenesis of Natural Compounds,
     Oxford, 1963, p. 641.
3   Y. Fujita, Kagaku (Kyoto), 11 (1956) 874.
4   R. Schoenheimer and D. Rittenberg, J. Biol. Chem., 114 (1936) 381.
5   R. Sonderhoff and H. Thomas, Ann. Chem., 530 (1937) 195.
6   D. Rittenberg and R. Schoenheimer, J. Biol. Chem., 121 (1937) 235.
7   K. Bloch and D. Rittenberg, J. Biol. Chem., 143 (1942) 297.
8   K. Bloch and D. Rittenberg, J. Biol. Chem., 145 (1942) 625.
9   D. Rittenberg and K. Bloch, J. Biol. Chem., 160 (1945) 417.
10   E.L. Tatum, R.W. Barratt, N. Fries and D. Bonner, Am. J. Bot., 37
     (1950) 38.
11   R.C. Ottke, E.L. Tatum, I. Zabin and K. Bloch, J. Biol. Chem., 189
     (1951) 429.
12   K. Bloch, E. Borek and D. Rittenberg, J. Biol. Chem., 162 (1946) 441.
13   H.N. Little and K. Bloch, J. Biol. Chem., 183 (1950) 33.
14   J. Wuersch, R.L. Huang and K. Bloch, J. Biol. Chem., 195 (1952).
15   J. Bonner and B. Arreguin, Arch. Biochem. Biophys., 21 (1949) 109.
16   E.C. Grob, G.G. Poretti, A. von Muralt and W.H. Schopfer, Experientia,
     7 (1951) 218.
17   R.O. Brady and S. Gurin, J. Biol. Chem., 189 (1951) 371.
18   I. Zabin and K. Bloch, J. Biol. Chem., 192 (1951) 261.
19   I. Zabin and K. Bloch, J. Biol. Chem., 192 (1951) 267.
20   J.M. Buchanan, W. Sakami and S. Gurin, J. Biol. Chem., 169 (1947)
     411.
21   R.W. Chen, D.D. Chapman and I.L. Chaikoff, J. Biol. Chem., 205
     (1953) 383.
22   G. Popjàk, in T.W. Goodwin (Ed.), British Biochemistry Past and Pres-
     ent, Biochemical Society Symposia No. 30, London and New York,
     1970.
22a G. Popjàk, Annu. Rev. Biochem., 27 (1958) 533.
23   J.W. Cornforth, G.D. Hunter and G. Popjàk, Biochem. J., 54 (1953)
     590.
24   J.W. Cornforth, G.D. Hunter and G. Popjàk, Biochem. J., 54 (1953)
     597.
25   J.W. Cornforth, G. Popjàk and I.Y. Gore, in Biochemical Problems of
     Lipids, Proceedings of the Second International Conference of Bio-
     chemical Problems of Lipids, Ghent, 1955, London, 1956, p. 216.
26   J.W. Cornforth, I.Y. Gore and G. Popjàk, Biochem. J., 65 (1957) 94.
27   K. Bloch, Helv. Chim. Acta, 36 (1953) 1611.
28   W.G. Dauben and K.H. Takemura, J. Am. Chem. Soc., 75 (1953) 6302.
28a R. Robinson, J. Soc. Chem. Ind., London, 51 (1932) 464.
29   H.J. Channon, Biochem. J., 20 (1926) 400.
30   H.J. Channon and G.R. Tristram, Biochem. J., 31 (1937) 738.

31 T. Kimizuka, J. Biochem. (Tokyo), 27 (1938) 469.
32 K. Täufel, K. Thaler and H. Schreyegg, Fettchem. Umschau, 43 (1936) 26.
33 J. Fitelson, J. Ass. Offic. Agr. Chem., 26 (1943) 499.
34 A. Dimter, Z. Physiol. Chem., 270 (1941) 247.
35 R.M.B. McKenna, V.R. Wheatley and A. Wormall, J. Invest. Dermatol., 15 (1950) 33.
36 H. Sobel, J. Invest. Dermatol., 13 (1949) 333.
37 R.G. Langdon and K. Bloch, J. Biol. Chem., 200 (1953) 129.
38 R.G. Langdon and K. Bloch, J. Biol. Chem., 200 (1953) 135.
39 J.W. Cornforth and G. Popjàk, Biochem. J., 58 (1954) 403.
40 N.L.R. Bucher, J. Am. Chem. Soc., 75 (1953) 498.
41 I.D. Frantz Jr. and N.L.R. Bucher, J. Biol. Chem., 206 (1954) 471.
42 N. Bucher and K. McGarrahan, J. Biol. Chem., 222 (1956) 1.
43 G.S. Boyd, Biochem. J., 55 (1953) 892.
44 W. Voser, M.W. Mijovic, H. Hensen, O. Jeger and L. Ruzicka, Helv. Chim. Acta, 35 (1952) 2414.
45 R.B. Woodward and K. Bloch, J. Am. Chem. Soc., 75 (1953) 2023.
46 T.T. Tchen, in D.M. Greenberg (Ed.), Metabolic Pathways (2nd ed. of Chemical Pathways of Metabolism), Vol. 1, New York and London, 1960, p. 389.
47 W.G. Dauben, S. Abraham, S. Hotta, I.L. Chaikoff, H.L. Bradlow and A.H. Soloway, J. Am. Chem. Soc., 75 (1953) 3038.
48 J.W. Cornforth, I. Youhotsky-Gore and G. Popjàk, Biochem. J., 64 (1956) 38 P.
49 W.G. Dauben and T.W. Hutton, J. Am. Chem. Soc., 78 (1956) 2647.
50 R.B. Clayton and K. Bloch, J. Biol. Chem., 218 (1956) 305.
51 R.B. Clayton and K. Bloch, J. Biol. Chem., 218 (1956) 319.
52 E. Schwenk and C.F. Baker, Arch. Biochem. Biophys., 45 (1953) 341.
53 E. Schwenk, D. Todd and C.A. Fish, Arch. Biochem. Biophys., 49 (1954) 187.
54 L. Ruzicka, Experientia, 9 (1953) 357.
55 A. Eschenmoser, L. Ruzicka, O. Jeger and D. Arigoni, Helv. Chim. Acta, 38 (1955) 1890.
56 H. Danielsson and T.T. Tchen, in D.M. Greenberg (Ed.), Metabolic Pathways, 3rd ed., Vol. 2, New York and London, 1968, p. 117.
57 E.J. Corey, W.E. Russey and P.R. Ortiz de Montellano, J. Am. Chem. Soc., 88 (1966) 4750.
58 T.T. Tchen and K. Bloch, J. Biol. Chem., 226 (1957) 921.
59 E.E. van Tamelen, J.D. Willett, R.B. Clayton and K.E. Lord, J. Am. Chem. Soc., 88 (1966) 4752.
60 S. Yamamoto and K. Bloch, J. Biol. Chem., 245 (1970) 1670.
61 P.D.G. Dean, P.R. Ortiz de Montellano, K. Bloch and E.J. Corey, J. Biol. Chem., 242 (1967) 3014.
62 D.E. Wolf, C.H. Hoffman, P.E. Aldrich, H.R. Skeggs, D.L. Wright and K. Folkers, J. Am. Chem. Soc., 78 (1956) 5273.
63 C.H. Shunk, B.O. Linn, J.W. Huff, J.L. Gilfillan, H.R. Skeggs and K. Folkers, J. Am. Chem. Soc., 79 (1957) 3294.

64 H. Rudney, J. Am. Chem. Soc., 76 (1954) 2595.
65 B.K. Bachawat, W.G. Robinson and M.J. Coon, J. Biol. Chem., 216 (1955) 727.
66 J.L. Rabinowitz and S. Gurin, J. Biol. Chem., 208 (1954) 307.
67 F.D. Florapearl, A. Cobey, J.V.B. Warms and S. Gurin, J. Biol. Chem., 221 (1956) 181.
68 L.F. Adamson and D.M. Greenberg, Biochim. Biophys. Acta, 23 (1957) 472.
69 P.A. Tavormina, M.H. Gibbs and J.W. Huff, J. Am. Chem. Soc., 78 (1956) 4498.
70 P.A. Tavormina and M.H. Gibbs, J. Am. Chem. Soc., 78 (1956) 6210.
71 J.W. Cornforth, R.H. Cornforth, G. Popjàk and I.Y. Gore, Biochem. J., 69 (1958) 146.
72 F. Dituri, S. Gurin and J.L. Rabinowitz, J. Am. Chem. Soc., 79 (1957) 2650.
73 G. Popjàk, L. Gosselin, I.Y. Gore and R.G. Gould, Biochem. J., 69 (1958) 238.
74 B.H. Amdur, H. Rilling and K. Bloch, J. Am. Chem. Soc., 79 (1957) 2547.
75 F. Lynen, Ciba Found. Symp., Biosynthesis of Terpenes and Sterols, London, 1959, p. 95.
76 H. Rilling, T.T. Tchen and K. Bloch, Proc. Natl. Acad. Sci. USA, 44 (1958) 167.
77 H. Rilling and K. Bloch, J. Biol. Chem., 234 (1959) 1424.
78 T.T. Tchen, J. Am. Chem. Soc., 79 (1957) 6344.
79 S. Chaykin, J. Law, A.H. Phillips, T.T. Tchen and K. Bloch, Proc. Natl. Acad. Sci. USA, 44 (1958) 998.
80 U. Henning, I. Kessel and F. Lynen, Abstr. Int. Cong. Biochem., 4th Congr., Vienna, 1958, p. 47.
81 A. de Waard and G. Popjàk, Biochem. J., 73 (1959) 410.
82 H.R. Levy and G. Popjàk, Biochem. J., 75 (1960) 417.
83 T.T. Tchen, J. Biol. Chem., 233 (1958) 1100.
84 K. Markley and E. Smallman, Biochim. Biophys. Acta, 47 (1961) 327.
85 B.W. Agranoff, H. Eggerer, U. Henning and F. Lynen, J. Am. Chem. Soc., 81 (1959) 1254.
86 U. Henning, E.M. Möslein and F. Lynen, Arch. Biochem. Biophys., 83 (1959) 259.
87 K. Bloch, S. Chaykin, A.H. Phillips and A. de Waard, J. Biol. Chem., 234 (1959) 2595.
88 F. Lynen, H. Eggerer, U. Henning and I. Kessel, Angew. Chem., 70 (1958) 738.
89 L.A. Witting and J.W. Porter, J. Biol. Chem., 234 (1959) 2841.
90 B.W. Agranoff, H. Eggerer, U. Henning and F. Lynen, J. Biol. Chem., 235 (1960) 326.
91 F.B. Lynen, B.W. Agranoff, H. Eggerer, U. Henning and E.M. Möslein, Angew. Chem., 71 (1959) 657.
92 DeW.S. Goodman and G. Popjàk, J. Lipid Res., 1 (1960) 286.
93 J.W. Cornforth and G. Popjàk, Tetrahedron Lett., No. 19 (1959) 29.

94 G. Popjàk and R.H. Cornforth, J. Chromatol., 4 (1960) 214.
95 D.H. Shah, W.W. Cleland and J.W. Porter, J. Biol. Chem., 240 (1965) 1946.
96 C.R. Benedict, J. Kett and J.W. Porter, Arch. Biochem. Biophys., 110 (1965) 611.
97 P.W. Holloway and G. Popjàk, Biochem. J., 104 (1967) 57.
98 G. Popjàk, D.S. Goodman, J.W. Cornforth, R.H. Cornforth and R. Ryhage, J. Biol. Chem., 236 (1961) 1934.
99 C.R. Childs Jr. and K. Bloch, J. Biol. Chem., 237 (1962) 62.
99a G. Popjàk and J.W. Cornforth, Biochem. J., 101 (1966) 553.
100 H.C. Rilling, J. Biol. Chem., 241 (1966) 3233.
101 W.W. Epstein and H.C. Rilling, J. Biol. Chem., 245 (1970) 4597.
101a J. Edmond, G. Popjàk, S.M. Wong and V. Williams, J. Biol. Chem., 246 (1971) 6254.
101b G. Popjàk, J. Am. Oil Chem. Soc., 54 (1977) 647A.
101c G. Popjàk, H.I. Ngan and W.S. Agnew, BioOrg. Chem., 4 (1975) 279.
101d T.W. Goodwin, Essays Biochem., 9 (1973) 103.
102 F. Gautschi and K. Bloch, J. Am. Chem. Soc., 79 (1957) 684.
103 A.A. Kandutsch and A.E. Russell, J. Biol. Chem., 235 (1960) 2253, 2256.
104 W.W. Wells and C.L. Lorah, J. Am. Chem. Soc., 81 (1959) 6089.
105 D.H. Neiderhiser and W.W. Wells, Arch. Biochem. Biophys., 81 (1959) 300.
106 I.D. Frantz, A.G. Davidson, E. Dulit and M.L. Mobberley, J. Biol. Chem., 234 (1959) 2290.
107 G.J. Schroepfer Jr. and I.D. Frantz, Jr., J. Biol. Chem., 236 (1961) 3137.
108 G.J. Schroepfer Jr., J. Biol. Chem., 236 (1961) 1668.
109 E.I. Mercer and J. Glover, Biochem. J., 80 (1961) 552.
110 W.M. Stokes and W.A. Fish, J. Biol. Chem., 235 (1960) 2604.
111 J. Avigan, D. Steinberg, J. Biol. Chem., 235 (1960) 3123.
112 D. Steinberg and J. Avigan, J. Biol. Chem., 235 (1960) 3127.
113 J. Avigan and D. Steinberg, J. Biol. Chem., 236 (1961) 2898.
114 G. Popjàk and J.W. Cornforth, Adv. Enzymol., 22 (1960) 281.
115 S. Dagley and D.E. Nicholson, An Introduction to Metabolic Pathways, Oxford and Edinburgh, 1970.
116 L.A. Witting, H.J. Knauss and J.W. Porter, Fed. Proc., 18 (1959) 353. Abs.
117 H. Rudney, Fed. Proc., 15 (1956) 342.
118 H. Rudney, J. Biol. Chem., 227 (1957) 363.
119 H. Rudney and J.J. Ferguson Jr., J. Am. Chem. Soc., 79 (1957) 5580.
120 J.J. Ferguson Jr. and H. Rudney, Fed. Proc., 16 (1957) 771.
121 H. Rudney and J.J. Ferguson Jr., J. Biol. Chem., 234 (1959) 1076.
122 F. Lynen, U. Henning, C. Bublitz, B. Sörbo and L. Kröplin-Rueff, Biochem. Z., 330 (1958) 269.
123 J.J. Ferguson Jr. and H. Rudney, J. Biol. Chem., 234 (1959) 1072.
124 J.J. Ferguson, I.F. Durr and H. Rudney, Fed. Proc., 17 (1958) 219.

125   F. Lynen, in G.E.W. Wolstenholme and M. O'Connor (Eds.), Ciba
      Foundation Symposium on the Biosynthesis of Terpenes and Sterols,
      London 1959, p. 95.
126   J. Knappe, E. Ringelmann and F. Lynen, Z. Physiol. Chem., 332 (1959)
      195.
127   I.F. Durr and H. Rudney, J. Biol. Chem., 235 (1960) 2572.
128   M.E. Kirtley and H. Rudney, Biochemistry, 6 (1967) 230.
129   T. Kawachi and H. Rudney, Biochemistry, 9 (1970) 1700.
130   R.B. Park and J. Bonner, J. Biol. Chem., 233 (1958) 340.
131   R.G.O. Kekwick, B.L. Archer, D. Barnard, G.M.C. Higgins, G.P. Mc-
      Sweeney and C.G. Moore, Nature (London), 184 (1959) 268.
132   F. Lynen and U. Henning, Angew. Chem., 72 (1960) 82.
133   B.L. Archer, G. Ayrey, E.G. Cockbain and G.P. McSweeney, Nature
      (London), 189 (1961) 663.
134   B.L. Archer, B.G. Audley, E.G. Cockbain and G.P. McSweeney, Bio-
      chem. J., 89 (1963) 565.
135   A.J. Birch, D. Boulter, R.I. Fryer, P.J. Thomson and J.L. Willis, Tetra-
      hedron Lett., 3 (1959) 1.
136   J.E. Graebe, D.T. Dennis, C.D. Upper and C.A. West, J. Biol. Chem.,
      240 (1965) 1847.
137   P. Karrer and C.H. Eugster, Helv. Chim. Acta, 33 (1950) 1172.
138   H.H. Inhoffer, F. Bohlman, K. Bartran, G. Rummert and H. Pommer,
      Ann. Chem., 570 (1950) 54.
139   N.A. Milas, P. Davis, I. Belic and D.A. Fles, J. Am. Chem. Soc., 72
      (1950) 4844.
140   L. Ruzicka, Proc. Chem. Soc., (1959) 341.
141   L. Ruzicka, Experientia, 9 (1953) 357.
142   E.C. Grob, Chimia, 10 (1956) 73.
143   C.C. Grob, Chimia, 11 (1957) 338.
144   G.D. Braithwaite, Ph.D. Thesis, University of Liverpool, quoted by
      Braithwaite and Goodwin (1960).
145   T.W. Goodwin, in Handbook of Plant Physiology, Vol. 10, p. 186.
      Heidelberg, 1958.
146   G. MacKinney, B.V. Chandler and A. Lukton, Abstr. Int. Cong. Bio-
      chem., 4th Congress, Vienna 1958, p. 130.
147   G.D. Braithwaite and T.W. Goodwin, Biochem. J., 67 (1957) 13 P.
148   G.D. Braithwaite and T.W. Goodwin, Biochem. J., 76 (1960) 1, 5, 194.
149   H. Yokoyama, C.O. Chichester and G. MacKinney, Nature (London),
      185 (1960) 687.
150   T.W. Goodwin, in T.W. Goodwin (Ed.), Chemistry and Biochemistry of
      Plant Pigments, London and New York, 1965, p. 127.
151   T.N.R. Varma and C.O. Chichester, Arch. Biochem. Biophys., 96
      (1962) 265.
152   H. Yamamoto, H. Yokoyama, K. Simpson, T.O.M. Nakayama and C.O.
      Chichester, Nature (London), 191 (1961) 1299.
153   D.G. Anderson and J.W. Porter, Arch. Biochem. Biophys. 97 (1962) 509.
154   L.W. Wells, W.J. Schelble and J.W. Porter, Fed. Proc., 23 (1964) 426
      Abstr.

155 E.C. Grob, K. Kirschner and F. Lynen, Chimia, 15 (1961) 308.
156 A.A. Kandutsch, H. Paulus, E. Levin and K. Bloch, J. Biol. Chem., 239 (1964) 2507.
157 E.C. Grob and A. Boschetti, Chimia, 16 (1962) 15.
158 B.H. Davies, D. Jones and T.W. Goodwin, Biochem. J., 87 (1963) 326.
159 E.I. Mercer and T.W. Goodwin, Biochem. J., 88 (1963) 46 P.
160 J.F. Pennock, F.W. Hemming and R.A. Morton, Biochem, J., 82 (1962) 11 P.
161 C.O. Chichester and T.O.M. Nakayama, in P. Bernfeld (Ed.), Biogenesis of Natural Compounds, Oxford, 1963, p. 475.
162 T.W. Goodwin and R.H.J. Williams, Proc. Roy. Soc., Ser. B, 163 (1966) 515.
163 J.W. Cornforth, R.H. Cornforth, C. Donniger and G.J. Popjàk, Proc. Roy. Soc., Ser. B, 163 (1966) 492.
164 J.W. Porter and R.E. Lincoln, Arch. Biochem., 27 (1950) 390.
165 H. Claes, Z. Naturforsch., 9b (1954) 461; 11b (1956) 260; 12b (1957) 401; 13b (1958) 222; 14b (1959) 4.
166 G. Suzue, Biochim. Biophys. Acta, 45 (1960) 616.
167 D. Beeler and J.W. Porter, Biochem. Biophys. Res. Commun., 8 (1962) 367.
168 G. MacKinney, in D.M. Greenberg (Ed.), Metabolic Pathways, 3rd ed., vol. II, New York and London, 1968, p. 221.
169 H. Yamamoto, C.O. Chichester and T.O.M. Nakayama, Arch. Biochem. Biophys., 96 (1962) 645.
170 W.W. Parson and H. Rudney, Proc. Natl. Acad. Sci. USA, 51 (1964) 444.
171 R.E. Olson, R. Bentley, A.S. Aiyar, G.H. Dialameh, P.H. Gold, V.G. Ramsey and M.C. Springer, J. Biol. Chem., 238 (1963) PC 3146.
172 A.S. Aiyar and R.E. Olson, Fed. Proc., 23 (1964) 425.
173 P. Friis, G.D. Daves Jr. and K. Folkers, J. Am. Chem. Soc., 88 (1966) 4754.
174 R.K. Olsen, J.L. Smith, G.D. Daves Jr., H.W. Moore, K. Folkers, W.W. Parson and H. Rudney, J. Am. Chem. Soc., 87 (1965) 2298.
175 W.W. Parson and H. Rudney, Proc. Natl. Acad. Sci. USA, 53 (1965) 599.
176 R.K. Olsen, G.D. Daves Jr., H.W. Moore, K. Folkers and H. Rudney, J. Am. Chem. Soc., 88 (1966) 2346.
177 D.L. Laidman, R.A. Morton, J.Y.F. Paterson and J.F. Pennock, Biochem. J., 74 (1960) 541.
178 J. Green, A.T. Diplock, J. Bunyan and D. McHale, Biochim. Biophys. Acta, 78 (1963) 739.
179 R.O. Brady, J. Biol. Chem., 193 (1951) 145.
180 A. Zaffaroni, O. Hechter and G. Pincus, J. Am. Chem. Soc., 73 (1951) 1390.
181 J.L. Rabinowitz and R.M. Dowben, Biochim. Biophys. Acta, 16 (1955) 96.
182 M. Levitz, G.P. Condon and J. Dancis, Fed. Proc., 14 (1955) 245.
183 E. Bloch, R.I. Dorfman and G. Pincus, J. Biol. Chem., 224 (1957) 737.

184   A. Salokangas, H.C. Rilling and L.T. Samuels, Biochemistry, 4 (1965) 1606.
185   H. Werbin, J. Plotz, G.V. Le Roy and E.M. Davis, J. Am. Chem. Soc., 79 (1957) 1012.
186   P.F. Hall and K.B. Eik-Nes, Biochim. Biophys. Acta, 63 (1962) 411.
187   N.R. Mason and K. Savard, Endocrinology, 75 (1964) 215.
188   H. Selye, Rev. Can. Biol., 1 (1942) 578.
189   L.T. Samuels, M.L. Helmreich, M.B. Lasater and H. Reich, Science, 113 (1951) 490.
190   N. Saba, O. Hechter and D. Stone, J. Am. Chem. Soc., 76 (1954) 3862.
191   E. Staple, W.S. Lynn Jr. and S. Gurin, J. Biol. Chem., 219 (1956) 845.
192   S. Solomon, R. Vande Wiele and S. Lieberman, J. Am. Chem. Soc., 78 (1956) 5453.
193   K. Shimizu, M. Hayano, M. Gut and R.H. Dorfman, J. Biol. Chem., 236 (1961) 695.
194   H. Levy, R.W. Jeanloz, R.P. Jacobsen, O. Hechter, V. Schenker and G. Pincus, J. Biol. Chem., 211 (1954) 867.
195   W.R. Slaunwhite and L.T. Samuels, J. Biol. Chem., 220 (1956) 341.
196   M.E. Davis and E.J. Plotz, Am. J. Obstet. Gynecol., 76 (1958) 939.
197   F.A. Hartman, K.A. Brownell, W.E. Hartman, G.A. Dean and C.G. Mac-Arthur, Am. J. Physiol., 86 (1928) 353.
198   W.W. Swingle and J.J. Pfiffner, Am. J. Physiol., 96 (1931) 153.
199   L.T. Samuels and K.B. Eik-Nes, in D.M. Greenberg (Ed.), Metabolic Pathways, 3rd ed., Vol. II, p. 169.
200   O. Hechter, A. Zaffaroni, R.P. Jacobsen, H. Levy, R.W. Jeanloz, V. Schenker and G. Pincus, Recent Progr. Hormone Research, 6 (1951) 215.
201   P.S. Hench, E.C. Kendall, C.H. Slocumb and H.F. Polley, Proc. Staff Meetings Mayo Clinic, 24 (1949) 181.
202   O. Hechter and G. Pincus, Physiol. Rev., 34 (1954) 459.
203   L.T. Samuels, Ciba Found. Coll. Endocrinol., Vol. 7, London 1953, p. 176.
204   D.H. Peterson and H.C. Murray, J. Am. Chem. Soc., 74 (1952) 1871.
205   M. Vogt, J. Physiol., 102 (1943) 341.
206   P.B. Raman, D.C. Sharma, R.I. Dorfman and J.L. Gabrilove, Biochemistry, 4 (1965) 1376.
207   L.F. Fieser and M. Fieser, Natural Products Related to Phenanthrene, New York, 1949.
208   P. Talalay and V.S. Wang, Biochim. Biophys. Acta, 18 (1955) 300.
209   W.S. Lynn Jr. and R.H. Brown, J. Biol. Chem., 232 (1958) 1015.
210   J.L. Rabinowitz, Arch. Biochem. Biophys., 64 (1956) 285.
211   B. Baggett, L.L. Engel, L. Balderas and G. Lanman, Endocrinology, 64 (1959) 600.
212   K.J. Ryan and O.W. Smith, Recent Progr. Horm. Res., 21 (1965) 367.
213   J. Hammerstein, B.F. Rice and K. Savard, J. Clin. Endocrinol. Metab., 24 (1964) 597.
214   M. Hayano, M.C. Lindberg, M. Wiener, H. Rosenkrantz and R.I. Dorfman, Endocrinology, 55 (1954) 326.

215　K.J. Ryan, Acta Endocrinol., 44 (1963) 81.
216　F.D. Maner, B.D. Saffan, R.A. Wiggins, J.D. Thompson and J.R.K. Preedy, J. Clin. Endocrinol. Metab., 23 (1963) 445.
217　G.F. Marrian, Biochem. J., 24 (1930) 435, 1021.
218　K.J. Ryan, J. Biol. Chem., 234 (1959) 2006.
219　A. Colas and W.L. Heinrichs, Steroids, 5 (1965) 753.
220　D.M. Cathro, K. Birchall, F.L. Mitchell and C.C. Forsyth, J. Endocrinol., 27 (1963) 53.
221　J.W. Reynolds, J. Clin. Endocrinol. Metab., 25 (1965) 416.
222　J.T. Lanman, Medicine, 32 (1953) 389.
223　B. Borgström, Biochim. Biophys. Acta, 106 (1965) 171.
224　J.R. Senior, J. Lipid Res., 5 (1964) 495.
225　J.M. Dietschy, H.S. Salomon and M.D. Siperstein, J. Clin. Invest., 45 (1966) 832.
226　N. Mani, Die historische Grundlagen der Leberforschung von Galen bis Claude Bernard, Basel, Stuttgart, 1967.
227　G.A.D. Haslewood, in M. Florkin and H.S. Mason (Eds.), Comparative Biochemistry, Vol. IIIA, New York, 1962, p. 205.
228　K. Bloch, B.N. Berg and D. Rittenberg, J. Biol. Chem., 149 (1943) 511.
229　H. Danielsson and K. Einarsson, Acta Chem. Scand., 18 (1964) 831.
230　D. Mendelsohn and E. Staple, Biochemistry, 2 (1963) 577.
231　H. Danielsson, Acta Chem. Scand., 14 (1960) 348.
232　H.M. Suld, E. Staple and S. Gurin, J. Biol. Chem., 237 (1962).
233　K. Okuda and H. Danielsson, Acta Chem. Scand., 19 (1965) 2160.
234　S. Bergström, H. Danielsson and B. Samuelsson, in K. Bloch (Ed.), Lipid Metabolism, New York, 1964, p. 291.
235　E. Staple and J.L. Rabinowitz, Biochim. Biophys. Acta, 59 (1962) 735.
236　J.B. Carey Jr., J. Clin. Invest., 43 (1964) 1443.
237　H. Danielsson, Arkiv Kemi, 17 (1961) 363.
238　H. Danielsson, Adv. Lipid Res., 1 (1963) 335.
239　O. Berséus, Acta Chem. Scand., 19 (1965) 325.
240　H. Danielsson and K. Einarsson, J. Biol. Chem., 241 (1966) 1449.
241　O. Berséus and H. Danielsson, Acta Chem. Scand., 17 (1963) 1293.
242　H.R.B. Hutton and G.S. Boyd, Biochim. Biophys. Acta, 116 (1966) 362.
243　S. Bergström, K. Pääbo and J.A. Rumpf, Acta Chem. Scand., 8 (1954) 1109.
244　S. Bergström, R.J. Bridgwater and U. Gloor, Acta Chem. Scand., 11 (1957) 836.
245　R.J. Bridgwater and S. Linstedt, Acta Chem. Scand., 11 (1957) 409.
246　G.A.D. Haslewood, in E. Schoffeniels (Ed.), Biochemical Evolution and the Origin of Life, Amsterdam, 1971.
247　G.A.D. Haslewood, Biol. Rev., 39 (1964) 537.
248　G.A.D. Haslewood, J. Lipid Res., 8 (1967) 535.
249　H. Danielsson and T. Kazuno, Acta Chem. Scand., 18 (1964) 1157.
250　S. Bergström and W. Gloor, Acta Chem. Scand., 8 (1954) 1373.
251　J. Bremer, Biochem. J., 63 (1956) 507.
252　W.H. Elliott, Biochem. J., 62 (1956) 433.

253   M.D. Siperstein and A.W. Murray, Science, 123 (1956) 377.
254   R.J. Bridgwater and D.A. Ryan, Biochem. J., 65 (1957) 24 P.
255   A. Butenandt and P. Karlson, Z. Naturforsch., 9b (1954) 389.
256   P. Karlson, Ann. Sci. Nat., 18 (1956) 125.
257   P. Karlson, Vitam. Horm., 14 (1956) 227.
258   P. Hocks and R. Wiechert, Tetrahedron Lett., (1966) 2989.
259   F. Hampshire and D.H.S. Horn, Chem. Commun., 1866 (1966) 37.
260   J. Hora, L. Làbler, A. Kasal, V. Černy, F. Sŏrm and K. Slàma, Steroids,
      8 (1966) 887.
261   P. Karlson and H. Hoffmeister, Z. Physiol. Chem., 331 (1963) 298.
262   D.S. King and J.B. Siddall, Nature (London), 221 (1969) 955.
263   J.A. Thomson, J.B. Siddall, M.N. Galbraith, D.H.S. Horn and E.J.
      Middleton, Chem. Commun., (1969) 669.
264   H. Roeller, K.H. Dahm, C.C. Sweeley and B.M. Tros, Angew. Chem., 79
      (1967) 190.
265   H. Roeller and K.H. Dahm, I.U.B. Tenth Int. Congr. of Biochemistry,
      Hamburg, July 25 to 31, 1976 (Abstracts), Hamburg, 1976, p. 642.
266   G.M. Happ and J. Meinwald, J. Am. Chem. Soc., 87 (1965) 2507.
267   J. Meinwald, G.M. Happ, J. Labows and T. Eisner, Science, 151 (1966)
      79.

Chapter 62

Biosynthesis of the purine nucleus

1. Isotopic experiments on whole animals

We have narrated, in Chapter 47, the discovery of uric acid by Scheele. In Chapter 49, we have retraced the discovery of "xanthic bases", later called purines as constituents of nucleic acids, the prosthetic groups of nucleoproteins. In Chapter 49 we have reported on a number of suggestions of the possible precursors of the purine nucleus. Urea, histidine, arginine and pyrimidines had been proposed. When the isotopic method became available, the problem was attacked again using whole organisms.

Barnes and Schoenheimer [1] observed, by feeding isotopic NH_4Cl to rats and pigeons that dietary ammonia nitrogen was rapidly incorporated into the purines of cellular nucleic acids as well as in the purine excretory products. The theory of Wiener according to which the purine ring was formed by the condensation of two moles of urea with one of tartronic acid, was discredited by experiments with ^{15}N-labelled urea.

We have reported in Chapter 49 the discussions concerning a possible derivation of the imidazole nucleus of allantoin from either histidine or arginine as suggested by Ackroyd and Hopkins [2], and the controversial results obtained by comparing the excretion of uric acid in man or allantoin in mammals other than primates when low or high amounts of arginine or histidine were given.

When isotopes were available, rats or pigeons were fed with [^{15}N]ammonium citrate. This led to incorporation of ^{15}N in higher concentrations in the purines of tissues or in the uric acid or allantoin of the excreta, than in histidine or arginine of tissue proteins (Barnes and Schoenheimer [1]).

Plate 218. John Machlin Buchanan.

Bloch [3] fed pigeons with [^{15}N]arginine and observed that the uric acid excreted did not contain significant amounts of ^{15}N.

Isotopic experiments of Plentl and Schoenheimer [4] also showed that pyrimidines do not participate in purine formation.

In contradistinction to these dismissals of suggested precursors, the first isotopic experiments (Barnes and Schoenheimer [1]) brought in positive information, for example the readily incorporated ammonium citrate, which led Sonne, Buchanan and Delluva [5] to conceive of the derivation of the uric acid ureide carbon from some simple metabolic unit. In this first paper of Buchanan's group on purine biosynthesis, the authors reported on the sources of the ureide carbon, tested by the use of isotopic compounds important in metabolism. They prepared $^{13}CO_2$, $H^{13}COOH$, $CH_3^{13}COOH$, $NH_2\text{-}CH_2\text{-}^{13}COOH$, CH_3CH $OH^{13}COOH$, and $^{13}CH_3^{13}CHOH$ $COOH$. These isotopic compounds were administered to pigeons and the uric acid excreted was degraded in order to gain information concerning the origin of both the ureide carbon atoms

$$\left(O\!=\!C\!\!\overset{\displaystyle N}{\underset{\displaystyle N}{\diagup\!\!\!\diagdown}} \right) .$$

From their results the authors discussed the reactions by which acetate and formate may enter into the ureide groups of uric acid (2 and 8 positions) as observed.

$$\begin{array}{c}
HN^1\!-\!C^6\!\!=\!\!O \quad H \\
O\!=\!C^2 \quad C^5\!\!-\!\!-\!\!N^7 \\
\mid \qquad \parallel \qquad \diagdown C^8\!\!=\!\!O \\
HN^3\!-\!C^4\!\!-\!\!-\!\!N^9 \diagup \\
\qquad\qquad H
\end{array}$$

They demonstrated that, in the rat, the carboxyl group of acetate is not a precursor of urea, from which the authors conclude that the carbon of urea and the ureide carbon of uric acid have different origins in metabolism.

These experiments led to other conclusions relating to the origin of the carbon chain 4-5-6 of uric acid. Buchanan, Sonne and Delluva [6] concluded that CO_2 is the source of carbon-6 and that the α-carbon of glycine is probably the source of carbon-5 while its carboxyl carbon is the source of carbon-4.

By indirect experiments with $^{15}NH_4Cl$, the same authors demonstrated that the amino nitrogen of glycine is probably the source of nitrogen-7.

The incorporation of CO_2 into carbon-6 of uric acid takes place, according to the authors, by a hitherto undescribed assimilation reaction.

In man, the nitrogen in position 7 was recognized as derived from glycine by Shemin and Rittenberg [7].

That the α-carbon of glycine is the precursor of carbon-5 in the pigeon was demonstrated by J.L. Karlsson and Barker [8]. That it can also be a precursor of C_2 and C_8 through the formation of formate from this α-carbon during metabolism was recognized in experiments on pigeons accomplished by Sakami [9] and by Siekewitz and Greenberg [10]. The incorporation of the α-carbon and amino nitrogen of serine into the ureide carbons of uric acid and into nitrogen in position 7 was explained by the conversion of serine to formate (Sakami [11]) and glycine (Shemin [12]).

In addition to the experiments on excretion products of pigeons or rats, it was recognized that similar glycine utilization was involved in the formation of the guanine of yeast nucleic acids (Abrams, Hammarsten and Shemin [13]). To quote the authors of this research accomplished in the department of Chemistry of the Karolinska Institute of Stockholm,

"The experiments reported indicate that in yeast glycine is used for the synthesis of nucleic acid purines in the same manner as for uric acid synthesis in humans and pigeons".

The same conclusion was reached by G.R. Greenberg [14] concerning the relation between glycine and the hypoxanthine of the nucleic acids of liver homogenates. It is clear that, at this time, the biosynthesis of nucleic acid purines and that of the excretory purines appearing as uric acid in the pigeon, which previously had been considered to arise by different pathways, now were recognized to be formed by similar reactions.

However, Heinrich and Wilson [15], as late as 1950, wrote:

"The metabolism of birds is different from that of mammals in that birds form uric acid as the chief end-product of nitrogenous metabolism, while mammals form urea. In mammals the excretory product of the purines, allantoin in most animals and uric acid in man, constitutes a very small part of the total nitrogen excretion. The differences between the relative extent of purine metabolism in birds and mammals suggest the possibility that the paths of metabolism and the precursors of purines may be different in these different animals''.

The authors described the carbon precursors of tissue purines in rats by administering a number of compounds tagged with ^{14}C carbon dioxide, carboxy-labelled acetate, carboxy-labelled glycine, doubly labelled glycine and formate. From tissue nucleic acids, adenine and guanine were isolated and degraded to determine the location of ^{14}C in their molecules. The authors found, as it has been demonstrated for pigeon's uric acid, that carbon-6 was derived of respiratory CO_2, carbons-4 and -5 from glycine and carbons-2 and -8 from formate. The experiments in vivo reported so far pointed to the origins of the carbons of the purine nucleus. The recognition of glycine as a source of carbons-4 and -5 derived from the experimental demonstration that C-4 derived from the carboxyl of lactate and C-5 from α- and β-carbons of this compound. Chemical reasoning led to assume that if the β-carbon of lactate was lost and if the carboxyl carbon remained attached to the α-carbon, glycine appeared as a likely intermediate. This was confirmed by the recognition that carbon-4 was highly labelled after carboxyl-labelled glycine was administered to pigeons. The chemical conceptual foundation of recognizing formate as precursor of carbons-2 and -8 came from the knowledge of benzimidazole condensation of formic acid with diamines. An example of such a condensation is represented by the synthesis of benzimidazole from orthophenylendiamine and formate. In these experiments in vivo, nitrogen-7 was recognized as originating from glycine. It was impossible to determine the origin of the other nitrogens. This resulted from interfering metabolic interactions of nitrogen compounds considered as possible precursors. It was still a common point of view to consider on the one hand the biosynthesis of uric acid excreted by birds as an endpoint of amino acid metabolism and on the other hand the biosynthesis of nucleic acids of tissues. It appeared nevertheless that the precursors of the purine ring were the same in both.

It was realized that experiments in vitro with homogenates were desirable in order to reduce the complexity of tissue metabolism. A first attempt had been made, as was stated in Chapter 49, with pigeon liver slices, a preparation lacking xanthine oxidase. When incubated with glutamine and oxaloacetate the liver slices accumulated hypoxanthine. But, since in the preparation of extracts there was considerable breakdown of cellular nucleotides yielding hypoxanthine, the biosynthesis de novo of this compound was obscured.

2. Hypoxanthine synthesis by pigeon liver homogenates

G.R. Greenberg [14] showed that carbon dioxide and formate are utilized in the synthesis of hypoxanthine by pigeon liver homogenates as they are in the pigeon as had been shown by Buchanan and his group.

Glycine is likewise converted into hypoxanthine in the homogenate as was observed by Schulman, Buchanan and Miller [16]. Schulman, Sonne and Buchanan [17] confirmed the hypoxanthine formation by pigeon liver homogenates, from glycine, CO_2 and formate.

G.R. Greenberg [18] provided evidence that in the de novo synthesis of hypoxanthine, the purine rings are completed after the introduction of a ribose and phosphate residue. In this system, inosine-5-phosphoric acid (inosinic acid) is formed prior to the formation of the free purine and is converted through inosine to hypoxanthine. These studies provided evidence which suggested that in the biosynthesis of hypoxanthine in pigeon liver, ring closures are effected after the introduction of a ribose moiety in the purine precursor base. According to these views, the immediate precursor of de novo synthesized hypoxanthine was a ribonucleotide having an incomplete purine structure.

3. Inosinic acid and purine biosynthesis

When G.R. Greenberg demonstrated that in liver homogenates, inosinic acid was formed before hypoxanthine the attention was called to the general problem of the biosynthesis of nucleotides from purine and pyrimidine bases. At the time, the precursors yielding the carbon atoms of the purine ring were identified but the identity of the intermediates in purine biosynthesis remained unknown. It was therefore necessary, in order to investigate possible ways of nucleotide formation, to resort to model systems.

Williams and Buchanan [19] observed that when they incubated radioactive hypoxanthine with non-isotopic inosinic acid or with inosine, and pigeon liver extract, either compound became strongly radioactive.

To account for such an observation, a possible explanation could be found in the nucleoside phosphorylase reaction proposed by Kalckar [20]. The reversible reaction could account for an incorporation of hypoxanthine into inosine.

inosine + HPO_4^{2-} $\rightleftharpoons$ hypoxanthine + ribose-1-P.

The conversion of inosine to inosinic acid could be explained by direct enzymatic phosphorylation

inosine + ATP $\rightleftharpoons$ inosinic acid + ADP

But these reactions were not supported by experimental data. Complete biosynthesis of inosinic acid was, under certain conditions, observed in the soluble fraction (extract) of pigeon liver homogenate. Proteins had been precipitated with alcohol to free these extracts of low molecular weight organic compounds among which possible substrates may exist. When these proteins were resuspended in buffered solutions containing ribose-5-P, ATP, glycine, formate, CO_2 and phosphoglyceric acid, a significant biosynthesis was obtained when glutamine and aspartic acid were added (Greenberg [21]; Schulman and Buchanan [22]; Schulman et al. [17]).

When experiments were performed with glutamine labelled in the amide group with ^{15}N and with unlabelled aspartic acid, the inosinic acid which was synthesized in vitro contained ^{15}N in nitrogen atoms 3 and 9 (Sonne, Lin and Buchanan [23,24]; Levenberg, Hartman and Buchanan [25]). To quote from the paper of Sonne et al. [24]:

"Although ammonium salts and the nitrogen of aspartic acid are readily incorporated into the purine ring in vivo [26] there is uncertainty that they are direct precursors because of the many side reactions which these compounds undergo. In studies in vitro, Orström, Orström and Krebs [27] have shown that glutamine, as well as pyruvate, oxaloacetate, and ammonium salts may stimulate hypoxanthine synthesis in pigeon liver slices, a fact that has been observed by Greenberg [21], using dialyzed pigeon liver extracts. However, a stimulatory effect of a given substance in a biosynthetic reaction does not definitively establish it as a precursor, while, on the other hand, the absence of such an effect in a crude metabolic system may indicate only that the compound is not a limiting factor. Because of this, we favored the more direct approach of comparing the simultaneous utilization of ^{15}N-labelled compounds suspected of being precursors and glycine-1-^{14}C for hypoxanthine synthesis by pigeon liver extracts''.

The authors showed that the nitrogen of glycine, glutamine, and aspartic and glutamic acids contribute in major proportions to the synthesis of the purine ring of hypoxanthine in pigeon liver extracts. While it was clear that N-7 and N-9 are respectively derived from glycine and from the amide nitrogen of glutamine, it remained difficult to define the origin of N-1 and N-3. Levenberg et al. [25] devised a new method, based on oxidation by H_2O_2 in alkaline solution, which in conjunction with other procedures permitted the separation of all the nitrogen atoms of the purine ring.

The authors found that the nitrogen of aspartic acid was the precursor of N-1, and that the amide nitrogen of glutamine was the precursor of N-3. They confirmed that the glutamine amide nitrogen is the precursor of N-9. By this work, the knowledge of the precursors of the atoms of the purine ring as shown in Fig. 50, was demonstrated.

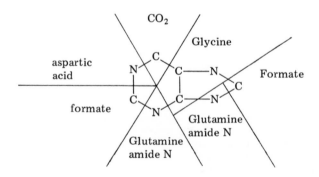

Fig. 50. Precursors of the atoms of the purine ring.

4. Experiments with 5-amino-4-imidazole carboxamide

In an entirely different field of studies, a diazotizable amine had been isolated from *Escherichia coli* cultures grown in the presence of sulphonamides (Stetten and Fox [28]) and identified as 4-amino-5-imidazole-carboxamide (AICA) (Shive, Ackermann, Gordon, Getzendaner and Eakin [29]).

5-amino-4-imidazole-carboxamide (AICA)

It appears immediately that it is only necessary to add a formate to AICA in order to complete a purine ring by adding the C-2. Schulman et al. [16] using [^{14}C]AICA showed that it was converted into uric acid by pigeon liver homogenates. Several studies showed that, while being an artificial precursor of the purine ring, there is evidence that AICA is not a normal intermediate product in the biosynthetic pathway (Schulman and Buchanan [29a]; Greenberg [18]).

5. Enzymatic synthesis and utilization of PRPP

After the important discovery, by G.R. Greenberg, in the Biochemical Department of Western Reserve University School of Medicine, in Cleveland (Ohio), that the biosynthesis of the purine ring proceeded through ribotide intermediates of which inosinic acid was the end-product, a fractionation of the enzymatic system leading from hypoxanthine to inosinic acid, was attempted in the laboratory of Buchanan and his associates. For the over-all reaction two enzymes were obtained. At the time, the nature of the ribose phosphate compound reacting with hypoxanthine to give inosinic acid was still unknown.

The work of Korn, Remy, Wasilejko and Buchanan [30] led to

the isolation of a phosphorylated ribose compound formed from ATP and ribose-5-P by Enzyme I and converted to the nucleotide by a direct reaction with the free base in the presence of Enzyme II. The compound was isolated. It was not ribose-1,5-diphosphate (Klenow [31]; Saffran and Scarano [32]). Remy, Remy and Buchanan [33] after Buchanan had moved from Philadelphia to the Division of Biochemistry of the Department of Biology of the Massachusetts Institute of Technology (MIT), in Cambridge, Mass., identified it as 5-phosphoribosyl-1-pyrophosphate and showed that in the presence of Enzyme I of pigeon liver (later called ribose phosphate pyrophosphokinase) ribose-5-P and ATP react to form AMP and the "active ribose", 5-phosphoribosylpyrophosphate which condenses with hypoxanthine in the presence of Enzyme II (from beef liver) to form inosinic acid.

As stated by Remy et al. [33]:

"The actual identification of the above compound as 5-phosphoribosyl-pyrophosphate was greatly facilitated by the report of Kornberg, Lieberman and Simms [34,35] that they had isolated and identified such a compound as an intermediate in their enzymatic systems which synthesized orotidylic and adenylic acids from orotic acid and adenine, respectively. The compound isolated from our system is undoubtedly identical with their product".

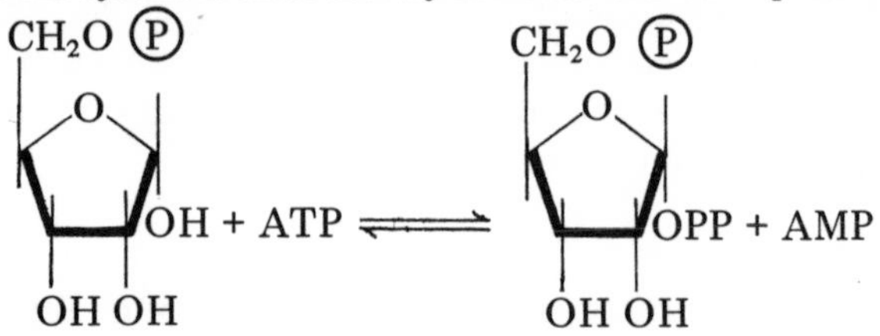

$$CH_2O \; \textcircled{P} \qquad\qquad CH_2O \; \textcircled{P}$$

OH + ATP $\rightleftharpoons$ OPP + AMP

OH OH   OH OH

5-phospho-α-D-ribose   5-phospho-α-D-ribosyl
                            pyrophosphate (PRPP)

The authors carried out the reaction of PRPP and hypoxanthine in the presence of Enzyme II (which they call inosinic acid pyrophosphorylase) with the formation of inosinic acid and pyrophosphate.

6. Origin of PRPP

The compound has been chemically synthesized by Tener and Khorana [36], in 1950. Khorana, Fernandes and Kornberg [37] accomplished the enzymatic synthesis in 1958. They showed that a direct transfer of a pyrophosphoryl group from ATP to 5-phospho-α-D-ribofuranose takes place. The enzyme involved, ribose-phosphate pyrophosphokinase, has been highly purified by Switzer [38]. It has received the systematic name [ATP : D-ribose-5-phosphate pyrophosphotransferase; 2.7.6.1]. It catalyses the reaction of Fig. 51.

7. Identification of two ribotide derivatives of glycinamide as intermediates in the biosynthesis of inosinic acid

After the recognition of inosinic acid as precursor of any purine base, including hypoxanthine, the attention focused on its biosynthesis in the presence of pigeon liver extracts.

In 1954, Goldthwait, Peabody and G.R. Greenberg [39] recognized an accumulation of two unknown compounds when glycine, formate, ribose-5-P, glutamine and ATP were incubated with pigeon liver extract. When radioactive glycine or formate was used, one of the compounds was labelled from radioactive glycine only. The second compound was labelled from both radioactive glycine and radioactive formate. When the enzymatic preparation was

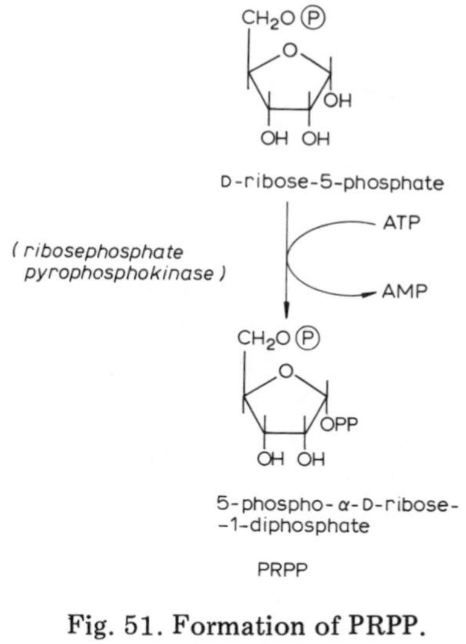

CH_2O ⓅP

D-ribose-5-phosphate

(ribosephosphate pyrophosphokinase)

ATP

AMP

CH_2O ⓅP

OPP

5-phospho-α-D-ribose-
-1-diphosphate

PRPP

Fig. 51. Formation of PRPP.

treated with activated charcoal, only the first compound was formed. Upon addition of tetrahydrofolic acid the second compound was also formed (Goldthwait, Peabody and Greenberg [41]). It was therefore concluded that the biosynthesis of the second compound was dependent upon the presence of a folic acid derivative (see Section 11 of this chapter).

The isolation of the second compound was made easier by the finding by Hartman, Levenberg and Buchanan [40] that azaserine, an antibiotic, inhibited the biosynthetic process and resulted in an accumulation of both compounds, and particularly of the second one.

The identity of both compounds was established by analysis and degradation. The first compound yielded glycine, ammonia nitrogen, ribose and phosphate approximately in the ratio 1 : 1 : 1 : 1, and in addition, the second compound yielded a formate residue. It was concluded that the compounds had the structures, 2-amino-N-ribosylacetamide-5'-phosphate (glycinamide ribotide) and 2-formamino-N-ribosylacetamide-5'-phosphate (formylglycinamide ribotide) (Goldthwait et al. [40]; Hartman et al. [41]).

8. From PRPP to FGAR

As stated above it was recognized that the path from ribose-5-P to GAR depended on the presence of two enzymes. This led to the consideration of two steps, the first leading from ribose-5-P and ATP to an unidentified ribose-containing compound and the second leading from this compound to GAR. As it was known (see above) that PRPP forms nucleotides with free bases it was supposed that it could replace the unknown compound. PRPP was indeed found to substitute for the first enzyme system and for ribose-5-P. But ATP remained necessary (Hartman et al. [40]). Purification of the second enzyme system showed that one of the enzymes it contains catalyses the reaction of PRPP with glutamine, yielding glutamic acid, pyrophosphate and a ribose compound which was recognized as phosphoribosylamine (PRA) (Goldthwait et al. [41]). Synthetic phosphoribosylamine replaced the new compound in the subsequent reaction with glycine and ATP to form GAR, ATP and P (Goldthwait [42]).

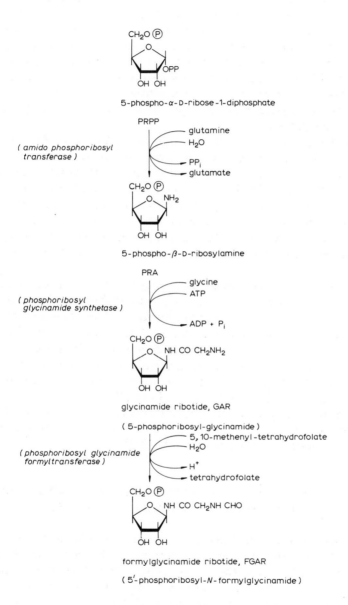

CH₂O ⓟ

OPP

OH OH

5-phospho-α-D-ribose-1-diphosphate

PRPP

(amido phosphoribosyl
transferase)

glutamine
H₂O

PP_i
glutamate

CH₂O ⓟ

NH₂

OH OH

5-phospho-β-D-ribosylamine

PRA

(phosphoribosyl
glycinamide synthetase)

glycine
ATP

ADP + P_i

CH₂O ⓟ

NH CO CH₂NH₂

OH OH

glycinamide ribotide, GAR

(5-phosphoribosyl-glycinamide)

(phosphoribosyl glycinamide
formyltransferase)

5,10-methenyl-tetrahydrofolate
H₂O

H⁺
tetrahydrofolate

CH₂O ⓟ

NH CO CH₂NH CHO

OH OH

formylglycinamide ribotide, FGAR

(5'-phosphoribosyl-N-formylglycinamide)

Fig. 52. From PRPP to FGAR.

References p. 353

As noted by Hartman [43], PRA has never been satisfactorily characterized, as a result of its instability. The enzyme catalysing the reaction between PRPP and glutamine, yielding phosphoribosylamine has been purified from pigeon liver by Hartman and Buchanan [44]. The authors called it 5-phosphoribosylpyrophosphate amidotransferase and it has also been designated as amidophosphoribosyltransferase. Its systematic name is [5-Phosphoribosylamine : pyrophosphate phosphoribosyltransferase (glutamate-amidating); 2.4.2.14].

The formation of GAR from glycine, ATP and PRA in the presence of magnesium ions was recognized as catalysed by an enzyme called glycinamide ribonucleotide kinosynthetase by Hartman and Buchanan [44]. The enzyme, also called GAR synthetase, or phosphoribosylglycinamide synthetase, has received the systematic name [5-Phosphoribosylamine : glycine ligase (ADP-forming); 6.3.4.13, formerly EC 6.3.1.3]. Studies with the enzymes mentioned have demonstrated the stoichiometry of the pathway of Fig. 52.

Data on the properties of both enzymes are given by Hartman [43]. The presumptive C-8 of the purine ring is introduced by the formation of formylglycinamide ribotide (FGAR) (see above). The reaction takes place by a transfer of the formyl group from a formylated derivative of tetrahydrofolate to GAR in the presence of an enzyme first called glycinamide ribonucleotide transformylase (G.R. Greenberg [45]; Greenberg, Jaenicke and Silverman [46]; Warren, Flaks and Buchanan [47]). This enzyme, also called phosphoribosylglycinamide formyltransferase has received the systematic name [5,10-Methenyltetrahydrofolate : 5'-phosphoribosylglycinamide formyltransferase; 2.1.2.2].

9. Conversion of FGAR to FGAM and then to AIRP, with formation of the imidazole ring (Fig. 53)

Levenberg, working in Buchanan's laboratory with a crude pigeon liver extract which had been precipitated with ethanol and washed free of organic substrates, observed that formylglycinamide ribonucleotide (FGAR) could be converted to inosinic acid provided glutamine, CO_2, aspartic acid, formate (or serine) ATP and a "high energy phosphate" regenerating source (such as phos-

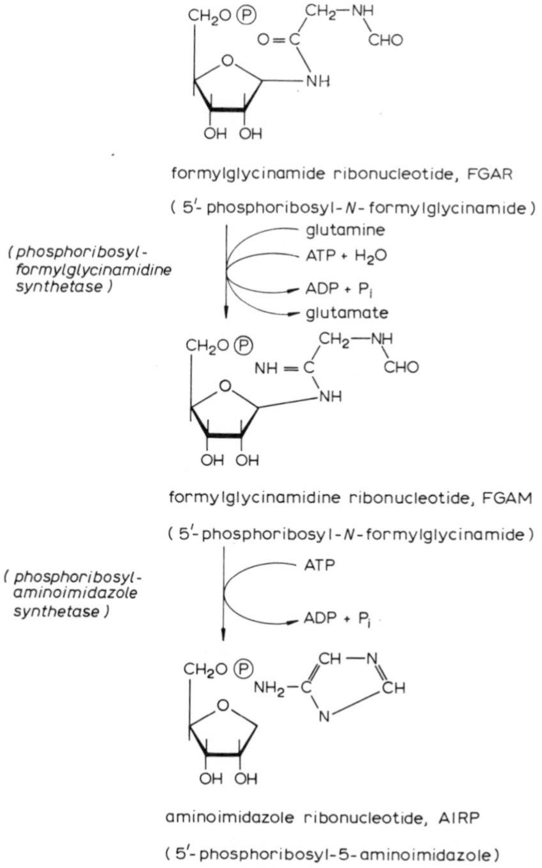
CH2O (P)
CH2—NH
O=C
CHO
NH
OH OH
formylglycinamide ribonucleotide, FGAR
(5'- phosphoribosyl-N- formylglycinamide)
(phosphoribosyl-formylglycinamidine synthetase)
glutamine
ATP + H2O
ADP + Pi
glutamate
CH2O (P)
CH2—NH
NH = C
CHO
NH
OH OH
formylglycinamidine ribonucleotide, FGAM
(5'- phosphoribosyl-N- formylglycinamide)
(phosphoribosyl-aminoimidazole synthetase)
ATP
ADP + Pi
CH2O (P)
CH — N
NH2—C
CH
N
OH OH
aminoimidazole ribonucleotide, AIRP
(5'- phosphoribosyl-5- aminoimidazole)

Fig. 53. From FGAR to AIRP.

phoglyceric acid) were present. If, in such an experiment, formate (or serine) was omitted, an "arylamine" (an amino derivative of a heterocyclic compound) accumulated. This compound proved to be 5-amino-4-imidazole ribonucleotide (Levenberg and Buchanan [48,49] which had already been isolated from natural sources and had been shown to be a precursor of inosinic acid (G.R. Greenberg [45]).

As shown by Levenberg and Buchanan, when CO_2, aspartic acid and serine were omitted from the reacting system, a second aryl-amine was formed, which could be distinguished on the basis of the absorption spectra of the dyes produced when the compounds were diazotized and treated with a coupling reagent according to the method of Bratton and Marshall [50].

To quote Hartman and Buchanan [51]:

"This fortunate difference in the properties of the two arylamines permitted the further segregation of the biosynthetic sequence of reactions into compo-nent steps each of which could be studied with more dispatch".

In 1957, Levenberg and Buchanan [49] had described the enzyme system of pigeon liver extract responsible for the conversion of formylglycinamide ribonucleotide (FGAR) to aminoimidazole ribonucleotide (AIRP). Studying this reaction further, Levenberg and Buchanan [52] noted that the enzymatic components of the system could readily be separated into two fractions. Incubating one of these fractions with FGAR, glutamine and ATP, the authors obtained a new compound which upon reaction with ATP and the second enzymatic compound, was converted to AIRP.

To quote Hartman and Buchanan [51]:

"Its non identity with formylglycinamide ribotide was shown by the fact that its conversion to amino-imidazole ribotide or inosinic acid was not inhibited by azaserine, whereas its formation from formylglycinamide ribotide was. Analysis of the hydrolysis products of the intermediate yielded results identical to those obtained in the case of aminoimidazole ribotide. However, the compound exhibited negative reactions in both the Bratton-Marshall and Pauly tests, indications that it was not yet cyclised to an imidazole com-pound. The new intermediate and formylglycinamide ribotide were compared by electrometric titrations. Whereas formylglycinamide ribotide contains only a secondary phosphate group dissociating in the range between pH 4 and 10 ($pK_1 = 6.4$) the new intermediate contains, in addition, a second group dis-sociating in the more alkaline region with a pK_2 of 9.2. On the basis of these properties, the new compound was shown to be 2-formamido-N-ribosyl-acetamidine-5'-phosphate (formyl-glycinamidine ribotide]."

The enzyme preparation involved in the conversion of formyl-glycinamide ribonucleotide to formylglycinamidine ribonucleotide was purified about 45-fold from pigeon liver by Melnick and Buchanan [53] who, by the use of this preparation, showed the following stoichiometry:

formylglycinamide ribonucleotide + ATP + glutamine + H_2O

→ formylglycinamidine ribonucleotide + ADP + H_3PO_4

+ glutamic acid

The enzyme involved, phosphoribosylformylglycinamidine synthetase, also called FGAR amidotransferase, has received the systematic name [5′-phosphoribosyl-formylglycinamide : L-glutamine
amido-ligase (ADP-forming); 6.3.5.3].

The cyclisation of formylglycinamidine derivative (FGAM) to
form 5-amino-4-imidazole ribonucleotide (AIRP), the first intermediate containing the imidazole nucleus, occurs in a dehydration
reaction in which the elements of water are removed to cleave
ATP to ADP and P_i according to the following reaction, catalysed
by phosphoribosyl-aminoimidazole synthetase:

ATP + 5′-phosphoribosylformylglycinamidine = ADP

+ orthophosphate + 5′-phosphoribosyl-5-aminoimidazole

(Levenberg and Buchanan [49,52]. The enzyme has received the
systematic name [5′-Phosphoribosylformylglycinamidine cycloligase (ADP-forming); 6.3.3.1].

10. From AIRP to AICRP

Levenberg and Buchanan [49], as stated above, had found that
AIRP was converted to inosinic acid, in the presence of aspartic
acid, bicarbonate, formate and ATP; by a partially purified fraction of pigeon liver extract. Since formate was necessary for the
completion of the purine ring, it was logical to perform an experiment without formate. This led to the formation of 5-amino-4-
imidazolecarboxamide ribonucleotide (AICRP) (Levenberg and
Buchanan [49]).

The crude enzyme system responsible for the conversion of
AIRP to AICRP was fractionated by Lukens and Buchanan [54],
with the view of identifying intermediate steps.

The authors obtained two enzyme fractions and observed that
one of these fractions (fraction I) catalysed the formation of a

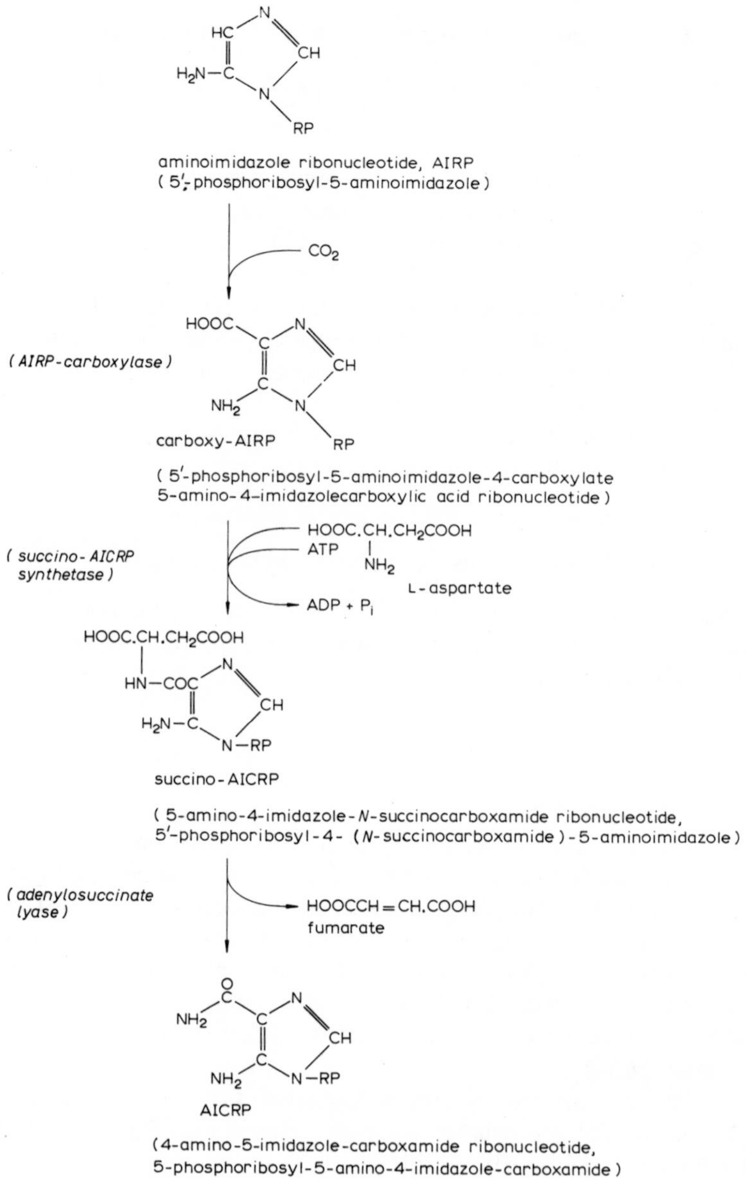
aminoimidazole ribonucleotide, AIRP
(5'-phosphoribosyl-5-aminoimidazole)
CO2
(AIRP-carboxylase)
carboxy-AIRP
(5'-phosphoribosyl-5-aminoimidazole-4-carboxylate
5-amino-4-imidazolecarboxylic acid ribonucleotide)
(succino-AICRP synthetase)
HOOC.CH.CH2COOH
ATP
NH2
L-aspartate
ADP + Pi
succino-AICRP
(5-amino-4-imidazole-N-succinocarboxamide ribonucleotide,
5'-phosphoribosyl-4- (N-succinocarboxamide)-5-aminoimidazole)
(adenylosuccinate lyase)
HOOCCH=CH.COOH
fumarate
AICRP
(4-amino-5-imidazole-carboxamide ribonucleotide,
5-phosphoribosyl-5-amino-4-imidazole-carboxamide)

Fig. 54. From AIRP to AICRP (RP represents the phosphoribosyl group).

new intermediate (in the presence of AIRP, bicarbonate, aspartic acid and ATP). The reaction mixture was heated to destroy the enzymatic fraction I. When fraction II was added to the resulting system, AICRP was formed. The intermediate resulting from the action of enzymatic fraction I was identified as 5-amino-4-imidazole-N-succinocarboxamide ribonucleotide (succino-AICRP) by Lukens and Buchanan [55].

When AIRP was incubated with enzymatic fraction I and with bicarbonate in high concentration but without ATP or aspartic acid, yet another intermediate was obtained, 5-amino-4-imidazole-carboxylic acid ribonucleotide (carboxy-AIRP) [55].

The reversible carboxylation of AIRP discovered by Lukens and Buchanan [55] in 1959 is catalysed by AIRP carboxylase which has received the systematic name [5'-Phosphoribosyl-5-amino-4-imidazole-carboxylate carboxy-lyase; 4.1.1.21].

When carboxy-AIRP reacts with aspartic acid and ATP to form succino-AICRP, the catalysing enzyme is succino-AICRP synthetase which was purified in 1962 by Miller and Buchanan [56] from chicken liver acetone powder. It has received the systematic name [5'-Phosphoribosyl-4-carboxy-4-amino-imidazole : L-aspartate ligase (ADP-forming); 6.3.2.6].

The enzyme catalysing the cleavage of succino-AICRP into AICRP and fumarate has been recognized as identical with the one which converts adenylosuccinic acid to adenylic acid and fumaric acid (Carter and Cohen [57]).

This has been convincingly confirmed by experiments with mutants of *Neurospora crassa* (Miller, Lukens and Buchanan [58]; Giles, Partington and Nelson [59]). It was observed in these experiments that the mutants lacking the enzyme for the splitting of adenylosuccinate are also inactive towards the succinocarboxamide ribonucleotide, while no such relation was observed with arginosuccinase.

The enzyme involved in the cleavage of succino-AICRP, adenylosuccinate lyase, or adenylosuccinase, has received the systematic name [Adenylosuccinate AMP-lyase; 4.3.2.2].

11. From AICRP to IMP

Flaks, Erwin and Buchanan [60] have observed that a purified enzymatic preparation from chicken liver catalyses the transfer of

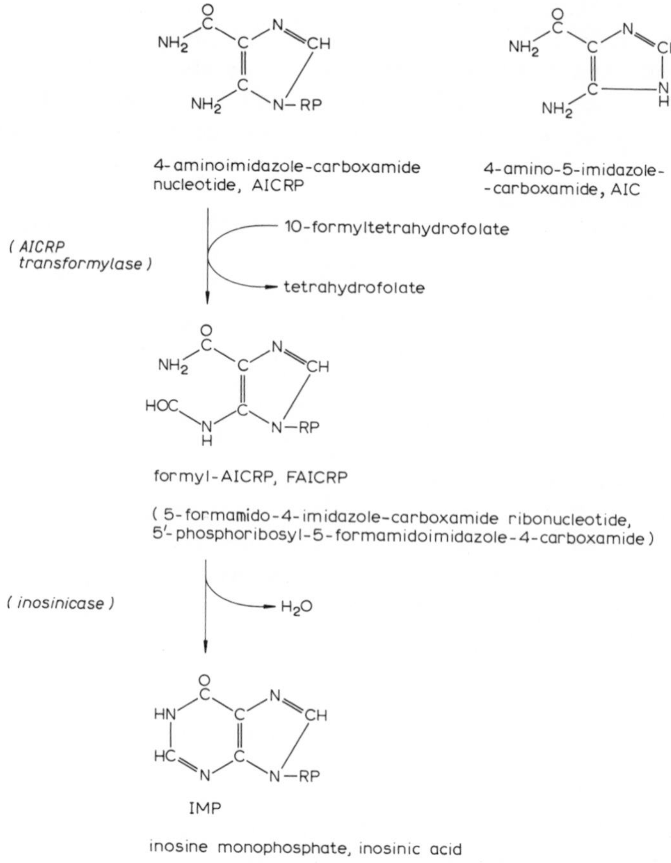

4-aminoimidazole-carboxamide   4-amino-5-imidazole-
nucleotide, AICRP   -carboxamide, AIC

(AICRP
transformylase)

10-formyltetrahydrofolate

tetrahydrofolate

formyl-AICRP, FAICRP

(5-formamido-4-imidazole-carboxamide ribonucleotide,
5'-phosphoribosyl-5-formamidoimidazole-4-carboxamide)

(inosinicase) H_2O

IMP

inosine monophosphate, inosinic acid

Fig. 55. From AICRP to IMP.

a formyl group from either methenyltetrahydrofolate or 10-for-
myltetrahydrofolate to AICRP with a formation of equivalent
amounts of IMP. In this paper, though not decisively resolving
them, the authors have presented some evidence that two enzymes
are involved. That FAICRP is a true intermediate is confirmed by
the observation that chemically synthesized preparations of this
compound are readily converted into IMP. That 10-formyltetra-
hydrofolate serves as the specific donor was shown by Hartman
and Buchanan [61]. The first enzyme involved, AICRP transformylase, or phosphoribosylglycinamide formyltransferase has

received the systematic name [5,10-Methenyltetrahydrofolate : 5'-phosphoribosylglycinamide formyltransferase; 2.1.2.2].

The second enzyme, inosinicase, or IMP-cyclohydrolase, catalyses the dehydration of FAICRP to yield IMP. It has received the systematic name [IMP 1,2-hydrolase (decyclizing); 3.5.4.10].

12. Interconversions among purine nucleotides

As was stated by Hartman [43] (p. 21):

"Inosinic acid is centrally located in the metabolic interrelationships of purines. Both of the primary nucleic acid purines, adenine and guanine, are derived in nucleotide form directly from IMP by amination reactions. The steps leading to synthesis of AMP were originally described by Lieberman [62] and by Carter and Cohen [57], the key finding being the intermediate role of adenylosuccinate. The route to GMP was shown to involve oxidation of IMP to xanthylic acid (XMP) followed by amination of this nucleotide intermediate. Three groups simultaneously described these processes in rabbit bone marrow (Abrams and Bentley [63]), pigeon liver (Lagerkvist [64]) and *Aerobacter aerogenes* (Gehring and Magasanik [65]). Subsequently, these sequences for AMP and GMP formation have been found to be of almost universal occurrence."

13. Formation of adenosine monophosphate

Lieberman [62] demonstrated that GTP specifically acts as the dehydrating agent in the condensation of IMP and L-aspartate. This reaction is catalysed by an enzyme which has been purified from wheat germ by Hatch [66]. This enzyme, adenylosuccinate synthetase, or adenylosuccinate synthase, has received the systematic name [IMP : L-aspartate ligase (GDP-forming); 6.3.4.4].

The enzyme catalysing the cleavage of adenylosuccinate to AMP and fumarate is the same as the enzyme described above as effecting a similar elimination from succino-AICRP (adenylosuccinase; 4.3.2.2).

14. Formation of xanthosine and guanosine phosphates

Greenberg [67] and Buchanan, Flaks, Hartman, Levenberg, Lukens and Warren [68], have established IMP as an intermediate

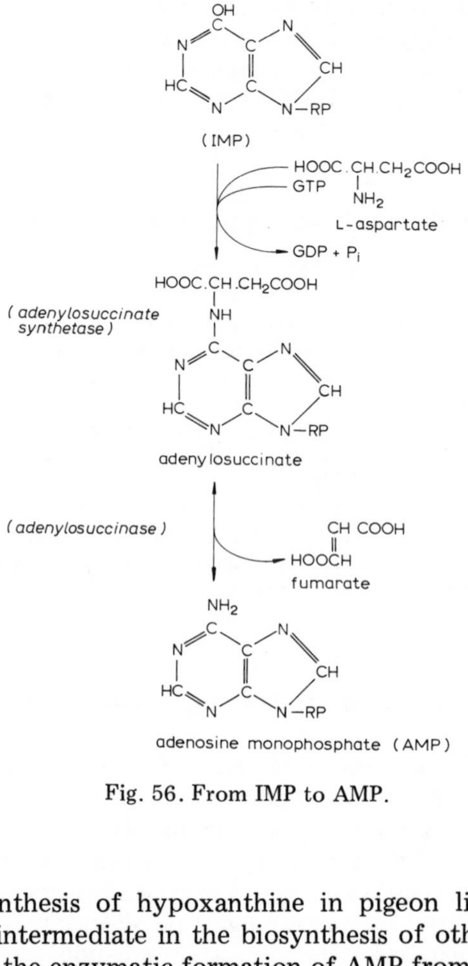

Fig. 56. From IMP to AMP.

in the biosynthesis of hypoxanthine in pigeon liver. That IMP could be an intermediate in the biosynthesis of other purines was suggested by the enzymatic formation of AMP from IMP, reported in the preceding section.

In order to investigate the possibility of the enzymatic formation of GMP from IMP, Lagerkvist [69] investigated the metabolism of IMP in extracts of pigeon liver.

To quote from Lagerkvist:

"It appeared probable that the introduction of an amino group in position 2 of IMP would be preceded by an oxidation in this position to give a xanthine nucleotide".

The hypothesis of Lagerkvist was strengthened by the discovery, by Magasanik and Brooke [70], of a guanineless auxotroph of *Aerobacter aerogenes* accumulating xanthosine. Lagerkvist incubated ^{14}C-labelled IMP with a crude extract of pigeon liver in the presence of oxidized NAD and obtained a new ^{14}C-labelled compound which was identified as XMP. When this compound was incubated with the crude extract together with ATP, Mg^{2+} and L-glutamine, a compound identified as GMP was formed. Independently, Abrams and Bentley [63,71] observed the same reaction in extracts of rabbit bone marrow. Gehring and Magasanik [65] and Magasanik, Moyed and Gehring [72] partially purified from *Aerobacter aerogenes* an enzyme catalysing the oxidation of IMP to XMP.

IMP dehydrogenase, catalysing the oxidation of IMP at posi-

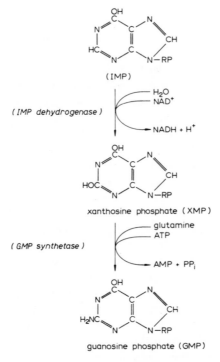

Fig. 57. From IMP to XMP and GMP.

tion 2 was described by McFall and Magasanik [73]. This enzyme has received the systematic name [IMP : NAD$^+$ oxidoreductase; 1.2.1.14]. The product of the oxidation, xanthosine phosphate is aminated concomitantly with a cleavage of ATP to AMP and pyrophosphate (Lagerkvist [69]; Abrams and Bentley [74]; Moyed and Magasanik [75]). The amine source can be ammonia, as has been observed with the enzyme from pigeon liver or mammalian tissues, but the preferred substrate is glutamine. With this substrate, the enzyme mediating the amination is GMP synthetase, which has received the systematic name [Xanthosine-5'-phosphate : glutamine amido-ligase (AMP-forming); 6.3.5.2].

15. From IMP to uric acid (Fig. 58)

As we shall see, along with pyrimidine nucleotides, the purine nucleotides participate in the biosynthesis of nucleic acids. It is in the nucleotide form that the main nucleic acid purines, adenine and guanine are derived from IMP by amination reactions as recalled above.

If we consider the nucleotide pool of animal tissues, we recognize an input flow resulting partly from the process of de novo biosynthesis of nucleotides described in this chapter.

There is also an output flow which involves dephosphorylation, deamination and oxidation of purines and their derivatives, leading as was shown in Chapter 49 to the purine end products uric acid, allantoin, allatoic acid, urea or even ammonia.

The enzymes catalysing the degradation of 5'-nucleotides are widely distributed. Little specificity is shown with respect to the base component of the substrate of 5-nucleotidase (systematic name: [5'-Ribonucleotide phosphohydrolase; 3.1.3.5]).

The conversion of IMP to hypoxanthine has always been conceived by Buchanan as taking place in two steps: from IMP to inosine in the presence of a phosphatase and from inosine to hypoxanthine and ribose-1-P (see Korn et al. [30]).

Because of the equilibrium constant, it should be considered that the enzyme hypoxanthine phosphoribosyltransferase, also called IMP pyrophosphorylase (systematic name: [IMP : pyrophosphate phosphoribosyltransferase; 2.4.2.8]), is involved in the reaction for the synthesis of IMP rather than its breakdown to

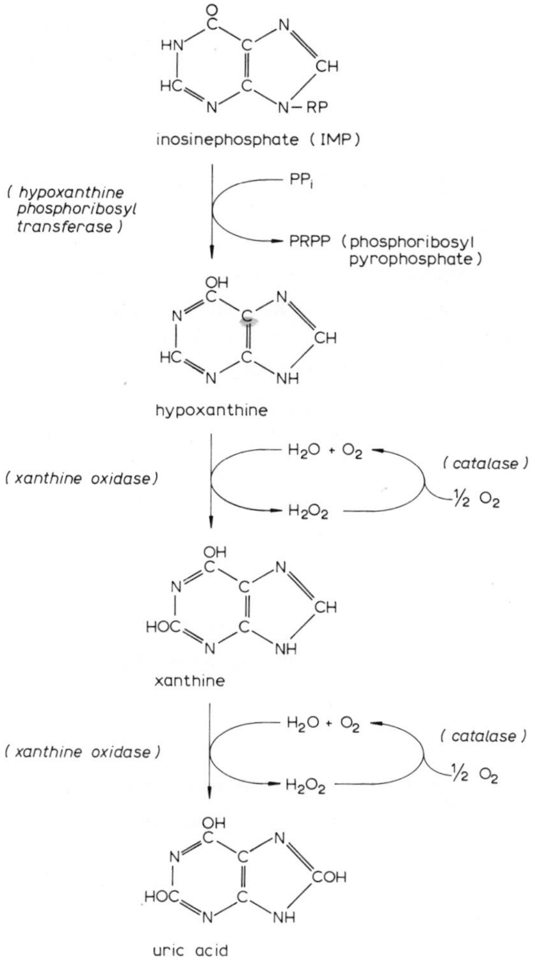

Fig. 58. From IMP to uric acid.

hypoxanthine and PRPP, as indicated in *Enzyme Nomenclature* (on this enzyme, see Kornberg, Lieberman and Simms [76]; Remy, Remy and Buchanan [77]; Lukens and Herrington [78]).

As was stated in Chapter 49 the enzyme catalysing the oxidation of hypoxanthine to xanthine and to uric acid was called xanthine oxidase by Burian. Xanthine oxidase has also been called hypoxanthine oxidase or aldehydrase. Its history has been reviewed by Bray [79]. It has received the systematic name [xanthine : oxygen oxidoreductase; 1.2.3.2].

16. Enzymes for the further degradation of uric acid

As retraced in Section 18 of Chapter 49, uricolysis may be accomplished in the presence of several enzymes (uricase, allantoinase; allantoicase, urease) (see Florkin and Duchâteau [80]). More information has been obtained concerning the uricolysis enzymes.

Uricase (also called urate oxidase or urico oxhydrase) is abundant in liver and kidney of mammals except primates. It has been purified from pig liver and recognized as a copper protein. (On the history of this enzyme, see Mahler [81].) The systematic name of this enzyme is [Urate : oxygen oxidoreductase; 1.7.3.3].

Allantoinase, catalysing the opening of the 5-membered purine ring (see Chapter 49) has received the systematic name (Allantoin amidohydrolase; 3.5.2.5].

Allantoicase, catalysing the conversion of allantoic acid to glyoxylic acid and urea, has received the systematic name [Allantoate amidinohydrolase; 3.5.3.4].

17. Persistence of the theory of Wiener in the case of molluscs and the demise of this theory

Because of the apparent discrepancy between the high level of arginase activity in snails, and their uricotelism, several authors have maintained that the main terminal product of amino acid metabolism in snails, uric acid, is biosynthesized in the hepatopancreas of these molluscs, from urea (Wolf [82]; Baldwin [83]; Grath [84]).

Snails, contrary to other animal forms, would accomplish the biosynthesis of the purine ring along the pathway proposed by Wiener, by which two molecules of urea are combined with a molecule of tartronic acid to form a molecule of uric acid. Here, the isotopic method has once more clarified the situation. According to Wiener's scheme, the labelled carbon of [^{14}C]urea should be found in the C-2 and C-3 of uric acid, as shown in Fig. 59. How-

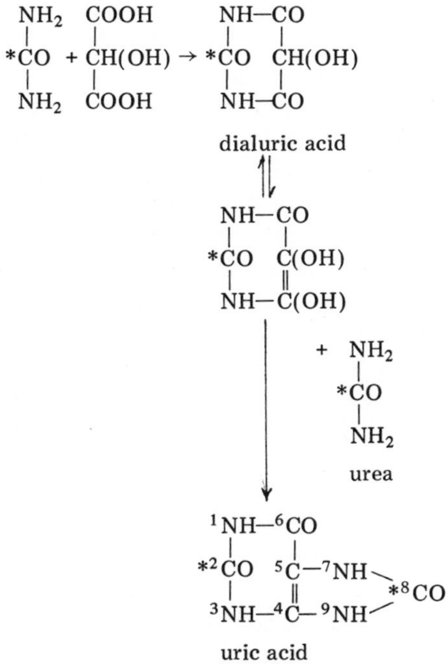

NH₂   COOH   NH—CO
  | | | |
*CO + CH(OH) → *CO CH(OH)
  | | | |
NH₂   COOH   NH—CO

dialuric acid

⇅

NH—CO
  | |
*CO C(OH)
  | ‖
NH—C(OH)

+ NH₂
     |
   *CO
     |
    NH₂

urea

¹NH—⁶CO
  | |
*²CO ⁵C—⁷NH
  | ‖   *⁸CO
³NH—⁴C—⁹NH

uric acid

Fig. 59. Wiener's theory.

ever, when [¹⁴C]urea was injected into the snail *Helix pomatia* by Bricteux-Grégoire and Florkin [85], activity appeared neither in C-2, nor in C-8 of uric acid. The greater part of the activity was found to be localized in C-6 and C-4, i.e. in the position chiefly labelled following the administration of [¹⁴C]bicarbonate to the pigeon (Buchanan et al. [6]). The interpretation given by Bricteux-Grégoire and Florkin [85] is that the urea injected is decomposed by the urease which has been identified in the kidney of the snails by Baldwin and Needham [86] in 1934 and by Heidermanns and Kirchner-Kühn [87], in 1952 and that the carbon dioxide resulting from this action takes part in the biosynthetic pathway, and becomes localized in C-6 and C-4 according to the classical scheme.

Subsequently, Jezewska, Gorzkowski and Heller [88] have in Heller's laboratory at Varshow, confirmed the utilization of C-1 of glycine for the biosynthesis of uric acid C-4 and C-5 in *Helix pomatia.* Lee and Campbell [89], in 1965, have confirmed the origin of C-6 from carbon dioxide in the snail *Otala lactea,* of C-4 and C-5 from C-1 and C-2 of glycine, and of C-2 and C-8 from the one-carbon pool. These results have been confirmed in *Helix pomatia* by Gorzkowski [90] in 1969. Lee and Campbell [89] also showed that N-3 and N-9 arise from glutamine and that 4-amino-5-imidazole carboxamide is a precursor of uric acid in *Otala.* These observations and those from their study of the inhibition of uric acid biosynthesis led them to consider that their biosynthesis involves ribosyl intermediates as it is accepted in the classical pathway of purine ring biosynthesis. The universality of this pathway in animals was also reinforced by the observation by Heller and Jezewska [91] of its utilization in an insect, the moth *Antheraea pernyi.*

18. The case of uricotelic animals

From all comparative studies it has been found that the excretion of uric acid in animals is always a result of the loss of purinolytic enzymes. According to our present knowledge, uric acid is always the result of the oxidation of the hypoxanthine resulting from inosinic acid in the course of purine catabolism.

Uricotelic reptiles and birds form a special case among vertebrates, as the uric acid excreted by these animals derives not only from purines, but also from amino acids.

These animals do not possess the complete enzymatic machinery for ureogenesis, and for instance they lack carbamoylphosphate synthase. The nitrogen of amino acids is, in these forms, as well as in uricotelic invertebrates, channeled into purine biosynthesis and excreted together with the products of purine metabolism (see Vol. 29, Part B, Chapter IV).

REFERENCES

1   F.W. Barnes and R. Schoenheimer, J. Biol. Chem., 151 (1943) 123.
2   H. Ackroyd and F.G. Hopkins, Biochem. J., 10 (1916) 551.
3   K. Bloch, J. Biol. Chem., 165 (1946) 497.
4   A.A. Plentl and R. Schoenheimer, J. Biol. Chem., 153 (1944) 203.
5   J.C. Sonne, J.M. Buchanan and A.M. Delluva, J. Biol. Chem., 173 (1948) 69.
6   J.M. Buchanan, J.C. Sonne and A.M. Delluva, J. Biol. Chem., 173 (1948) 81.
7   D. Shemin and D. Rittenberg, J. Biol. Chem., 167 (1947) 875.
8   J.L. Karlsson and H.A. Barker, J. Biol. Chem., 177 (1949) 597.
9   W. Sakami, J. Biol. Chem., 176 (1948) 995.
10   P. Siekewitz and D.M. Greenberg, J. Biol. Chem., 180 (1949) 845.
11   W. Sakami, J. Biol. Chem., 178 (1949) 519.
12   D. Shemin, J. Biol. Chem., 162 (1946) 297.
13   R. Abrams, E. Hammarsten and D. Shemin, J. Biol. Chem., 173 (1948) 429.
14   G.R. Greenberg, Arch. Biochem., 19 (1948) 337.
15   M.R. Heinrich and D.W. Wilson, J. Biol. Chem., 186 (1950) 447.
16   M.P. Schulman, J.M. Buchanan and C.S. Miller, Fed. Proc., 9 (1950) 225.
17   M.P. Schulman, J.C. Sonne and J.M. Buchanan, J. Biol. Chem., 196 (1952) 499.
18   G.R. Greenberg, J. Biol. Chem., 190 (1951) 611.
19   W.J. Williams and J.M. Buchanan, J. Biol. Chem., 203 (1953) 583.
20   H.M. Kalckar, J. Biol. Chem., 167 (1947) 477.
21   G.R. Greenberg, Fed. Proc., 10 (1951) 192.
22   M.P. Schulman and J.M. Buchanan, Fed. Proc., 10 (1951) 244.
23   J.C. Sonne, I. Lin and J.M. Buchanan, J. Am. Chem. Soc., 75 (1953) 1516.
24   J.C. Sonne, I. Lin and J.M. Buchanan, J. Biol. Chem., 220 (1956) 369.
25   B. Levenberg, S.C. Hartman and J.M. Buchanan, J. Biol. Chem., 220 (1956) 379.
26   U. Lagerkvist, Ark. Kemi, 5 (1953) 569.
27   A. Orström, M. Orström and H.A. Krebs, Biochem. J., 33 (1939) 990.
28   M.R. Stetten and C.L. Fox Jr., J. Biol. Chem., 161 (1945) 33.
29   W. Shive, W.W. Ackermann, M. Gordon, M.E. Getzendaner and R.E. Eakin, J. Am. Chem. Soc., 69 (1947) 725.
29a M.P. Schulman and J.M. Buchanan, J. Biol. Chem., 196 (1952) 513.
30   E.D. Korn, C.N. Remy, H.C. Wasilejko and J.M. Buchanan, J. Biol. Chem., 217 (1955) 875.
31   H. Klenow, Arch. Biochem. Biophys., 46 (1953) 186.
32   M. Saffran and E. Scarano, Nature (London), 172 (1953) 949.
33   Ch.N. Remy, W.T. Remy and J.M. Buchanan, J. Biol. Chem., 217 (1955) 885.

34   A. Kornberg, I. Lieberman and E.S. Simms, J. Am. Chem. Soc., 76 (1954) 2027.
35   A. Kornberg, I. Lieberman and E.S. Simms, J. Biol. Chem., 215 (1955) 389.
36   G.M. Tener and H.G. Khorana, J. Am. Chem. Soc., 80 (1950) 1999.
37   G. Khorana, J.T. Fernandes and A. Kornberg, J. Biol. Chem., 230 (1958) 941.
38   R.L. Switzer, Biochem. Biophys. Res. Commun., 32 (1968) 320.
39   D.A. Goldthwait, R.A. Peabody and G.R. Greenberg, J. Am. Chem. Soc., 76 (1954) 5258.
40   S.C. Hartman, B. Levenberg and J.M. Buchanan, J. Biol. Chem., 221 (1956) 1057.
41   D.A. Goldthwait, R.A. Peabody and G.R. Greenberg, J. Biol. Chem., 221 (1956) 555.
42   D.A. Goldthwait, J. Biol. Chem., 222 (1956) 1051.
43   S.C. Hartman, in D.M. Greenberg (Ed.), Metabolic Pathways, 3rd ed., Vol. 4, Nucleic Acids, Protein-Synthesis and Coenzymes, New York and London, 1970, p. 1.
44   S.C. Hartman and J.M. Buchanan, J. Biol. Chem., 233 (1958) 456.
45   G.R. Greenberg, J. Am. Chem. Soc., 76 (1954) 1458.
46   G.R. Greenberg, L. Jaenicke and M. Silverman, Biochim. Biophys. Acta, 17 (1955) 589.
47   L. Warren, J.G. Flaks and J.M. Buchanan, J. Biol. Chem., 229 (1957) 1812.
48   B. Levenberg and J.M. Buchanan, J. Am. Chem. Soc., 78 (1956) 504.
49   B. Levenberg and J.M. Buchanan, J. Biol. Chem., 224 (1957) 1005.
50   A.C. Bratton and E.K. Marshall Jr., J. Biol. Chem., 128 (1939) 537.
51   S.C. Hartman and J.M. Buchanan, Erg. Physiol., 50 (1959) 75.
52   B. Levenberg and J.M. Buchanan, J. Biol. Chem., 224 (1957) 1019.
53   I. Melnick and J.M. Buchanan, J. Biol. Chem., 225 (1957) 157.
54   L.N. Lukens and J.M. Buchanan, J. Am. Chem. Soc., 79 (1957) 511.
55   L.N. Lukens and J.M. Buchanan, J. Biol. Chem., 234 (1959) 1791.
56   R.W. Miller and J.M. Buchanan, J. Biol. Chem., 237 (1962) 485.
57   C.E. Carter and L.H. Cohen, J. Biol. Chem., 222 (1956) 17.
58   R.W. Miller, J.N. Lukens and J.M. Buchanan, J. Am. Chem. Soc., 79 (1957) 1513.
59   N.H. Giles, C.W.H. Partington and N.J. Nelson, Proc. Natl. Acad. Sci. USA, 43 (1957) 305.
60   J.G. Flaks, M.J. Erwin and J.M. Buchanan, J. Biol. Chem., 229 (1957) 603.
61   S.C. Hartman and J.M. Buchanan, J. Biol. Chem., 234 (1959) 812.
62   I. Lieberman, J. Biol. Chem., 223 (1956) 327.
63   R. Abrams and M. Bentley, J. Am. Chem. Soc., 77 (1955) 4179.
64   U. Lagerkvist, Acta Chem. Scand., 9 (1953) 1028.
65   L.B. Gehring and B. Magasanik, J. Am. Chem. Soc., 77 (1955) 4685.
66   M.D. Hatch, Biochem. J., 98 (1966) 148.
67   G.R. Greenberg, Fed. Proc., 12 (1953) 651.

68   J.M. Buchanan, J.G. Flaks, S.C. Hartman, B. Levenberg, L.N. Lukens and L. Warren, in Chemistry and Biology of Purines, CIBA Found. Symp., London, 1957, p. 233.

69   U. Lagerkvist, J. Biol. Chem., 233 (1958) 138.

70   N. Magasanik and M.S. Brooke, J. Biol. Chem., 206 (1954) 83.

71   M. Bentley and R. Abrams, Fed. Proc., 15 (1956) 218.

72   B. Magasanik, H.S. Moyed and L.B. Gehring, J. Biol. Chem., 226 (1957) 339.

73   E. McFall and B. Magasanik, J. Biol. Chem., 235 (1960) 2103.

74   R. Abrams and M. Bentley, Arch. Biochem. Biophys., 79 (1959) 91.

75   H.S. Moyed and B. Magasanik, J. Biol. Chem., 241 (1966) 351.

76   A. Kornberg, I. Lieberman and E.S. Simms, J. Biol. Chem., 215 (1955) 417.

77   C.N. Remy, W.T. Remy and J.M. Buchanan, J. Biol. Chem., 217 (1955) 885.

78   L.N. Lukens and K.A. Herrington, Biochim. Biophys. Acta, 24 (1957) 432.

79   R.C. Bray, in P.D. Boyer, H. Lardy and K. Myrbäck (Eds.), The Enzymes, 2nd ed., Vol. 7, New York, 1963, p. 533.

80   M. Florkin and Gh. Duchâteau, Arch. Int. Physiol., 53 (1943) 267.

81   H.R. Mahler, in P.D. Boyer, H. Lardy and K. Myrbäck (Eds.), The Enzymes, 2nd ed., Vol. 8, New York, 1963, p. 285.

82   G. Wolf, Z. Vergl. Physiol., 19 (1933) 1.

83   E. Baldwin, Biochem. J., 29 (1935) 1538.

84   H. Grath, Zool. Jahrb., Abt. Allg. Zool. Physiol., 57 (1937) 355.

85   S. Bricteux-Grégoire and M. Florkin, Arch. Int. Physiol. Biochim., 70 (1962) 144.

86   E. Baldwin and J. Needham, Biochem. J., 28 (1934) 1372.

87   C. Heidermanns and I. Kirchner-Kühn, Z. Vergl. Physiol., 34 (1952) 166.

88   M.M. Jezewska, B. Gorzkowski and J. Heller, Acta Biochim. Polon., 11 (1964) 135.

89   T.W. Lee and J.W. Campbell, Comp. Biochem. Physiol., 15 (1965) 457.

90   B. Gorzowski, Acta Biochim. Polon., 16 (1969) 193.

91   J. Heller and M.M. Jezewska, Bull. Acad. Polon. Sci., Classe II, 7 (1959) 1.

Chapter 63

Extensions on the pathway of purine biosynthesis: riboflavin, pterin, pteridine. Tetrahydrofolic acid as coenzyme of C_1 transfer in biosynthesis

1. Pterins and pteridines

In 1889, F.G. Hopkins [1] studied the pigments of the wings of butterflies. These pigments, first taken to be purines and now known as pterins, contained — as shown by Purrmann [1a] in 1940 — a heterocyclic ring which was called pteridine by Weygand [2]. Many pterin pigments were studied: xanthopterin, leucopterin, isoxanthopterin, chrysopterin, erythropterin, ekapterin, rhodopterin, etc. The chemistry of pterins and pteridines is treated by W. Shive in Vol. 11 (Chapter VII) of this Treatise.

2. Riboflavin

In the course of their studies on vitamin B_2, Kuhn, György and Wagner-Jauregg [3] isolated in crystalline form a water-soluble, growth-promoting compound which was called riboflavin, and the photolysis of which yielded lumiflavin. The chemistry of riboflavin and related flavins, derivatives of the alloxazine nucleus, is treated by J.P. Lambooy in Vol. 11 (Chapter II) of this Treatise.

It has been demonstrated that the ring system of a purine precursor is incorporated into the pteridine and flavin skeleton (except the C-8 of the purine ring, extruded as formate). The carbons of the pteridine skeleton not derived from purine are supplied by the C-1 and C-2 of a pentose (Watt [4]). The incorporation of radioactive formate, glycine and CO_2 occurs in similar positions into purines, riboflavin and pteridine.

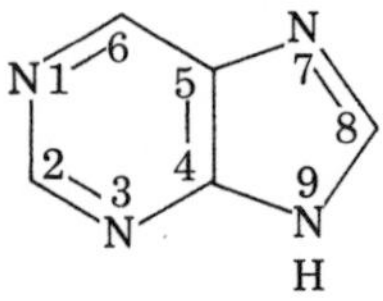

Purine skeleton

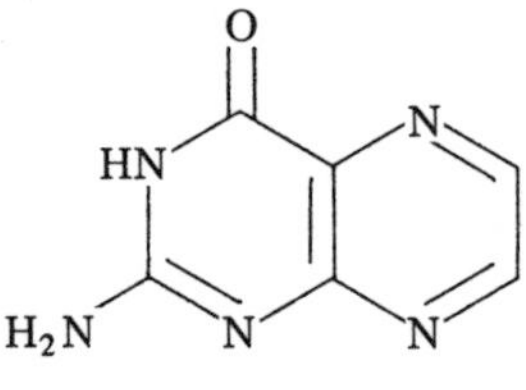

Pterine skeleton

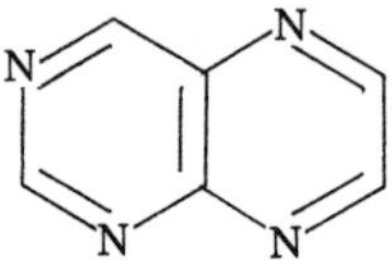

Pteridine skeleton

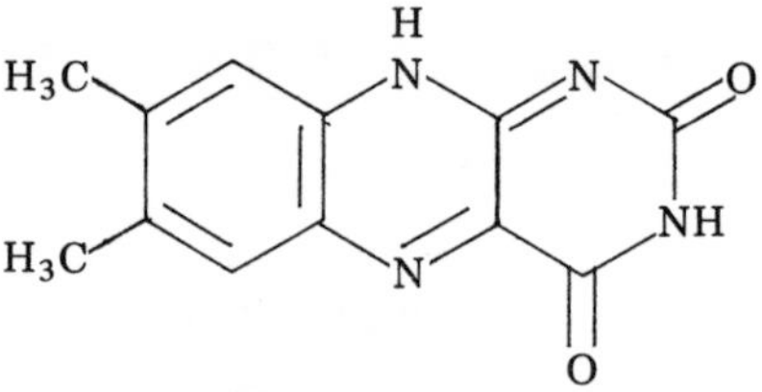

Flavin skeleton

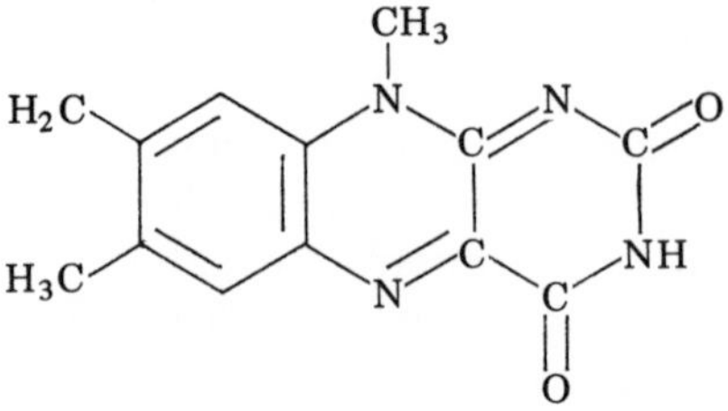

Lumiflavin

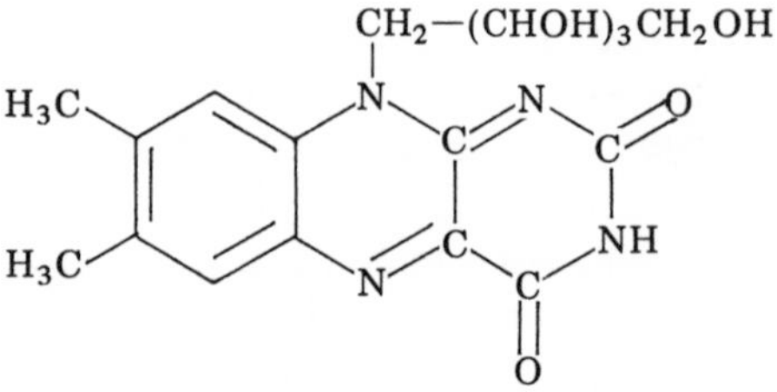

Riboflavin

3. Biosynthesis of riboflavin

The detailed mechanism of the biosynthesis of riboflavin has been retraced from a historical viewpoint by G.W.E. Plaut in Vol. 21 (Chapter I, Section b) of this Treatise. Riboflavin is obtained, in the presence of the enzyme riboflavin synthetase, from 6,7-dimethyl-8-ribityllumazine. The green fluorescent 6,7-dimethyl-8-ribityllumazine was first isolated from cultures of *Eremothecium ashbyii* and has been chemically synthesized in forms labelled in various positions in the molecule. Direct evidence of its conversion into riboflavin, in the presence of riboflavin synthetase, came from the demonstration by Maley and Plaut [5] of the conversion of radioactive 6,7-dimethyl-8-ribityllumazine into radioactive ribo-

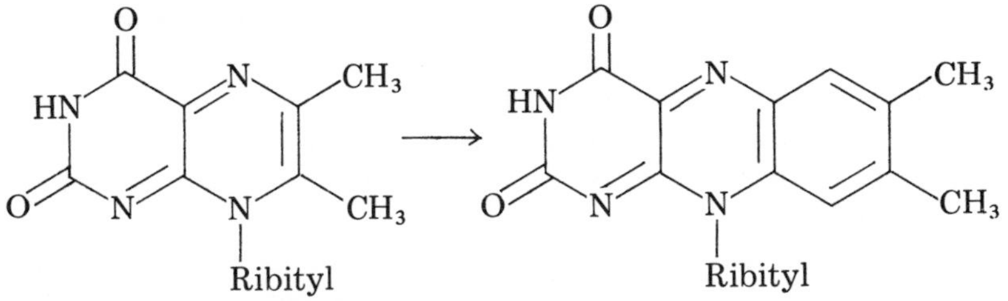

O
HN
O   N
    Ribityl
6,7-Dimethyl-8-ribityllumazine

→

O
HN
O   N
    Ribityl
Riboflavin

CH₃
CH₃

CH₃
CH₃

+

O
HN
O   N
    H
    Ribityl

NH₂
NH

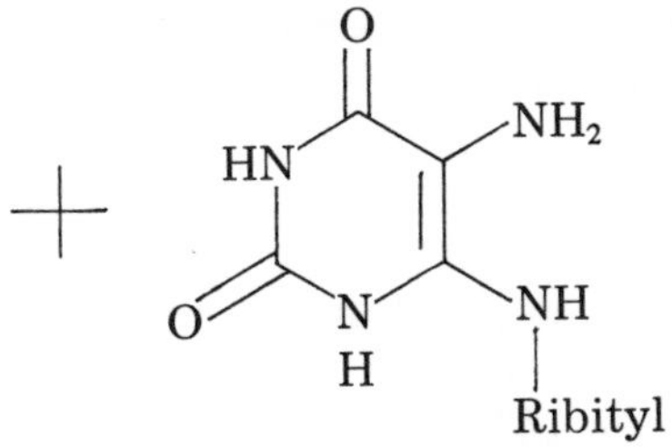

4-Ribitylamino-5-amino-2,6-dioxotetrahydropyrimidine

flavin by cell-free extracts of *Ashbya gossypii*, as well as of other organisms (literature in Vol. 11, Chapter I, section b).

4. Biosynthesis of pteridines

That purines may be precursors of pteridine compounds was suggested by Albert [6], in 1954. He based this on his observations of the chemical transformation of purines into pteridines under mild conditions in the presence of glyoxal. Using the isotopic method, Weygand and Waldschmidt [7] injected [¹⁴C]glycine and [¹⁴C]-formate into butterfly larvae. They observed the incorporation of these compounds in the leucopterin isolated from the wings of the adult butterflies. The pattern of labelling was in favour of purines and pyrimidines being formed by similar biosynthetic pathways. By a number of subsequent observations it was substantiated that purines can be directly used to form pteridines.

   Ziegler-Günder, Simon and Wacker [8] showed the incorporation of [2-¹⁴C]guanine into a pteridine by *Xenopus* larvae.

Brenner-Holzach and Leuthardt [9,10] gave arguments in favour of a purine as a precursor of pteridine pigments in *Drosophila melanogaster*. Weygand, Simon, Dahms, Waldschmidt, Schliep and Wacker [12] gave arguments in favour of guanine or guanosine-5'-phosphate as direct precursor of xanthopterin and leucopterin of the cabbage butterfly *Pieris brassicae*.

Vieira and Shaw [11], using whole cells of a *Corynebacterium* observed that the administration of [2-^{14}C]adenine resulted in the formation of radioactive teropterin (pteroyltriglutamic acid) while the administration of [8-^{14}C]adenine had no such effect. This indicated that the imidazole ring portion of the purine is opened and that the carbon-8 of the purine is removed. This would leave, in the case of the utilization of guanine, 2,4,5-triamino-6-hydroxypyrimidine. Using the 5-^{14}C form of this compound, Baugh and Shaw [13] observed its incorporation by *Corynebacterium*, into pteridine compounds. Previously, Weygand and Waldschmidt [7] had already observed the incorporation of the 2-^{14}C form of this compound in the xanthopterin biosynthesized in butterflies.

Reynolds and Brown [14,15] reported that cell-free extracts of *Escherichia coli* can catalyse the conversion of guanosine into the pteridine moiety of folic acid (see Sections 5 and 6). Besides demonstrating for the first time the enzymatic synthesis of the pteridine moiety of folic acid, Reynolds and Brown observed that radioactive dihydrofolate was biosynthesized from uniformly labelled [^{14}C]guanosine and from guanine plus [1-^{14}C]ribose and that no radioactivity was detected in the dihydrofolate synthesized from [8-^{14}C]guanosine or from guanine plus [5-^{14}C]ribose-5-P. This confirms the requirement of a ring-opening reaction as a step prior to pteridine formation.

With the exception of the experiments of Ziegler-Günder, Simon and Wacker [8] on *Xenopus* larvae, vertebrates had not been used as experimental material in studies of pteridine biosynthesis. Levy [16] showed that when sections of skin were taken from bullfrog tadpoles and immersed in solutions of [2-^{14}C]-guanine or [2,4-^{14}C]guanine, incorporation of the purine occurred into at least three highly fluorescent compounds, identified as the pteridines: 2-amino-4-hydroxypteridine-6-carboxylic acid, isoxanthopterin and biopterin.

From available knowledge, a biosynthetic pathway was formulated (Reynolds and Brown [15]; Schliep and Wacker [17];

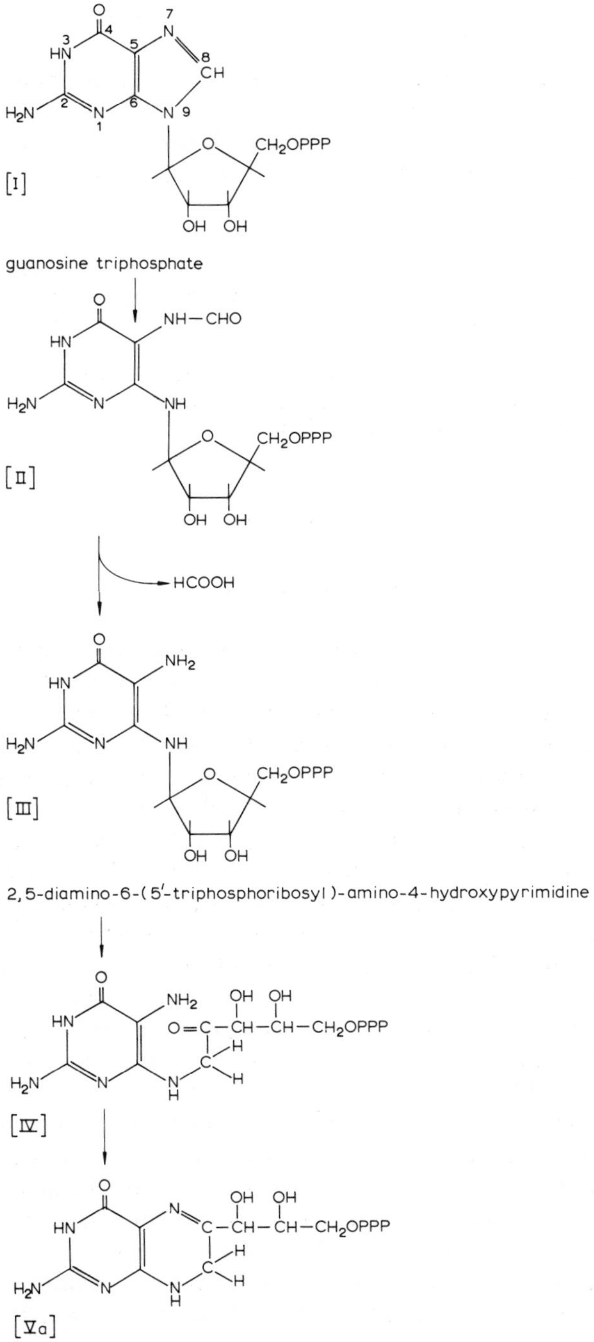

guanosine triphosphate

[I]

[II]

→ HCOOH

[III]

2,5-diamino-6-(5'-triphosphoribosyl)-amino-4-hydroxypyrimidine

[IV]

[Va]

Fig. 60. Pathway of pteridine biosynthesis. (It is established that the pteridine structure in solution is almost exclusively the tautomer indicated.)

Krumdieck, Shaw and Baugh [18]) starting from guanosine tri-phosphate (GTP), Recognized as the most efficient substrate. T. Shiota, in section f of Chapter I of Volume 21 of this Treatise proposed the unified pathway of Fig. 60, and comments on it as follows.

"A scission of the imidazole bond of a guanine nucleotide (guanosine tri-phosphate, GTP [I]) ultimately results in the elimination of carbon-8 and the formation of 2,5-diamino-6-(5'-triphosphoribosyl)-amino-4-hydroxypyrimi-dine [III].

The ribose moiety then undergoes an Amadori-type rearrangement result-ing in a deoxypentulose derivative [IV] and contributes its carbon atoms 1' and 2' to form the pteridine ring [Va]" (p. 112).

5. Folic acid

(a) Introduction

Folic acid is a broadly distributed vitamin of the B-group which was first detected in the course of studies on tropical macrocytic anaemia (tropical sprue), a disease cured by liver extracts.

Working on these extracts, and with the help of microbiological studies and of growth studies on chickens, a yellow sparely soluble compound was obtained. It was also called vitamin Bc on account of its action on the growth of chicken, or *Sfr factor* on account of its action on the growth of *Streptococcus faecalis* R. or *Lc factor*, on account of its action on the growth of *Lactobacillus casei*. The Lc factor was also designated as "norite eluate factor" by Snell and Peterson [19], and Mitchell, Snell and Williams [20] called the Sfr factor folic acid, indicating its acidity and its abundance in spinach leaves. The vitamin was isolated in crystalline form from liver by Pfiffner, Binkley, Bloom, Brown, Bird, Emmett, Hogan and O'Dell [21].

This yellow crystalline product contained one glutamate residue and was called pteroylglutamic acid or folic acid. The structure of folic acid was established by SubbaRow, who had made a name for himself through his work with C. Fiske on creatine phosphate (see Chapter 23), and a number of researchers of the Lederle Laboratories (Stokstad, Hutchings, Mowat, Boothe, Waller, Angier, Semb and SubbaRow [22]; Mowat, Boothe, Hutchings,

$$\longleftarrow \text{pteridine} \longrightarrow \longleftarrow p\text{-AB} \longrightarrow \longleftarrow \text{glutamic acid} \longrightarrow$$
$$\longleftarrow \text{pteroic acid} \longrightarrow \qquad n = 3 \text{ or } 7$$
$$\longleftarrow p\text{-Aminobenzoylglutamic acid} \longrightarrow$$
$$\longleftarrow \text{folic acid} \longrightarrow$$

(pteroylglutamic acid)

Stokstad, Waller, Angier, Semb, Cosulich and SubbaRow [23]).

The structure of folic acid was confirmed by the total chemical synthesis (see Huennekens and Osborn [24] for scientific and patent literature on chemical synthesis, as well as for literature on the properties of folic acid).

It was recognized that p-amino-benzoic acid (p-AB) was incorporated into folic acid (literature in Woods [25]). It was reported that several pteridines stimulated folic acid production by bacteria but the possibility of interconversions in the complex cellular medium rendered the observations inconclusive. It was only when extracts of cells able to catalyse the formation of folic acid were available that the problem could be tackled efficiently.

(b) Biosynthesis

A first step in the study of the biosynthesis of folic acid was accomplished when resting cell suspensions of bacteria were recognized as synthesizing folate-like compounds from p-AB. The conclusion was that a pteridine precursor of folic acid was available in the cells (Nimmo-Smith, Lascelles and Woods [26]). From chemical studies on the nature of this precursor (literature in Shiota, Chapter 1, Section f, Vol. 21 of this Treatise and in Brown [27]) revealed it as being 2-amino-4-hydroxy-6-hydroxymethyldihydro-

Plate 219. Donald Devereux Woods.

pteridine. The enzymatic synthesis of dihydropteroate or dihydro-folate from this precursor by bacterial extracts has been shown to require ATP (Shiota [28]; Brown, Weisman and Molnar [29]; Merola and Koft [30]).

The requirement of ATP led to the consideration that a phos-phorylated pteridine intermediate is involved. It was suggested by Jaenicke and Chan [31] that the pyrophosphate derivative of dihydropteridine was active in the enzymatic synthesis of folate compounds. Shiota, Disraely and McCann [32] and Shiota, McCann and Disraely [33] synthesized the mono- and diphosphate esters of 2-amino-4-hydroxy-6-hydroxymethylpteridine and found that, in the absence of ATP, and after reduction to the dihydro compound, the diphosphate ester was, in the presence of extracts of *Lactobacillus plantarum* and of *p*-aminobenzoylglutamic acid, converted into dihydrofolate.

But Brown et al. [29] postulated that dihydropteroic acid, instead of *p*-aminobenzoylglutamic acid was the normal inter-mediate in dihydrofolic acid synthesis in *E. coli*. These authors suggested the following sequence of enzymatic reactions:

2-amino-4-hydroxy-6-hydroxymethyldihydropteridine

+ *p*-aminobenzoic acid $\xrightarrow{\text{ATP}}$ dihydropteroic acid

Dihydropteroic acid + glutamic acid $\xrightarrow{\text{ATP}}$ dihydrofolic acid.

Griffin and Brown [34] formulate as follows the reasons in favour of this sequence:

"(a) in experiments with *E. coli* extracts, *p*-aminobenzoic acid is used as sub-strate at least 10 times more effectively than is *p*-aminobenzoylglutamic acid, (b) an enzyme that catalyses the formation of dihydrofolate from glutamic acid and dihydropteroate was found in extracts of *E. coli*, whereas no enzyme for the conversion of *p*-aminobenzoic acid and glutamic acid into *p*-amino-benzoylglutamic acid and glutamic acid into *p*-aminobenzoylglutamic acid could be detected (Brown et al. [29])" (p. 310).

The notion that the pteridine substrate is utilized in the dihydro form is, as stated by Weisman and Brown [35],

"based on the observation that dihydropteroic acid was produced even when 2-amino-4-hydroxy-6-hydroxymethyltetrahydropteridine was added as sub-

Plate 220. Gene M. Brown.

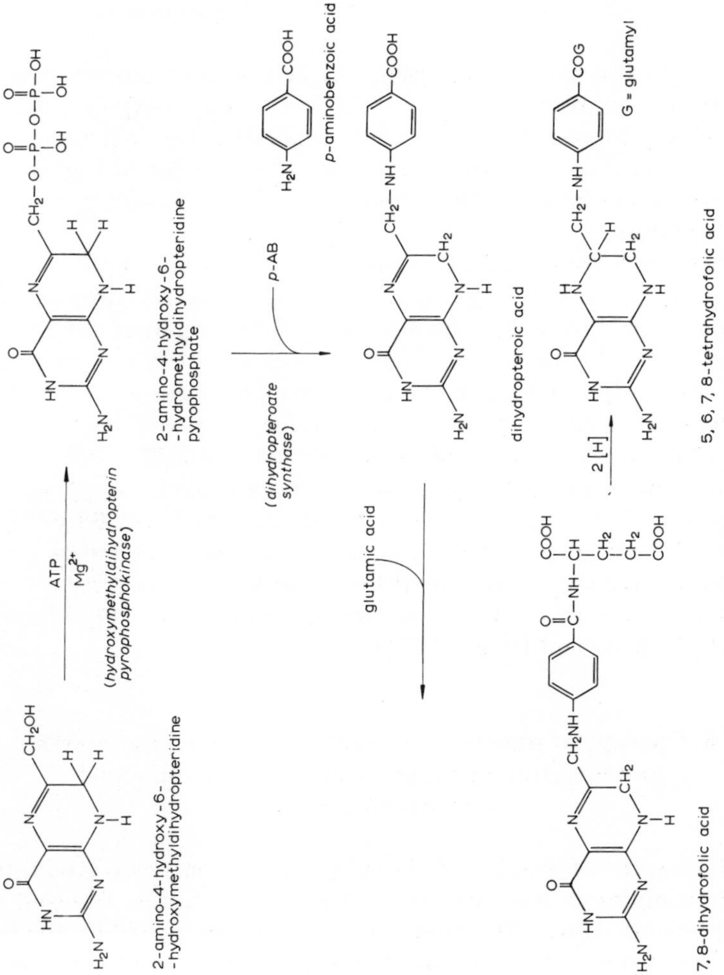

Fig. 61. Biosynthesis of tetrahydrofolic acid.

strate (Brown et al. [29]). It was thought that the tetrahydro compound was oxidized to the dihydro derivative before being utilized" (p. 326).

Fig. 61 relates the biosynthetic pathway of the coenzyme form. It appears that it is as a result of the oxidation of the reduced (dihydro and tetrahydro) forms of the vitamin folic acid that this vitamin occurs in natural products.

(c) Enzymatic aspects of folic acid biosynthesis

The enzymatic system catalysing the synthesis of dihydropteroic acid from p-aminobenzoic acid and 2-amino-4-hydroxy-6-hydroxy-methyldihydropteridine has been separated by Weisman and Brown [35] into two protein fractions (fraction A and fraction B) by chromatography of cell-free extracts of E. coli on diethylamino-ethylcellulose.

Fraction A contains a heat-stable enzyme catalysing the formation of 2-amino-4-hydroxy-6-hydroxymethyldihydropteridine pyrophosphate from 2-amino-4-hydroxy-6-hydroxymethyldihydro-pteridine in the presence of ATP and Mg^{2+}. It has been called hydroxymethyldihydropterin pyrophosphokinase. The second enzyme, utilizing the pyrophosphate ester along with p-aminoben-zoic acid for the production of dihydropteroic acid is commonly called dihydropteroate synthase or dihydropteroate pyrophos-phorylase. Its systematic name is [2-Amino-4-hydroxy-6-hydroxy-methyl-7,8-dihydropteridine-diphosphate : 4-aminobenzoate 2-amino-4-hydroxydihydropteridine-6-methenyltransferase; 2.5.1.15]. (On this enzyme, see Richey and Brown [36]; Shiota, Baugh, Jackson and Dillard [37]).

6. Concept of transfer of C_1 units as catalysed by pteroin proteins containing, as coenzymes, derivatives of tetrahydrofolic acid

In the course of studies of the inhibition of bacterial growth by sulphonamides it was discovered that this effect is reversed by p-AB (Woods [38]). The mode of action was suggested by the fact that the amine which accumulated in sulphanilamide-treated E.

coli (Stetten and Fox [39]), was recognized by Shive, Ackerman, Gordon, Getzendaner and Eakin [40] as 5-amino-4-imidazole carboxamide.

$$\begin{array}{c} NH_2-C=O \\ | \\ C-N \\ \parallel \quad \diagdown CH \\ NH_2-C-NH \diagup \end{array}$$

5-amino-4-imidazole carboxamide

To convert this amine to a purine, only a single carbon atom is necessary and the authors suggested that *p*-AB worked as a cofactor of a 1-carbon fragment, and that sulphanilamide had blocked the enzymatic addition of this 1 carbon fragment. It appeared that a common metabolic step in the biosynthesis of the various metabolites which relieve sulphonamide inhibition or which stimulate growth in *p*-AB-requiring mutant strains of *E. coli* was a reaction involving formate or one of its derivatives. It was known from isotopic experiments (see Chapter 62) that the carbon of formate gets fixed into carbon atoms 2 and 8 of the purine uric acid. Sakami [41] had shown that serine which is effective in the partial reversal of sulphanilamide inhibition was formed in the rat from formate and glycine (see Chapter 73). Furthermore, it was known that serine biosynthesis is promoted in bacteria by "folic acid" (Holland and Meinke [42]).

Lampen, Jones and Roepke [43] showed that the growth of *p*-AB-less mutants of *E. coli* was stimulated by glycine, in which the α-carbon had been shown by Sakami [41] to give rise to formate.

The concept of the transfer of C_1 units as catalysed by pteroprotein enzymes, containing derivatives of folic acid as coenzymes, was confirmed by the establishment, by tracer techniques, of a close metabolic relationship between formate, formaldehyde and potential donors or acceptors of formyl and hydroxymethyl groups. When the isotopic method became available, both in animals and bacteria, it was established that a deficiency of folic acid depressed the incorporation of [14C]formate in a number of systems: nucleic acid purines (Berg [44,45]; Skipper, Mitchell and Bennett [46]; Drysdale, Plaut and Lardy [47]), serine (Lascelles, Cross and Woods [48]; Woods [49]) and methionine (Stekol,

Plate 221. T. Shiota.

Weiss, Smith and Weiss [50]). A similar decrease of C_1 incorporation was observed after the administration of folic acid antagonists such as aminopterin (Goldthwait and Bendich [51]; Broquist [52]). But it was not clear whether these metabolic interconversions were accomplished through "activated" forms. Experimental data were in favour of the latter. For instance, Siegel and Lafaye [53] observed that the addition of unlabelled formaldehyde to a homogeneous system incorporating $H^{14}COOH$ into serine did not significantly reduce the radioactivity of the serine produced. This problem could only be resolved after the identification, characterization, and enzymatic relationships of the naturally occurring formylated derivatives of folic acid.

7. Enzymatic production of tetrahydrofolate

Folic acid has no coenzyme activity. It is converted by reduction to 5,6,7,8-tetrahydrofolate, the coenzyme form of folic acid. In vivo, the formation of tetrahydrofolic acid from folic acid has been observed in tissue slices by Nichol and Welch [54] and in microorganisms by Nichol, Anton and Zakrzewski [55] and by Broquist, Kohler, Hutchison and Burchenal [56].

It was not realized that the cells convert folic acid to tetrahydrofolic acid in a two-stage reduction involving dihydrofolic acid as intermediate. Apparently, the reducing agent can be different in different systems. For instance, Miller and Waelsch [57] found that in liver preparations folic acid was reduced by NADPH, whereas in a bacterial system studied by Wright and Anderson [58], reduction of folic acid to dihydrofolic acid required in addition the operation of the pyruvate oxidase system. Dihydrofolic acid had first been obtained as a colourless derivative of folic acid by the catalytic hydrogenation of folic acid in alkaline solution, over platinum (O'Dell, Vandenbelt, Bloom and Pfiffner [59]). Chemical evidence for the assignment of the 7,8-dihydrostructure (literature in Huennekens and Osborn [24]) has been confirmed by enzymatic evidence (Osborn and Huennekens [60]). For the reduction of dihydrofolic to tetrahydrofolic acid by pigeon liver preparations G.R. Greenberg [61] observed a requirement for reduced pyridine nucleotide. A partial purification of the enzyme system was accomplished, from chicken liver, by Futterman [62].

This enzyme system carries out both an NADPH-dependent reduction of folic acid to dihydrofolic acid and an NADPH or NADH-dependent reduction of dihydrofolic acid to tetrahydrofolic acid. From acetone powder extracts of chicken liver, Zakrzewski and Nichol [63] isolated an NADPH-linked enzyme system converting folic acid to tetrahydrofolic acid.

Osborn and Huennekens [60] have partially purified from acetone powder extracts of chicken liver, by means of protamine treatment, fractionation with ethanol, and adsorption and elution from calcium phosphate gel, an NADPH-linked dehydrogenase which they called dihydrofolic acid reductase and which catalyses the reaction

7,8-dihydrofolic acid + NADPH $\rightleftharpoons$ 5,6,7,8-tetrahydrofolic

acid + NADP

The equilibrium was far to the right at pH 7.5.

8. Formylated natural derivatives related to folic acid and their interconversions

Several substances related to folic acid were recognized as for-mylated compounds. It was the case with the *Streptococcus lactis* R. (SLR) factor (Wolf, Anderson, Kaczka, Harris, Arth, South-wick, Mozingo and Folkers [64] and with the citrovorum factor (CF) (Shive, Bardos, Bond and Rogers [65]). In formylations, the CF was generally active in lieu of folic acid, unlike the SLR factor, but it was not inhibited by folic acid inhibitor (Welch and Heinle [66]).

(a) Rhizopterin

From cultures of *Rhizopus nigricans*, Keresztesy, Rickes and Stokes [67] isolated rhisopterin, a growth factor for *S. faecalis*. The chemical structure was established and confirmed by chemical synthesis (Rickes, Trenner, Conn and Keresztesy [68]; Wolf et al. [64]). The formyl group was found to be located on N^{10}.

N^{10}-formylpteroic acid (rhizopterin)

(b) Teropterin

Hutchings, Stokstad, Bohonos, Sloane and Y. SubbaRow [69] isolated from a *Corynebacterium* another crystalline folic acid compound, teropterin, with three additional glutamic residues. Pfiffner, Calkins, Bloom and O'Dell [70,104] had previously recognized a pteroylheptaglutamate in material isolated from yeast.

(c) N^5-Formyltetrahydrofolic acid (folinic acid, citrovorum factor, leucovorin)

Sauberlich and Baumann [71] observed that *Leuconostoc citrovorum* required a growth factor present in liver and yeast. This factor, CF, was found to be present at low level in the urine of rats maintained on a folic acid-deficient diet, but at high level when folic acid was fed again (Sauberlich [72]). Historically the elucidation of the structure of the CF has not followed the usual degradative approach but has proceeded by synthetic steps. The CF was found to be more effective than folic acid in reversing the toxicity of aminopterin in rats, and the same effect was observed with N^{10}-formylfolic acid compared to folic acid (literature in Huennekens and Osborn [24]). This suggested that CF could be a formyl derivative of folic acid. On this trail of research, considerable CF activity was found when N^{10}-formylfolic acid was submitted to catalytic hydrogenation followed by autoclaving in alkaline solution a process which, as was later known, generated folinic acid (literature in Huennekens and Osborn [24]).

If folinic acid was first obtained by chemical procedures, it was concurrently isolated in the crystalline form from liver by Keresztesy and Silverman [73] and by Sauberlich [74].

References p. 385

The structure (location of formyl) was established conclusively by the classical methods of organic chemistry (see Huennekens and Osborn [24]).

(d) N^{10}-Formyltetrahydrofolic acid

This compound was first obtained as an unstable intermediate in the chemical synthesis of folinic acid from folic acid, by May,

N^{10}-formyl-5,6,7,8-tetrahydrofolic acid

N^5-formyl-5,6,7,8-tetrahydrofolic acid
(folinic acid, leucovorin, citrovorum factors)

Bardos, Barger, Lansford, Ravel, Sutherland and Shive [75]. G.R. Greenberg, Jaenicke and Silverman [76] have demonstrated the presence of an enzyme in pigeon liver homogenates which they called formyltetrahydrofolate synthetase, catalysing the reaction

ATP + formate + tetrahydrofolate

= ADP + orthophosphate + 10-formyltetrahydrofolate

The enzyme, also called formate-activating enzyme, has received the systematic name [Formate : tetrahydrofolate ligase (ADP-forming); 6.3.4.3]. It was later found in a number of bacteria, including *Clostridium cylindrosporum* (Rabinowitz and Pricer [77]) and *Micrococcus aerogenes* (Whiteley, Osborn and Huennekens [78]). The enzyme was purified from acetone powders of pigeon liver (Greenberg and Jaenicke [79]), from *Micrococcus aerogenes* (Whiteley, Osborn and Huennekens [80]) and obtained in crystalline form by Rabinowitz and Pricer [81] from *Clostridium cylindrosporum.*

The participation of N^{10}-formyltetrahydrofolic acid as an intermediate in the enzymatic biosynthesis of folinic acid was proven by Nichol, Anton and Zakrzewski [82] through their experiments on microorganisms.

The enzymatic conversion of folinic acid (N^5-formyltetrahydrofolic acid) to N^{10}-formyltetrahydrofolic acid was first demonstrated by G.R. Greenberg [79,83] in pigeon liver extracts as requiring ATP and formulated as follows:

N^5-formyltetrahydrofolate + ATP

$\rightarrow N^{10}$-formyltetrahydrofolate + ADP + P_i (1)

The product was identified by its utilization in purine biosynthesis (see Chapter 62, and Sections 9 and 11 of this chapter) and by its conversion to N^{10}-formylfolate by air oxidation. Both groups have presented evidence for the participation of N^5, N^{10}-methenyltetrahydrofolic acid as an intermediate. This compound, (also called anhydroleucovorin or isoleucovorin) had been obtained in the course of the studies on the structure of folic acid (literature in Huennekens and Osborn [24]) as a product of the acid treatment of 5-formyltetrahydrofolate.

On account of their findings, G.R. Greenberg [61,83], J.M. Peters and D.M. Greenberg [84,85] have suggested replacing reaction (1) with two steps:

N^5-formyltetrahydrofolate + ATP + H^+

$\rightarrow (N^5,N^{10}$-methenyltetrahydrofolate$)^+$ + ADP + P_i (2)

$(N^5,N^{10}$-methyltetrahydrofolate$^+$) + H_2O

$\rightarrow N^{10}$-formyltetrahydrofolate + H^+ (3)

Peters and D.M. Greenberg called cyclohydrase the enzyme

Plate 222. Lothar Jaenicke.

catalysing reaction (2), and Rabinowitz and Pricer [86] called cyclohydrolase the enzyme catalysing reaction (3).

Nevertheless, in extracts of chicken liver acetone powder, Kay, Osborn, Hatefi and Huennekens [87] have not identified anhydroleucovorin in the course of the conversion.

As stated by Huennekens and Osborn [24],

"the mechanism, the intermediates, and even the endproducts of ATP remain in doubt" (p. 412).

As we have stated in Chapter 35, during the catabolism of histidine by liver and bacterial preparations, formimino glutamic acid was formed. The possible role of folic acid coenzyme in the further metabolism of formiminoglutamic acid was indicated by an increased urinary excretion of this compound by folic acid-deficient rats (Bakerman, Silverman and Daft [88]). That formiminoglutamic acid was a very effective formylating agent in the enzymatic synthesis of formylfolic acid in liver preparations was demonstrated by Slavik and Matoulkova [89]. Miller and Waelsch [90] have described liver enzymes which carried out the overall reaction

formimino-L-glutamic acid + tetrahydrofolic acid

→ L-glutamic acid + 10-formyltetrahydrofolic acid + NH_3

Tabor and Wyngarden [91] have demonstrated that this overall reaction represents the sum of three separate reactions. The first is catalysed by formiminotransferases. In the case of formiminoglutamate the reaction can be formulated as represented in Fig. 62.

The formimino residue can also be transferred from formiminoglycine, the product of uric acid degradation in *Clostridia* [81,86]. The enzymes involved are now called glutamate formiminotransferases (systematic name: [5-Formiminotetrahydrofolate : L-glutamate N-formiminotransferase; 2.1.2.5] and glycine formiminotransferase (systematic name: [5-Formiminotetrahydrofolate : glycine N-formiminotransferase; 2.1.2.4]).

The second reaction of Tabor and Wyngarden can be formulated as consisting in a deamination of 5-formiminotetrahydrofolate leading to 5,10-methenyltetrahydrofolate. The authors named the enzyme involved formiminotetrahydrofolate cyclodeaminase. This

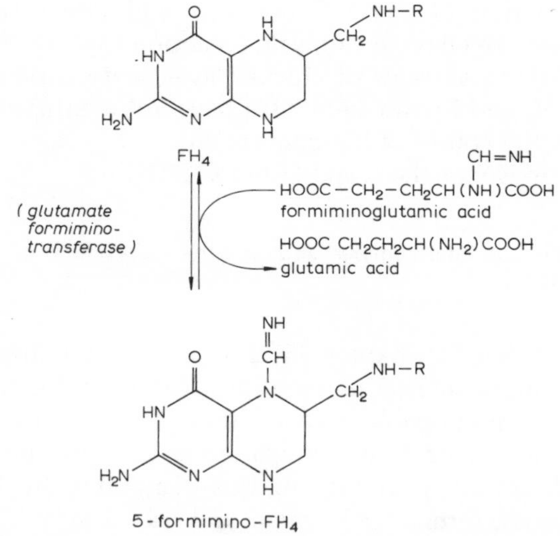

Fig. 62. Example of reaction catalysed by a formiminotransferase.

enzyme has received the systematic name [5-Formiminotetrahy-drofolate ammonia-lyase (cyclizing); 4.3.1.4].

The third reaction formulated by Tabor and Wyngarden is the one already shown by Peters and D.M. Greenberg [85] as being catalysed by the enzyme they have called methenyltetrahydro-folate cyclohydrolase (5,10-methenyltetrahydrofolate → 10-formyl-tetrahydrofolate).

This enzyme has been given the systematic name [5,10-Methenyl-tetrahydrofolate 5-hydrolase (decyclizing); 3.5.4.9].

9. "Active formate" and "active formaldehyde"

N^5-Formyltetrahydrofolic acid (folinic acid, CF) was the first com-pound tentatively proposed as accomplishing the function of "formyl" donor. This was based on its higher efficiency, compared with folic acid, in reversing aminopterin toxicity (see above).

As retraced in Chapter 62, the presumptive C-8 of the purine ring is introduced by the formation of formylglycinamide ribotide (FGAR). This is the result of a transfer to glycine amide ribotide

(GAR) of the "formyl" donated by a formyl derivative of folic acid, acting as coenzyme of the enzyme phosphoribosylglycinamide formyltransferase. The specific donor was identified by G.R. Greenberg [83] as being 5,10-methenylhydrofolate.

In the step from AICRP to IMP, Flaks, Erwin and Buchanan [92] showed that either 5,10-methenylhydrofolate or 10-formyl-tetrafolate acts as the "formyl" donor. This was confirmed by Warren, Flaks and Buchanan [93]. In most other pteroprotein reactions, "active formate" also corresponds to the N^5,N^{10}- or the N^{10}-formyl isomer rather than the N^5-formyl isomer (Huennekens, Osborn and Whiteley [94].

Nevertheless, though incompletely clarified, the 5-formimino-FH_4 isomerase reaction provides an entrance for folinic acid in C_1 transfers.

As it had been found that formaldehyde could be utilized for purine biosynthesis without passing through free formate (G.R. Greenberg [61,79]), Welch and Nichol [95], in a review article in *Annual Review of Biochemistry* postulated, besides "active formate" the existence of "active formaldehyde" which was identified after the unravelling of the relevant enzymatic systems.

Its nature was clarified by Kisliuk and Sakami's [96,97] and Blakley's [98] discovery that FH_4 is a cofactor required in the interconversion of serine and glycine (see Chapter 73).

Active formaldehyde was identified as the bridge compound N^5,N^{10}-methylenetetrahydrofolic acid (literature in Huennekens and Osborn [24]).

In Chapter 66 we shall retrace the discovery of the production of methyl groups, de novo, from formate in the course of the formation of methionine from homocysteine, the donor being

N^5,N^{10}-Methenyltetrahydrofolic acid

N^5,N^{10}-Methylenetetrahydrofolic acid (active formaldehyde)

N^5-Methyltetrahydrofolic acid

5-methyltetrahydrofolate. Using a mutant of *E. coli* requiring either methionine or vitamin B_{12} for growth, Mangum and Scrimgeour [99]; Cathou and Buchanan [100]; and Rosenthal and Buchanan [101] have identified the enzyme system involved. 5-Methyltetrahydrofolate is derived from 5,10-methylenehydrofolate by the reductive process

5-methyltetrahydrofolate + NAD

= 5,10-methylenetetrahydrofolate + NADH

This reaction is catalysed by 5,10-methylenetetrahydrofolate reductase. This enzyme has received the systematic name [5-Methyltetrahydrofolate : NAD^+ oxidoreductase; 1.1.1.68].

We have described in Chapter 62 the action of the enzyme phosphoribosylglycinamide formyltransferase which catalyses the transfer of "formyl" from the specific "formyl donor" 5,10-methenylhydrofolate to the C-8 of the purine ring. Other transformylases are described in the reviews of Huennekens and Osborn [24], Friedkin [102] and Brown [27].

10. Polyglutamates of pteroic acid

Rabinowitz and Himes [103] noticed that almost all of the folate compounds found in microorganisms occur as polyglutamates of pteroic acid. While the first glutamyl is added to dihydropteroate as mentioned above, the other glutamyl residues are added at the tetrahydro state of reduction. Diglutamic acid cannot be used for a direct formation of tetrahydropteroyltriglutamic acid (Griffin and Brown [34]). In tetrahydropteroyltriglutamic acid, the extra glutamyl residues are bound in peptide linkage as γ-glutamyl components.

As many as seven glutamyl components have been shown to occur in tetrahydrofolate compounds (Pfiffner, Calkins, O'Dell, Bloom, Brown, Campbell and Bird [104]). The scanty knowledge so far acquired concerning the occurrence and biosynthesis of polyglutamates and the possibility of their replacement by monoglutamate derivatives of folic acid has been epitomized by Shiota in section f, Chapter I, Vol. 21 of this Treatise.

11. Recognition of specific participation of folate cofactors in C_1 transfers

According to our present views, the introduction of C_1 units in biosynthesized compounds results from a transfer of C_1 units fixed to folate cofactors (donors) to C_1 acceptors.

The pool of C_1 donors consists of free formate and of the combinations, with folic coenzymes, of C_1 groups such as methyl, hydroxymethyl, formyl or formimino groups.

Free formic acid as was stated in Section 8d of this Chapter, can be converted, in the presence of ATP and of formyltetrahydrofolate synthetase, into N^{10}-formyltetrahydrofolic acid. Other tetrahydrofolate derivatives are produced in the course of catalytic degradations. For instance (see Chapter 35, p. 332) the α-carbon of glycine appears in $N^{5,10}$-methylene-FH_4.

In the course of the pathway of histidine catabolism to glutamic

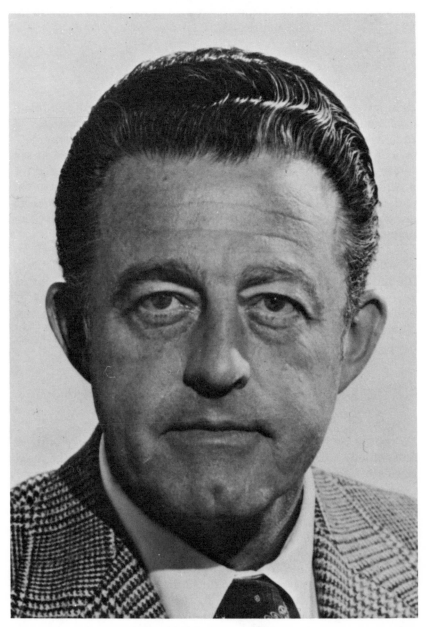

Plate 223. F.M. Huennekens.

acid (Chapter 35, p. 335) the C-2 appears in 5-formimino-FH_4. Concerning the biosynthetic uses of "active formates" we have retraced the history of their unravelling. In the case of purines, two steps involve the participation of N^{10}-formyl-FH_4 and of $N^{5,10}$-methenyl-FH_4 respectively (Chapter 62). As is retraced in Chapter 73 it was shown that $N^{5,10}$-methylene-FH_4 is used for the biosynthesis of serine from glycine. It has also been shown that FH_4 is involved in the biogenesis of the methyl groups of methionine and thymine (Chapter 66).

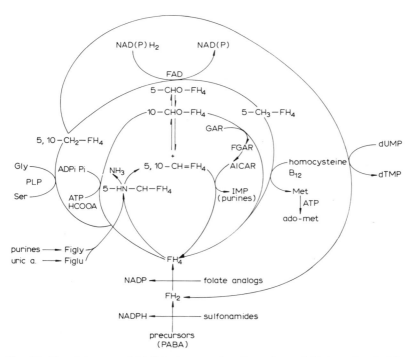

Fig. 63. (Jaenicke, unpubl.) Scheme showing the nature of C_1 transfers in biosynthesis.
PABA, paraminobenzoate; FH_2, dihydrofolate; FH_4, tetrahydrofolate; PLP, pyridoxalphosphate; B_{12}, cobalamine; ado-met, 5-adenosyl methionine; GAR, glycinamide ribotide; FGAR, formylglycinamide ribotide; AICAR, 5-aminoimidazole-4-carboxamide ribotide; Figlu, formiminoglutamate; Figly, formiminoglycinate.

References p. 385

These demonstrations of the role of donor and acceptor of C_1, in combination with the knowledge of the interconversions of functional forms of folic acid, have led to the synthetic scheme of Fig. 63.

REFERENCES

1   F.G. Hopkins, Nature (London), 40 (1889) 335.
1a R. Purrmann, Ann. Chem., 544 (1940) 182.
2   F. Weygand, Öster. Chem. Z., 44 (1941) 254.
3   R. Kuhn, P. György and T. Wagner-Jauregg, Chem. Ber., 66 (1933) 576.
4   W.B. Watt, J. Biol. Chem., 242 (1967) 565.
5   G.F. Maley and G.W.E. Plaut, J. Am. Chem. Soc., 81 (1959) 2025.
6   A. Albert, Biochem. J., 57 (1954) X.
7   F. Weygand and M. Waldschmidt, Angew. Chem., 67 (1955) 328.
8   I. Ziegler-Günder, H. Simon and A. Wacker, Z. Naturforsch., Pt. b, 11 (1956) 82.
9   O. Brenner-Holzach and F. Leuthardt, Helv. Chim. Acta, 42 (1959) 2254.
10 O. Brenner-Holzach and F. Leuthardt, Helv. Chim. Acta, 44 (1961) 1480.
11 E. Vieira and E. Shaw, J. Biol. Chem., 236 (1961) 2507.
12 F. Weygand, H. Simon, G. Dahms, M. Waldschmidt, H.J. Schliep and H. Wacker, Angew. Chem., 73 (1961) 402.
13 C.M. Baugh and E. Shaw, Biochem. Biophys. Res. Commun., 10 (1963) 28.
14 J.J. Reynolds and G.M. Brown, J. Biol. Chem., 237 (1962) PC 2713.
15 J.J. Reynolds and G.M. Brown, J. Biol. Chem., 239 (1964) 317.
16 C.C. Levy, J. Biol. Chem., 239 (1964) 560.
17 H.J. Schliep and H. Wacker, Angew. Chem., 73 (1961) 402.
18 C.L. Krumdieck, E. Shaw and C.M. Baugh, J. Biol. Chem., 241 (1966) 383.
19 E.E. Snell and W.H. Peterson, J. Bacteriol., 39 (1940) 273.
20 H.K. Mitchell, E.E. Snell and R.J. Williams, J. Am. Chem. Soc., 63 (1941) 2284.
21 J.J. Pfiffner, S.B. Binkley, E.S. Bloom, R.A. Brown, O.D. Bird, A.D. Emmett, A.G. Hogan and B.L. O'Dell, Science, 97 (1943) 404.
22 E.L.R. Stokstad, B.L. Hutchings, J.H. Mowat, J.H. Boothe, C.W. Waller, R.B. Angier, J. Semb and Y. SubbaRow, J. Am. Chem. Soc., 70 (1948) 5.
23 J.H. Mowat, J.H. Boothe, B.L. Hutchings, E.L.R. Stokstad, C.W. Waller, R.B. Angier, J. Semb, D.B. Cosulich and Y. SubbaRow, J. Am. Chem. Soc., 70 (1948) 14.
24 F.M. Huennekens and M.J. Osborn, Adv. Enzymol., 21 (1959) 369.
25 D.D. Woods, Ciba Found. Symp. Chem. Biol. Pteridines, (1954) 220.
26 R.H. Nimmo-Smith, J. Lascelles and D.D. Woods, Br. J. Exp. Pathol., 29 (1948) 264.
27 G.M. Brown, in D.M. Greenberg (Ed.), Metabolic Pathways, 3rd ed., Vol. IV, New York and London, 1970, p. 383.
28 T. Shiota, Arch. Biochem. Biophys., 80 (1959) 155.
29 G.M. Brown, R.A. Weisman and D.A. Molnar, J. Biol. Chem., 236 (1961) 2534.

30   A.J. Merola and B.W. Koft, Bacterial Proceedings; Soc. Am. Bacteriol.,
     Baltimore, 1960, p. 175.
31   L. Jaenicke and P.C. Chan, Angew. Chem., 72 (1960) 752.
32   T. Shiota, M.N. Disraely and M.P. McCann, Biochem. Biophys. Res.
     Commun., 7 (1962) 194.
33   T. Shiota, M.P. McCann and M.N. Disraely, Fed. Proc., 22 (1963) 204.
34   M.J. Griffin and G.M. Brown, J. Biol. Chem., 239 (1964) 310.
35   R. Weisman and G.M. Brown, J. Biol. Chem., 239 (1964) 326.
36   D.P. Richey and G.M. Brown, J. Biol. Chem., 244 (1969) 1582.
37   T. Shiota, C.M. Baugh, R. Jackson and R. Dillard, Biochemistry, 8
     (1969) 5022.
38   D.D. Woods, Br. J. Exp. Pathol., 21 (1940) 74.
39   M.R. Stetten and C.L. Fox Jr., J. Biol. Chem., 161 (1945) 333.
40   W. Shive, W.W. Ackermann, M. Gordon, M.E. Getzendaner and R.E.
     Eakin, J. Am. Chem. Soc., 69 (1947) 725.
41   W. Sakami, J. Biol. Chem., 176 (1948) 995.
42   B.R. Holland and W.W. Meinke, J. Biol. Chem., 178 (1949) 7.
43   J.O. Lampen, M.J. Jones and R.R. Roepke, J. Biol. Chem., 180 (1949)
     423.
44   P. Berg, Fed. Proc., 11 (1952) 186.
45   P. Berg, J. Biol. Chem., 205 (1953) 145.
46   H.E. Skipper, J.H. Mitchell Jr. and L.L. Bennett, Cancer Res., 10
     (1950) 510.
47   G.R. Drysdale, G.W.E. Plaut and H.A. Lardy, J. Biol. Chem., 193
     (1951) 533.
48   J. Lascelles, M.J. Cross and D.D. Woods, Biochem. J., 49 (1951) Lxvi.
49   D.D. Woods, American Chemical Society, 120th Meeting, New York,
     1951, Abstracts, p. 14.
50   J.A. Stekol, S. Weiss, P. Smith and K. Weiss, J. Biol. Chem., 201 (1953)
     299.
51   D.A. Goldthwait and A. Bendich, Fed. Proc., 10 (1951) 190.
52   H.P. Broquist, Fed. Proc., 11 (1952) 191.
53   I. Siegel and J. Lafaye, Proc. Soc. Exp. Biol. Med., 74 (1950) 620.
54   C.A. Nichol and A.D. Welch, Proc. Soc. Exp. Biol. Med., 74 (1950) 52.
55   C.A. Nichol, A.H. Anton and S.F. Zakrzewski, Science, 121 (1955) 275.
56   H.P. Broquist, A.R. Kohler, D.J. Hutchison and J.H. Burchenal, J. Biol.
     Chem., 202 (1953) 59.
57   A. Miller and H. Waelsch, Arch. Biochem. Biophys., 63 (1956) 263.
58   B.E. Wright and M.L. Anderson, J. Am. Chem. Soc., 79 (1957) 2027.
59   B.L. O'Dell, J.M. Vandenbelt, E.J. Bloom and J.J. Pfiffner, J. Am.
     Chem. Soc., 69 (1947) 250.
60   M.J. Osborn and F.M. Huennekens, J. Biol. Chem., 233 (1958) 969.
61   G.R. Greenberg, Fed. Proc., 13 (1954) 745.
62   S. Futterman, J. Biol. Chem., 228 (1957) 1031.
63   S.F. Zakrzewski and C.A. Nichol, Biochim. Biophys. Acta, 27 (1958)
     425.
64   D.E. Wolf, R.C. Anderson, E.A. Kaczka, S.A. Harris, G.E. Arth, P.L.
     Southwick, P.L. Mozingo and K. Folkers, J. Am. Chem. Soc., 69 (1947)
     2753.

65 W. Shive, T.H. Bardos, T.J. Bond and L.L. Rogers, J. Am. Chem. Soc., 72 (1950) 2817.
66 A.D. Welch and R.W. Heinle, Pharmacol. Rev., 3 (1951) 345.
67 J.C. Keresztesy, E.L. Rickes and J.L. Stokes, Science, 97 (1943) 465.
68 E.L. Rickes, N.R. Trenner, J.B. Conn and J.C. Keresztesy, J. Am. Chem. Soc., 69 (1947) 2751.
69 B.L. Hutchings, E.L.R. Stokstad, N. Bohonos, N.H. Sloane and Y. SubbaRow, J. Am. Chem. Soc., 70 (1948) 1.
70 J.J. Pfiffner, D.G. Calkins, E.S. Bloom and B.L. O'Dell, J. Am. Chem. Soc., 68 (1946) 1392.
71 H.E. Sauberlich and C.A. Baumann, J. Biol. Chem., 176 (1948) 165.
72 H.E. Sauberlich, J. Biol. Chem., 181 (1949) 467.
73 J.C. Keresztesy and M. Silvermann, J. Am. Chem. Soc., 73 (1951) 5510.
74 H.E. Sauberlich, J. Biol. Chem., 195 (1952) 337.
75 M. May, T.J. Bardos, F.L. Barger, M. Lansford, J.M. Ravel, G.L. Sutherland and W. Shive, J. Am. Chem. Soc., 73 (1951) 3067.
76 G.R. Greenberg, L. Jaenicke and M. Silverman, Biochim. Biophys. Acta, 17 (1955) 589.
77 J.C. Rabinowitz and W.E. Pricer Jr., J. Am. Chem. Soc., 78 (1956) 589.
78 H.R. Whiteley, M.J. Osborn and F.M. Huennekens, J. Am. Chem. Soc., 80 (1958) 757.
79 G.R. Greenberg and L. Jaenicke, in E.W. Wolstenholme and C.M. O'Connor (Eds.), Ciba Foundation Symposium on the Chemistry and Biology of Purines, London, 1957, p. 204.
80 H.R. Whiteley, M.J. Osborn and F.M. Huennekens, J. Biol. Chem., 234 (1959) 1538.
81 J.C. Rabinowitz and W.E. Pricer Jr., J. Biol. Chem., 237 (1962) 2898.
82 C.A. Nichol, A.H. Anton and S.F. Zakrzewski, Science, 121 (1955) 275.
83 G.R. Greenberg, J. Am. Chem. Soc., 76 (1954) 1458.
84 J.M. Peters and D.M. Greenberg, J. Biol. Chem., 226 (1957) 329.
85 J.M. Peters and D.M. Greenberg, J. Am. Chem. Soc., 80 (1958) 2719.
86 J.C. Rabinowitz and W.E. Pricer Jr., J. Am. Chem. Soc., 78 (1956) 5702.
87 L.D. Kay, M.J. Osborn, Y. Hatefi and F.M. Huennekens, J. Biol. Chem., 235 (1960) 195.
88 H.A. Bakerman, M. Silverman and F.S. Daft, J. Biol. Chem., 188 (1951) 117.
89 K. Slavik and V. Matoulkova, Coll. Czech. Chem. Com., 19 (1954) 1032.
90 A. Miller and H. Waelsch, J. Biol. Chem., 228 (1957) 397.
91 H. Tabor and L. Wyngarden, J. Biol. Chem., 234 (1959) 1830.
92 J.G. Flaks, M.J. Erwin and J.M. Buchanan, J. Biol. Chem., 229 (1957) 603.
93 L. Warren, J.G. Flaks and J.M. Buchanan, J. Biol. Chem., 229 (1957) 613.
94 F.M. Huennekens, M.J. Osborn and H.R. Whiteley, Science, 128 (1958) 120.
95 A.D. Welch and C.A. Nichol, Annu. Rev. Biochem., 21 (1952) 633.

96   R.L. Kisliuk and W. Sakami, J. Am. Chem. Soc., 76 (1954) 1456.
97   R.L. Kisliuk and W. Sakami, Fed. Proc., 13 (1954) 242.
98   R.L. Blakley, Biochem. J., 58 (1954) 448.
99   J.H. Mangum and K.G. Scrimgeour, Fed. Proc., 21 (1962) 242.
100 R.E. Cathou and J.M. Buchanan, J. Biol. Chem., 238 (1963) 1746.
101 S. Rosenthal and J.M. Buchanan, Acta Chem. Scand., 17 (1963) S 288.
102 M. Friedkin, Annu. Rev. Biochem., 32 (1963) 185.
103 J.C. Rabinowitz and R.H. Himes, Fed. Proc., 19 (1960) 963.
104 J.J. Pfiffner, D.G. Calkins, B.L. O'Dell, E.S. Bloom, R.A. Brown, C.J. Campbell and O.D. Bird, Science, 102 (1945) 228.

Subject Index

Acetaldehyde, 175, 177
—, as intermediate between carbohydrates and fats, 171
—, oxidation by permanganate, 125
Acetate, carboxy-labelled, 175,
205, 207, 208
—, conversion to mevalonic acid,
273
—, incorporation in squalene, 259
—, in the isoprenoid pathway, 240
—, methyl-labelled, 205, 207, 208
Acetate activation, mechanism, 38
Acetic thiokinase, 175
Acetoacetate, 241
—, conversion to mevalonic acid,
259
—, incorporation into cholesterol,
243, 259
Acetoacetyl-ACP, 180, 181, 182,
184
Acetoacetyl-CoA, 259
—, condensation with acetyl-CoA,
274
—, in biosynthesis of mevalonic acid,
274
Acetoacetyl-enzyme, 179
Acetoacetyl-synthase, 180
Acetobacter xylinum, 158
Aceto-CoA-kinase reaction, 38, 41
Acetokinase, 38
Acetyl-ACP, 180, 181, 182—184
Acetylcholine, biosynthesis, 31, 33,
35, 41
—, role in nerve activity, 33

Acetylcholine cycle, 24
Acetylcholinesterase activity, 33
Acetyl-CoA, 22, 23, 30, 38, 39, 41,
131, 173—175, 177, 178, 179,
181, 183, 184, 208, 257, 259
—, in biosynthesis of mevalonic acid,
274
—, labelled, 287
Acetyl-CoA-ACP- transacylase, 181,
184
Acetyl-CoA: carbon dioxide ligase
(ADP-forming), 179
Acetyl-CoA carboxylase, 178, 184
N-Acetyl-D-glucosamine, 165
N-Acetyl-D-glucosamine units, in
chitin, 164
Acetylphosphates, 31, 35, 36, 38,
87, 173
—, in cell metabolism, 32
Acetylsulphanilamide, 36
Acetyl synthase complex, 179
"Active acetate", 22
"Active dihydroxyacetone", 71
"Active formaldehyde", 378, 379
"Active formate", 378, 379, 383
"Active glycoaldehyde", 70
"Active ribose", 334
"Active succinate", 211, 213, 217,
218
Acyl-ACP, 181, 183
Acyl carrier proteins, 180
Acyl coenzyme A = Acyl-CoA, 181,
185, 186, 188
Acyl CoA: 1,2-diacylglycerol

O-acyltransferase, 189
Acyl-CoA: Sn-glycerol-3-phosphate
    O-acyltransferase, 186
Acyl-CoA groups, 260
Acyl-malonyl-ACP condensing en-
    zyme, 181, 182, 184
Acylphosphates, 32, 57
—, formation, 32
ACP = Acyl carrier protein, 180,
    181
Adenine, 165, 334, 345, 348, 360
—, from tissue nucleic acids, 329
Adenosine, phosphorolysis, 67
Adenosine diphosphate = ADP, see
    also under ADP, 31
Adenosine monophosphate, see also
    under AMP, 58, 345, 346
Adenosine triphosphate, see also
    under ATP, 31
S-Adenosylethionine, 230
S-Adenosyl-L-homocysteine, 229
5-Adenosyl methionine, 383
S-Adenosyl-L-methionine, 229, 230
S-Adenosyl-L-methionine: magne-
    sium-protoporphyrin-O-methyl-
    transferase, 229
Adenylacetate, 39, 41
Adenylbenzoate, 57, 58
Adenylic acid, 36, 37, 55, 146, 147,
    334, 343
Adenylosuccinase, 343, 345, 346
Adenylosuccinate, 343, 345, 346
Adenylosuccinate AMP-lyase, 343
Adenylosuccinate lyase, 342, 343
Adenylosuccinate synthase, 345
Adenylosuccinate synthetase, 345,
    346
Adenylpyrophosphate, 30, 32, 36
Adonitol, 116
ADP = Adenosine diphosphate, 31,
    38, 95, 130, 131, 134, 136, 153,
    262, 341, 375
Adrenal cortex, hormonal secretion,
    289
Adrenocortical hormones, biosyn-
    thesis, 292
Aerobacter aerogenes, 345, 347
Aetioporphyrin III, 214

Agnosterol, 252
AICA = 5-Amino-4-imidazole car-
    boxamide, 333
AICAR = 5-Amino-4-imidazole car-
    boxamide ribotide, 383
AICRP = 5-Amino-4-imidazole-car-
    boxamide ribonucleotide, 341
—, conversion to IMP, 343, 344
—, from AIRP, 341, 342
AICRP transformylase, 344
AIRP = Aminoimidazole ribonu-
    cleotide, 339, 340, 341, 342,
    343
—, conversion to AICRP, 341, 342
—, from FGAR, 338, 339
AIRP-carboxylase, 342, 343
Alanine, 51
—, ^{14}C carbon dioxide fixation, 86
—, conversion to glucose, 109
—, distribution of ^{14}C, 92
Alanylalanine, 51
Albumin, 51
Aliphatic aldehydes, condensation
    with malonyl CoA, 177
β-Aldehydes, reaction with glycine,
    197
Aldehydrase, 350
Aldolase, 76, 99, 100, 102, 104
—, activity in leaves, 93
Aldose-l-phosphate, 143, 155
Aldosterone, 290, 291
—, from corticosterone, 293
Allantoate amidinohydrolase, 350
Allantoic acid, 348, 350
Allantoicase, 350
Allantoin, 325, 348
Allantoin amidohydrolase, 350
Allantoinase, 350
Alloxazine, 357
Allyl pyrophosphates, 264
Amido phosphoribosyl transferase,
    337, 338
Amino acids, 22
—, acylphosphates, 32
—, biosynthesis, 15, 23
—, ^{14}CO$_2$ fixation, 86
—, conversion to glucose, 109
—, —, to uric acid, 352

—, equilibrium with peptides, 52
—, incorporation by peptide linkage, 45, 46
—, —, in proteins, 46, 51
—, in protein synthesis, 49, 51
—, in ripening pea seeds, 53
—, labelled, 47, 51, 53
—, metabolic activity, 12
—, metabolism, 329
—, origin of glucose, 109
—, reaction with dibenzoylphosphate, 57
—, separation by paper chromatography, 91
—, ^{14}C-tagged, 91
p-Aminobenzoic acid, 55, 363, 365, 367, 368, 369, 383
p-Aminobenzoylglutamic acid, 363, 365
p-Aminohippuric acid, 55, 57
2-Amino-4-hydroxy-6-hydroxymethyldihydropteridine, 363, 365, 367, 368
2-Amino-4-hydroxy-6-hydroxymethyl-7,8-dihydropteridine-diphosphate: 4-aminobenzoate 2-amino-4-hydroxydihydropteridine-6-methenyltransferase, 368
2-Amino-4-hydroxy-6-hydroxymethyldihydropteridine pyrophosphate, 367
2-Amino-4-hydroxy-6-hydroxymethylpteridine, 365, 367
2-Amino-4-hydroxypteridine-6-carboxylic acid, 360
4-Amino-5-imidazole-carboxamide, 344, 352
5-Amino-4-imidazole carboxamide = AICA, 333, 369
4-Aminoimidazole-carboxamide nucleotide, 344
4-Amino-5-imidazolecarboxamide ribonucleotide, 342
5-Amino-4-imidazolecarboxamide ribonucleotide = AICRP, 341
5-Aminoimidazole-4-carboxamide-ribotide = AICAR, 383
5-Amino-4-imidazolecarboxylic acid ribonucleotide, 342, 343

Aminoimidazole ribonucleotide = AIRP, 339, 340, 341, 342, 343
5-Amino-4-imidazole-N-succinocarboxamide ribonucleotide, 342, 343
α-Amino-β-ketoadipic acid, 218, 220
δ-Amino-β-ketoadipic acid, 214
Aminolaevulinate dehydratase, 221
5-Aminolaevulinate hydro-lyase (adding 5-aminolaevulinate and cyclysing), 221
δ-Aminolaevulinate synthase, 220
δ-Aminolaevulinic, acid, 23, 216, 229, 231, 245
—, as precursor for heam, 216
—, biosynthesis, 214, 217, 218
—, conversion to porphobilinogen, 220
—, —, to protoporphyrin, 218, 219
—, —, to protoporphyrin IX, 227
—, —, to vitamin B$_{12}$, 232
—, —, synthesis, 214
γ-Aminolaevulinic acid, 211
λ-Aminolaevulinic acid, 227
δ-Aminolaevulinic acid synthetase, 218
Aminolaevulinic acid dehydratase, 220, 221
α-Amino-β-oxoadipic acid, 216, 217
o-Aminophenol, 257
o-Aminophenyl β-D-glucuronic acid, 157
Aminopterin, 371
Aminopterin toxicity, 378
2-Amino-N-ribosylacetamide-5'-phosphate, 336
Ammonia, precursor of porphyrins, 197
Ammonium citrate, 327
Ammonium oxoglutarate, 117
Ammonium pyruvate, 117
AMP = Adenosine monophosphate, 38, 39, 41, 58, 345, 346, 348
Amylase, 159
α-Amylase, 159
β-Amylase, 149, 151
Amylo-1,6-glucosidase, 149
Amylopectin, 149—151
—, structure, 149, 150

Amylose, 148, 149, 151
—, conversion to branched amylo-
pectin, 151
—, in starch, 149
Anabolic studies, methodology, 17
Anabolism, 1
Androgens, 294
Androstenedione, 289
Δ^4-Androstenedione, 294, 295
—, biosynthetic pathway from preg-
nenolone, 295
—, —, to oestrone and oestradiol,
296, 297, 298
Androst-5-ene-3,17-dione, 295
Anhydroleucovorin, 375, 377
Aniline, 49
Animal metabolism, 21
Anion-exchange resins, 86
Antheraea pernyi, 352
Arabinose-5-phosphate = A5P, 39,
41, 63
Arabinose-5-phosphate-deaminase,
41
Arabinose-5-phosphate 5'-nucleo-
tidase, 41
L-Araboketose, 151
Arbutin, 157
Arginase, 19
— activity, in snails, 350
Arginine, 325
— ^{15}N, 327
Arginosuccinase, 343
Arthropod hormones, 239
Arylamines, 340
Ashbya gossypii, 359
Asparagine, conversion to glucose,
109
Aspartic acid, 20, 51, 131, 197,
331, 332, 338, 339, 340, 341,
343, 345
Aspergillus niger, 131
Atherosclerosis, 5, 7, 9
ATP = Adenosine triphosphate, 21,
31, 38, 39, 41, 42, 49, 53, 55,
57, 83, 85, 95, 100, 130, 131,
134, 136, 137, 146, 153, 155,
174, 175, 177, 178, 186, 217,
260, 261, 262, 264, 311, 331,

335, 336, 338, 339, 340, 341,
348, 365, 375, 377, 381
—, as acetylation reagent, 33, 35,
37
—, energy release during hydrolysis,
30
—, hydrolysis, 33, 35
—, —, as source of energy of muscular contraction, 29
—, in biosynthesis reactions, 30, 31,
32
—, in synthesis of hexose phosphate,
102, 103
ATP-ase, 33
ATP-dependent carboxylative phase,
85, 103
ATP-dependent reductive phase, 97
ATP: 5-diphosphomevalonate carboxylyase (dehydrating), 263
ATP: glycerol-3-phosphotransfer-
ase, 186
ATP: mevalonate 5-phosphotrans-
ferase, 261
ATP: 5-phosphomevalonate phosphotransferase, 262
ATP: D-ribose-5-phosphate pyrophosphotransferase, 335
ATP: D-ribulose-5-phosphate
1-phosphotransferase, 95
ATP → Ru-5-P transphosphatase, 95
Autoradiography, 91, 98
Auxotroph mutants, 23
Azaserine, 336, 340
Azotobacter vinelandii, 65

Bacillus coli, 119
Bacterial metabolism, 21, 112
Bacteriochlorin, 230
Bacteriochlorophyll, biosynthesis,
229, 230
Bacteroides vulgatus, 125
Barium carbonate, 115
Barium propionate, pyrolysis, 116
Benzimidazole, condensation formic
acid with diamines, 329
—, synthesis from o-phenylenedia-
mine and formate, 329
Benzoic acid, in synthesis of hip-

puric acid, 54, 55, 57
Benzoylation coenzyme, 57
Benzoyl-CoA, 57, 58
Benzoylglycinamide, 49
Benzoylglycinanilide, 49
Benzoylglycine, 49
Benzoylphosphate, 55, 57
Bicarbonate, 81, 343
—, ^{11}C, 127
—, ^{13}C, 129
—, ^{14}C, 351
Bile acids, 239
—, biosynthesis, 300, 301, 302, 303
—, C_{24}, 301
—, conjugation, 311
—, from coprostanol, 307
—, in Amphibia and fishes, 309
Bile alcohols, 239
—, biosynthesis, 300, 302
—, conjugation, 311
—, from coprostanol, 307
—, in synthesis of cholic acid, 310
Bile salts, 300
Bioenergetics, 29
Biogenetic isoprene rule, 240, 255, 276, 277
Biological isoprene units, 277
—, identification, 263
Biological pathways, 2
Biopterin, 360
Biose phosphates, 92
Biosynthetic pathways, 2, 3
—, determination by chemical methods, 15, 16
—, sequencing into a chain of enzymatic separate links, 18
Biosynthetic relationship, sedoheptulose and ribulose, 93
Biosynthesis, energetics, 29
Biotin, 177, 178
Bombyx, 311
Branching enzyme, 151
p-Bromophenacetylglycerate, 89
Büchner's zymase, 14
5β-Bufol, 302, 306, 309, 310
Butyric acid, 174, 241
—, anaerobic synthesis, 171
— hydroxamate, 175

Butyryl-ACP, 182, 183, 184
Butyryl-CoA, 173, 184
Butyryl-synthase, 180

Calcium carbonate, 112
Calliphora, 313
Caproic acid, 174
Capryl-CoA, 183
Carbamoylglutamate, 20
Carbamoylphosphate synthase, 352
Carbohydrates, 240
—, biosynthesis, 81, 156
—, —, from CO_2, 102
—, conversion to fats, 171
—, gluconeogenesis, formation from non-carbohydrate, 129
—, from carbon dioxide, 21, 22
—, labelled, 81
—, synthesis, 93
Carbohydrate cycle, 21
Carbohydrate metabolism, 76
—, tracer studies with ^{11}C, 122
Carbon, ^{11}C, 21, 22, 81, 83, 85, 86, 117, 119
—, ^{13}C, 12, 115, 116
—, ^{14}C, 13, 85, 86
Carbon dioxide, as universal metabolic oxidizing agent, 123
—, assimilation in photosynthesis, 96
—, bacterial reduction to methane, 113
—, biosynthetic action, 21, 22
—, ^{11}C labelled, 22, 81, 83, 122, 124, 126, 127, 129
—, ^{11}C, uptake by propionic acid bacteria, 113, 115
—, ^{13}C, 113, 115, 129
—, ^{14}C, 13, 98
—, —, fixation, 86
—, —, uptake by algae, 89
—, fixation, 81
—, —, by plants 122
—, —, by propionic bacteria, 112, 113, 116
—, —, in animal tissues, 117, 119
—, —, in heterotrophs, 109
—, —, in heterotrophic bacteria, 111

—, —, in mammalian tissues, 123, 124
—, in biosynthesis of α-oxoglutarate, 121
—, in metabolic studies, 21, 22
—, in synthesis of hypoxanthine, 330
—, in urea synthesis, 111
—, photosynthetic dark fixation, 124
—, reduction to formate, 124
—, — propionic acid, 115
Carbonic acid, 81
Carbon reduction, kinetics, 85
—, photosynthetic cycle, 81
Carboxy-AIRP, 342, 343
α-Carboxy-β-carboxymethyl-β'-carboxyethylpyrrole, 208
Carboxydismutase, 96, 104
Carboxylation enzyme, 96
β-Carotene, 243, 282
—, biosynthesis, 279
—, synthesis, 277
δ-Carotene, 282
γ-Carotene, 282
ζ-Carotene, 282, 283
Carotenoids, 230, 239
—, biosynthetic pathways, 277
Catabolic studies, methodology, 17
Catabolism, 1
Cellobiose glucosyltransferase, 151
Cellular respiration, 22
Cellulose, biosynthesis from UDPG, 158
—, in plants, 158
—, from sugar nucleotides, 165
Chemosynthesis, 111
Chemosynthetic autotrophs, 81
Chemosynthetic bacteria, 111
Chenodeoxycholic acid, 301, 302, 303
—, biosynthesis, 306
Chimaera, 307
Chimaerol, 302
5β-Chimaerol, 309, 310
Chitin, biosynthesis, 164
Chlorella, 83, 86, 91, 95, 223, 227

Chlorella mutants, 229, 231, 282
Chlorella vulgaris, 225, 284
Chlorophyll, 193
—, biosynthesis, 229
—, pyrrole skeleton, 229
Chlorophyllide a, 231
Chloroplasts, as seat of sugar synthesis, 105
5β-Cholanoic acids, 301
5β,25ζ-Cholestane-3α,7α,12α-tetrol, 310
Cholestanol, conversion to bile acids, 303
Cholesterol, acetoacetate in the biosynthesis, 243, 297
—, biosynthesis, 7
—, —, by cell free preparations, 250
—, —, through isoprene pathway, 240, 241, 243
—, chemical synthesis, 245
—, conversion to bile acids, 301, 302, 303
—, —, to cholic acid, 303, 305, 310
—, —, to ecdysone, 313
—, —, to progesterone, 298
—, —, to steroid hormones, 287
—, deposition in the arteries, 5
—, deuterium labelled, 240, 241
—, distribution of acetate carbons, 245, 247, 251
—, enzymes in the conversion to pregnenolone, 300
—, from labelled acetate, 245, 247, 249, 250, 251
—, from lanosterol, 252, 253, 255, 273
—, from mevalonic acid, 274
—, from squalene, 249, 250, 251, 253, 255
—, from sterol intermediates, 273
—, metabolism, 5, 7, 9
—, precursors, 241
—, relation with mevalonic acid, 259
—, role of coenzyme A in biosynthesis, 250
—, structure, 250, 251, 252, 253
—, to pregnenolone, 287

Cholic acid, 301, 302
—, conjugation with taurine, 311
—, from cholesterol, 303, 305, 310
—, from coprostane, 307
Choline, acetylation, 33, 35, 37
—, enzymatic acetylation, 36
Choline acetylase, 35, 37, 41
Cholyl-CoA, 307
Chromanes, from mevalonic acid, 284
Chrysanthemols, 270
Chrysanthemum monocarboxylic acid, 270
Chrysopterin, 357
Cineole, 276
Citral, 315
Citric acid, 87, 121, 205
—, biosynthesis, 117
—, labelled, 212, 213
—, synthesis by avian tissue, 112
—, — by moulds, 113
Citric acid cycle, 209, 217, 247
Citronellal, 276, 315
Citrovorum factor (CF), 372, 373, 374, 375
Citrulline, 19, 20
[^{14}C]Citrulline, 20
Clostridia, 377
Clostridium, 76
Clostridium cylindrosporum, 375
Clostridium kluyveri, 171, 173, 174
Cobalamine, 383
Cobinamides, 233
Cobyric acid, 233
Cobyrinic acid, 232, 233
Cobyrinic acid amides, 233
CoA = Coenzyme A. (Coenzyme of Acetylation)
Coenzyme A, 22, 31, 37, 38, 39, 41, 55, 57, 58, 174, 175, 177, 188, 209, 211, 212, 217, 218, 311
—, in biological acetylation, 250
—, in cholesterol biosynthesis, 250
Coenzymes Q, 284
Complex saccharides, biosynthesis from monosaccharides, 141
—, from sugar nucleotides, 164, 165

Copper protein, 350
Coproporphyrins, 195, 216, 217, 230
—, from uroporphyrinogen, 225, 227
Coproporphyrin I, 196, 223, 225, 228
Coproporphyrin III, 209, 221, 223, 225
Coproporphyrin isomer III, 221, 225, 227
Coproporphyrinogen, 223
Coproporphyrinogen III, 225, 227, 228
Coproporphyrinogenase, 227
Coproporphyrinogen oxidase, 227
Coproporphyrinogen: oxygen oxidoreductase (decarboxylating), 227
Coprostane, conversion to cholic acid, 307
Coprostane-3α,7α,12α,24,26-pentol, 307
Coprostane-3α,7α,12α,26-tetrol, 303, 307
Coprostane-3α,7α,12α-triol, 303, 307
Coprostanol, conversion to bile acids, 303
Corrin, 232
Corrinoids, biosynthesis, 22, 193, 231
Corticosteroids, biological sources, 293
Cortin, 289
Cortisol, 289, 290, 291
Cortisone, commercial synthesis, 292, 293
Corynebacterium, 360, 373
Cozymase, 147
Creatine phosphate, 362
Crotonase, 182
Crotonyl-ACP, 182, 183, 184
Crustacean ecdysones, 311
CTP, 261
Cyanocobalamin, 232
C$_{30}$ Cyclic triterpenes, 255
Cyclobutyl pyrophosphates, 272

Cyclohydrase, 375
Cyclohydrolase, 377
5β-Cyprinol, 302, 306, 309, 310
Cysteine, 174
Cysteine-activated papain, 49
Cytidine, 165

Dakin-West reaction, 216, 217
Decanoic acid, 175
2-Decaprenyl-5-hydroxy-6-methoxy-
    3-methyl-1,4-benzoquinone, 286
2-Decaprenyl-6-methoxy-1,4-benzo-
    quinone, 285, 286
2-Decaprenyl-6-methoxy-3-methyl-
    1,4-benzoquinone, 285, 286
2-Decaprenyl-6-methoxyphenol,
    286
2-Decaprenylphenol, 286
(C₃-C₄)Decarboxylation   cycle   of
    Szent-Gyorgyi, 89
Degenerative diseases, 7
Dehydroagnosterol, 252
Dehydroepiandrosterone, 295, 297,
    298
15,15′-Dehydrolycopersene, 280
Dehydropeptidase, 14
Dehydrosqualene, 266
Deoxycholic acid, 301, 302, 311
11-Deoxycorticosterone, 291,
    292
11-Deoxycortisol, 291
Deoxypentulose derivative, 362
Deuterated linseed oil, 11
Deuterated organic molecules, opti-
    cal properties, 5
Deuteroacetate, 23, 204
—, in sterol biosynthesis, 240, 241,
    243
Deuterobutyric acid, 11
Deuterocholesterol, 11, 240, 241,
    243
Deuterocoprostane, 11
Deuterohaemin, 197, 204
Deuterohexanoic acid, 11
Deuteropalmitic acid, 11
Deuterostearic acid, 11
Deuterium, 12
—, as molecular tracer, 9

—, as molecular tracer in cholesterol
    metabolism, 5
—, biological applications, 4
—, in metabolic studies, 1
—, in tracer studies, 3
Dextrins, 143
Diabetes, origin of glucose, 109
1,2-Diacylglycerol, 188
Diacylglycerol acyltransferase, 188
Diacylglycerol phosphate, 186
Diacylphosphates, 57
β,β′-Dialkyl-α-aminomethylpyrrole,
    213
Dialuric acid, 351
2,5-Diamino-6-(5′-triphosphoribo-
    syl)-amino-4-hydroxypyrimidine,
    361, 362
Dibenzoylphosphate, 57
Diethylaminoethylcellulose, 368
Diethyl ketone, 115
D-α,β-Diglyceride, 186, 187, 188
Diglyceride acyltransferase, 188
1,2-Diglyceride-3-phosphate, 187
Dihydrofolic acid, 360, 365, 367,
    371, 372, 383
Dihydrofolic acid reductase, 372
Dihydrolanosterol, 252
Dihydropteridine phosphate, 365
Dihydropteroate, 381
Dihydropteroate   pyrophosphoryl-
    ase, 368
Dihydropteroate synthase, 367,
    368
Dihydropteroic   acid,   365,   367,
    . 368, 381
Dihydrotriterpenes, 247
Dihydroxyacetone, 93
Dihydroxyacetone phosphate, 99,
    100, 136, 186, 187
3α,7α-Dihydroxy-5β-cholestane,
    306
3α,7α-Dihydroxy-5β-cholestan-26-
    oic acid, 306
7α,12α-Dihydroxy-5β-cholestan-3-
    one, 305
7α,12α-Dihydroxycholest-4-en-3-
    one, 305, 309, 310
Dimethylallyldiphosphate: isopen-

tenyldiphosphate dimethylallyltransferase, 265
Dimethylallyl pyrophosphate, 264,
266, 267, 277, 281
—, from isopentenyl pyrophosphate,
263, 264
—, from mevalonic acid, 260, 262
Dimethylallyltransferase, 264, 265,
266
6,7-Dimethyl-8-ribityllumazine,
358, 359
3,17-Dioxoandrost-4-en-19-al, 296
Dipeptides, 51
Diphenylamine, 281
Diphosphoglucose pyrophosphorylase, 155
1,3-Diphosphoglycerate, 136
5-Diphosphomevalonate, 261
Disaccharides, 143, 157, 159
—, glycosidic linkages, 156
Diterpenes, 276
—, biosynthesis through isoprene
pathway, 240
2,4-Divinylphaeoporphyrin a_5 monomethyl ester, 231
Dolichodial, 315
Dolichols, 239
Dowex-1 ion exchange resin, 57,
174
Drosophila melanogaster, 360

Ecdysone, 311, 312
—, biosynthesis from cholesterol,
313
β-Ecdysone, 311
Ehrlich reagent, 213
Ekapterin, 357
Energetics, of biosynthesis, 29
—, of living cells, 29
Energy, of food transformation, 29
—, of metabolic processes, 31
Energy-rich chemical compounds,
29
Energy-rich phosphate derivatives,
31
trans-2-Enoyl-ACP, 183, 185
Enoyl-ACP hydrase, 182, 184, 185
Enoyl-ACP reductase, 183, 184,
185

Enzymatic acetylation, sulphanilamide, 36
Enzymes, systematic nomenclature,
23
Enzymes, activity, 14
—, acyl-malonyl-ACP condensing
181, 182
—, branching, 151, 160
—, extraction by solvents, 14
—, in biosynthesis, 17
—, — of complex saccharides, 164,
165
—, — of steroid hormones, 300
—, isolation, 13, 14
—, nomenclature, 14, 23, 24
—, of glycolysis, 76
—, of the oxidative pathway, 64
—, of the reductive phase, 100
—, β-oxidation, 22
—, purification, 14, 174
—, reductive carbohydrate cycle,
105
—, specific activity, 14
Enzyme-catalysed replacement, 49
Epimerase, 65, 67, 69, 70
4-Epimerases, 165
Epinephrine, 159
Eremothecium ashbyi, 358
Ergosterol, bioreaction with deuteroacetate, 240, 241
—, from labelled acetates, 251
Erythritol, 116
Erythropterin, 357
Erythrose, in biosynthetic reactions, 77
Erythrose-4-phosphate, 21, 61, 71,
73, 104
Escherichia coli, (= E. coli), 53, 64,
67, 76, 105, 113, 124, 180, 181,
182, 183, 333, 360, 365, 368,
369, 380
Essential oils, 239
Ethanolamine, 199
Ethionine, 230
Eucalyptus citriodora, 276
Eucalyptus globulus, 276
Euglena, 225
Exacting mutants, in biosynthetic
studies, 15

Farnesol, 276, 277
Farnesyl pyrophosphate, 267, 277, 279, 281
—, biosynthesis, 266, 284
—, condensation, 264, 265
—, conversion to presqualene pyrophosphate, 270, 271
—, — to squalene, 268
—, enzymic dimerisation to squalene, 265
—, from isoprene units, 263, 264
—, in conversion of isopentenyl pyrophosphate to squalene, 263, 264
—, in squalene biosynthesis, 264, 267, 269
Farnesyl pyrophosphate synthetase, 264, 265
Fats, biosynthesis in animals, 171
—, metabolism, 9
Fatty acids, acylphosphates, 32
—, association with bile salts, 300
—, biosynthesis, 22, 171, 177, 180, 185
—, —, in cell-free preparations, 173
—, —, in the presence of β-oxidation enzymes, 173
—, $^{14}CO_2$ fixation, 86
—, catabolism, 86
—, degradation, 11
—, desaturation, 185
—, deuterated, 11
—, elongation in animal tissues, 183
—, from the glucose pathway, 75
—, metabolism, 11
—, β-oxidation, 9, 171
Fatty acid synthase, 179
Fatty acid synthetases, 180
Ferrochelatase, 229
FGAM = Formylglycinamidine ribonucleotide, 339
—, from FGAR, 338, 339
FGAR = Formylglycinamide ribotide, 336, 337, 338, 378, 383
—, conversion to AIRP, 338—340
—, — to FGAM, 338
—, — to inosinic acid, 338, 339, 340

—, from PRPP, 336, 337
FGAR amidotransferase, 341
FH_2 = Dihydrofolate, 360, 383
FH_4 = Tetrahydrofolate, 338, 344, 383
Fixed air, 1
Flavins, 357, 358
Folate cofactors, in C_1 transfers, 381
Folic acids, 360, 369, 378
—, as coenzymes, 369
—, biosynthesis, 363
—, —, enzymatic aspects, 368
—, conversion to tetrahydrofolic acid, 371, 372
—, formylated natural derivatives, 371, 372, 373, 374, 375, 377
—, glutamate derivatives, 381
—, structure, 362, 363
Folic acid coenzymes, 381
Folinic acid, 373, 374, 375, 378, 379
—, enzymatic biosynthesis, 375
Folin's theory of protein metabolism, 47
Formaldehyde, in purine biosynthesis, 379
Formaldehyde theory, 81, 83
5-Formamido-4-imidazole-carboxamide ribonucleotide, 344
2-Formamido-N-ribosylacetamidine-5'-phosphate, 340
2-Formamino-N-ribosylacetamide-5'-phosphate, 336
Formate, 111, 113, 327, 329, 331, 338, 339, 369
—, in biosynthesis of leucopterin, 359
—, in synthesis of hypoxanthine, 330
—, precursor in uric acid synthesis, 329
Formate-activating enzyme, 374
Formate: tetrahydrofolate ligase (ADP-forming), 375
5-Formimino-FH_4, 378, 383
Formiminoglutamic acid, 337, 378, 383

Formiminoglycinate, 383
Formiminoglycine, 377
5-Formiminotetrahydrofolate ammonia-lyase (cyclizing), 378
Formiminotetrahydrofolate cyclodeaminase, 377
5-Formiminotetrahydrofolate: L-glutamate N-formiminotransferase, 377
5-Formiminotetrahydrofolate: glycine N-formiminotransferase, 377
Formiminotransferases, 377, 378
Formylacetone, reaction with glycine, 213
Formyl-AICRP, 344
Formyl-donor, 378
N^{10}-Formyl-FH$_4$, 383
Formylfolic acid, enzymatic synthesis, 377
N^{10}-Formylfolic acid, 373, 375
Formylglycinamide ribonucleotide, 339, 341
Formylglycinamide ribotide, see also under FGAR, 336, 378, 383
Formylglycinamidine ribonucleotide, see under FGAM
Formylglycinamidine ribotide, 339—341
N^{10}-Formylpteroic acid, 373
10-Formyltetrafolate, 379
10-Formyltetrahydrofolate, 344
Formyltetrahydrofolate synthetase, 374, 381
N^5-Formyltetrahydrofolic acid = Folinic acid, 373, 374, 375, 378
N^{10}-Formyltetrahydrofolic acid, 374, 375, 377, 379, 381.
Free energy in organisms, 2, 31
Fructose, 93, 143, 150, 151, 158
D-Fructose-1,6-biphosphate 1-phosphohydrolase, 134
Fructose diphosphatase, 71, 73, 99, 100, 134, 137
Fructose-1,6-diphosphate, 73, 99, 100, 104, 134, 135, 136, 137, 153
Fructose-6-phosphate, 70, 71, 72, 101, 104, 134, 135, 136, 158, 163
—, from phosphoglyceric acid, 99

2-Fructose-6-phosphate, 73
5-Fructose-6-phosphate, 73
Fumaric acid, 86, 89, 116, 121, 207, 343
—, $^{14}CO_2$ fixation, 86
—, labelled, 89
—, paper chromatography, 91

Galactose, utilization by yeast, 153
D-Galactose, 165
—, distribution of ^{14}C in lactose, 162, 163
[^{14}C]D-Galactose, 163, 164
Galactose-1-phosphate, 146, 153, 155
[^{14}C]α-D-Galactose-1-phosphate, 163
Galactose-1-phosphate uridyl transferase, 156
β-Galactosidase, 53
Galactosyltransferase, 161, 164
Galactowaldenase, 154
GAR = Glycinamide ribotide, 336, 337, 383
—, formation from glycine, 338
Geranylgeranyl pyrophosphate, 271, 281
GAR synthetase, 338
GDP, 130
GDP-glucose, 158
Geraniol, 277
Geranylgeraniol, 276, 277
Geranylgeranyl phosphate, 279, 281
Geranylgeranyl pyrophosphate, 271, 281
Geranylgeranyl pyrophosphate synthetase, 281
Geranyl pyrophosphate, 263, 264, 266, 267, 268, 277, 279, 281
Geranyltransferase, 264, 265
Globin synthesis, 52
Glucagon, 159
Glucan, structure, 148
β-1,3-Glucan, from sugar nucleotides, 165
α-1,4-Glucan:α-1,4-glucan-6-glucosyl-transferase, 160
Glucocorticoid activity, 292
D-Gluconate-6-phosphate, 61, 63, 64

Gluconate 6-phosphate dehydrogenase, 64
D-Gluconate-6-phosphate δ-lactone, 65
Gluconeogenesis, 22, 109, 124, 128, 129, 130, 133, 134, 153, 165
—, intermediate stages, 135
—, participation of pyruvate, 109
—, reactions from lactate, 136
Gluconolactonase, 65, 68
δ-Gluconolactone, 65
D-Glucono-δ-lactone hydrolase, 65
D-Glucono-δ-lactone-6-phosphate, 64, 65, 68
β-D-Glucopyranose-6-phosphate, 65
D-Glucopyranose units, in starch, 148
Glucosazones, 126, 129
Glucose, 93, 129, 143, 151, 154, 159, 165
—, biosynthesis, 109, 111
—, —, from non-carbohydrate precursors, 109
—, conversion to lactose, 161
—, —, to maltose, 141
—, — to ribulose-5-phosphate, 64
—, degradation by *Lactobacillus casei*, 76
—, direct oxidation, 63
—, from glucose-6-phosphate, 135, 136
—, labelled, 83
—, origins in diabetes, 109
—, phosphorylation, 147
—, synthesis from carbon dioxide by plants, 122
—, units in starch, 149, 150
[^{14}C]D-Glucose, 163
—, distribution of ^{14}C in lactose, 162, 163
[1-^{14}C]Glucose, metabolism by glycolysis, 75
[6-^{14}C]Glucose, 163
—, metabolism by glycolysis, 75
[2,6-^{14}C]Glucose, 163
Glucose diphosphate, 154
Glucose-6-phosphatase, 135, 137
Glucose-1-phosphate, 135, 136,

137, 147, 153, 154, 155, 158, 161, 174
—, conversion to glucose-6-phosphates, 145, 146
α-D-Glucose-1-phosphate, 143, 144, 148, 150, 158, 159, 165
β-D-Glucose-1-phosphate, 151
Glucose-6-phosphate (= G-6-P), 61, 63, 64, 67, 71, 72, 75, 77, 136, 153, 157, 165
—, conversion to D-gluconate-6-phosphate, 61
—, — to glucose, 135, 137
—, — to ribulose-5-phosphate, 61
—, from glucose-1-phosphate, 145, 146
—, oxidative sequence to pentosephophates, 68
5-Glucose-6-phosphate, 73
6-D-Glucose-6-phosphate, 68
Glucose-6-phosphate dehydrogenase, 61, 63, 64, 65, 68
D-Glucose-6-phosphate: NADP$^+$ 1-oxidoreductase, 64
D-Glucose-6-phosphate phosphohydrolase, 135
1,4-α-Glucoside linkages, biosynthesis, 144, 145
Glucosidic bonds, formation by transglycosylation, 156
α-1,6-Glucosidic bonds, in starch, 148, 149
4-O-α-D-Glucosyl-D-glucose, 151
Glycosyltransferase, 150, 151
Glutamate formimino transferase, 377, 378
Glutamic acid, 47, 117, 332, 336, 339, 341, 363, 365, 377
—, from histidine, 381
—, precursor of porphyrins, 196, 197
Glutamine, 20, 117, 330, 331, 332, 335, 336, 338, 339, 340, 341, 352
Glutamyl components, in tetrahydrofolic acids, 381
γ-Glutamylcysteine, 53
Glutathione, 53, 174

Glutathione synthetase, 53
Glyceraldehyde, 100
[2,3-^{14}C]Glyceraldehyde phosphate, 163
D-Glyceraldehyde-3-phosphate, 70, 71, 73, 100, 136
Glyceraldehyde phosphate dehydrogenase, 99, 100
D-Glyceraldehyde-3-phosphate ketol isomerase, 100
D-Glyceraldehyde-3-phosphate : NADP$^+$ oxidoreductase, 100
Glyceric acid, 100
Glycerides, association with bile salts, 300
—, biosynthesis, 171
—, emulsification, 300
Glycerol, 32, 116
—, fermentation by propionic acid bacteria, 112, 113
—, phosphorylation, 186
Glycerol kinase, 186
Glycerol phosphate pathway, 189
L-α-Glycerol phosphate, 186, 187
Glycerol phosphate acyltransferase, 186
Glycerol-3-phosphate dehydrogenase, 186, 187
Glycerophosphate, 134
Glycinamide, ribotide derivatives, 335
Glycinamide ribonucleotide kinosynthetase, 338
Glycinamide ribonucleotide transformylase, 338
Glycinamide ribotide, see also GAR, 336, 337, 383
Glycine, 51, 53, 218, 301, 331, 332, 335, 336, 369, 379
—, combination with α-ketoglutaric acid, 208
—, condensation with succinyl-CoA, 218
—, conjugates with bile acids, 307, 309, 311
—, conversion to glucose, 109
—, — into hypoxanthine, 330
—, — to porphobilinogen, 220

—, — to serine, 199
—, formation of aminolaevulinic acid, 212, 218
—, in biosynthesis of uric acid, 327, 328, 329, 351, 352
—, in porphyrin formation, 207
—, in synthesis of hippuric acid, 54, 55, 57
—, — of protoporphyrin, 215, 216
—, labelled, 52
—, labelled carbons in protoporphyrin synthesis, 202
—, ^{13}N-labelled, 197
—, ^{15}N-labelled, 205
—, precursor of porphyrins, 197, 199, 201
—, — of pyrrole rings, 205
—, pyridoxal phosphate derivative, 218
—, reaction with formylacetone, 213
—, — with β-ketoaldehydes, 197, 199
—, — with succinic acid, 215
—, — with succinyl compounds, 213, 214
—, succinylation, 216—218
[2-^{14}C]Glycine, 201
Glycinamide ribonucleotide kinosynthetase, 338
Glycinamide ribonucleotide transformylase, 338
Glycinamide ribotide, 337, 378, 383
Glycine formiminotransferase, 377
Glycoaldehyde, 93
Glycogen, 125, 126, 127, 128, 129, 135, 136, 143
—, action of branching enzymes, 151
—, biosynthesis, 141, 144, 156
—, — by phosphorylase, 150
—, — in vitro, 147
—, — using UDPG, 159, 160
—, from sugar nucleotides, 165
—, phosphorolytic breakdown, 144
—, preparation from liver or muscle, 146

—, reaction with inorganic phosphates, 145
—, structure, 149, 150
Glycogenolysis, 147
Glycogen synthase, 160
Glycogen synthetases, 160
Glycolic acid, paper chromatography, 91
Glycolysis, 14, 75, 76, 129, 135
—, by liver and kidney cortex, 109
—, catalysed by phosphofructokinase, 134
—, in glucose catabolism, 76
Glycosides, 143, 157
—, biosynthesis from monosaccharides, 141
Glycoside β-o-aminophenol glucuronide, 157
1,4-Glycosidic linkages, 148, 151
1,6-Glycosidic linkages, 151
Glycosylation reactions, 141
Glyoxal, 359
Glyoxylic acid, 350
GMP, 347
—, synthesis, 345
GMP synthetase, 348
G6P, see under Glucose-6-phosphate
GTP, see also Guanosine triphosphate, 130, 133, 137, 158, 261, 345, 361, 362
GTP: oxaloacetate carboxy-lyase (transphosphorylating), 130
Guanine, 165, 345, 348
—, from tissue nucleic acids, 329
—, incorporation into pteridines, 359, 360
—, of yeast nucleic acids, 328
Guanosine, 360
Guanosine diphosphate D-glucose, 158
Guanosine phosphate, synthesis, 345, 347
Guanosine-5′-phosphate, 360
Guanosine triphosphate, see also GTP, 130, 158, 361, 362

Haem, 193, 197, 206, 221
—, biosynthesis, 245
—, — by fowl erythrocytes, 203

—, — from PBG, 223
Haematinic acid, 201, 203
Haemin, 201
—, biosynthesis from carboxy-labelled succinate, 210
—, degradation, 210
—, precursors, 197, 199
—, relation with aminolaevulinic acid, 214
—, synthesis from methylene-labelled succinate, 211
—, ^{14}C-labelled, 205
Haemoglobin, 52
HDP, 49
Heavy isotopes, 12
—, use in biosynthesis, 3
Heavy water, 3
—, in fluid transport studies, 4
Heavy water biology, 3
Helix pomatia, 351, 352
Hemin, 61
Heptose phosphate, 70
Heptulose, 101
Heterotrophic bacteria, 111
Heterotrophic synthesis, 111
Hevea brasiliensis, 275
Hexanoic acids, 11
Hexanoyl-ACP, 183, 185
Hexanyl-CoA, 183
Hexapeptides, 51
Δ_2-trans-Hexenoyl-ACP, 185
Hexokinase, 161
Hexose, formation from two trioses, 101, 102
Hexose diphosphatase, 134, 135
Hexose diphosphate, 91, 134
Hexose monophosphate, 72, 77, 91
Hexose phosphates, 89, 92, 98, 143, 150, 159
—, formation from pentose derivatives, 67
—, — from pentose-5-phosphate, 71, 72, 101
High group potential, 31
Hippuric acid, 49, 203
—, synthesis by liver slices, 55
—, — from benzoic acid and glycine, 54, 55, 57, 58

Histidine, 197, 325
—, catabolism to glutamic acid, 381
Homocysteine, 230, 379
3'-Homomevalonate, 313
Homoserine, 230
Hydrogen, radio-isotopes, 13
—, ^{2}H see Deuterium, 12
Hydroquinone, 157
Hydroquinone β-D-glucopyranoside, 157
Hydroxamic acid, 64
3-Hydroxyacyl-ACP, 185
β-Hydroxyacyl-ACP, 182, 183
Hydroxyacyl-CoA, 182
Hydroxyacyl esters, 182
16α-Hydroxyandrostenedione, 298, 299
19-Hydroxyandrosterone, 296
Hydroxyaspartic acid, 208
p-Hydroxybenzoic acid, precursor for ubiquinones, 284, 285
β-Hydroxybutyrate hydroxamate, 175
β-Hydroxybutyryl-ACP, 182, 184
7α-Hydroxy-5β-cholestan-3-one, 306
7α-Hydroxycholest-4-en-3-one, 305, 306
7α-Hydroxycholesterol, 301, 305
20-β-Hydrocholesterol, 287
18-Hydroxycorticosterone, 291
16α-Hydroxydehydroepiandrosterone, 298, 299
20-Hydroxyecdysone, 311, 312
3-Hydroxyhexanoyl-ACP, 185
3-Hydroxyisovaleric acid, 259
11β-, 17α-, 21-Hydroxylases, 300
Hydroxymethyldihydropterin pyrophosphokinase, 367, 368
3-Hydroxy-3-methylglutaric acid, 257, 259, 274
β-Hydroxy-β-methylglutaric acid coenzyme A ester condensing enzyme, 274
Hydroxymethylglutaryl-CoA, 274, 275
3-Hydroxy-3-methylglutaryl-CoA-acetoacetyl-CoA-lyase (CoA-acetylating), 274

Hydroxymethylglutaryl-CoA reductase (NADPH), 274, 275
Hydroxymethylglutaryl-CoA syntase, 275
3-Hydroxy-3-methylpentano-5-lactone, 257
β-Hydroxy-β-methyl reductase, 274
β-Hydroxyoctanoate hydroxamate, 175
17β-Hydroxy-3-oxoandrost-4-en-19-al, 296
17-Hydroxypregn-5-ene-3,20-dione, 295
17-Hydroxypregnenolone, 295, 297
20α-Hydroxypregn-4-en-3-one, 297
11β-Hydroxyprogesterone, 291
17-Hydroxyprogesterone, 291, 294, 295, 297
8-Hydroxyquinoline, 230, 231
Δ^5-3β-Hydroxysteroids, 298
3α-Hydroxysteroid dehydrogenase, 300
3β-Hydroxysteroid dehydrogenase, 294, 295, 300
17β-Hydroxysteroid dehydrogenase, 300
19-Hydroxytestosterone, 296
Hypoxanthine, 144, 330, 331
—, biosynthesis, 346
—, conversion to inosinic acid, 333, 334
—, from IMP, 348, 349
—, from inosine, 331
—, of nucleic acids, 328
—, oxidation to uric acid, 352
—, oxidation to xanthine, 349
—, synthesis by pigeon liver homogenates, 330
—, — in pigeon liver slices, 331, 332
Hypoxanthine oxidase, 350
Hypoxanthine phosphoribosyltransferase, 348, 349

IMP = Inosine monophosphate, 344, 349
—, conversion to adenine and guanine, 345
—, conversion to hypoxanthine, 348
—, — to uric acid, 348, 349

—, from AICRP, 343, 344
—, oxidation to XMP, 347
—, reaction with L-aspartate, 345
IMP: L-aspartate ligase (GDP-forming), 345
IMP-cyclohydrolase, 345
IMP dehydrogenase, 347
IMP 1,2-hydrolase (decyclizing), 345
IMP: NAD$^+$ oxidoreductase, 348
IMP: pyrophosphate phosphoribosyltransferase, 348
IMP pyrophosphorylase, 348
Inorganic phosphates, 36
Inosine, 330, 331
Inosine phosphate = IMP, 344, 349
Inosine-5-phosphoric acid, 330
Inosine triphosphate = ITP, 130, 261
Inosinic acid, 146, 330, 333, 339, 340, 341, 344
—, biosynthesis, 331, 335
—, from FGAR, 338
—, from hypoxanthine and PRPP, 333, 334
—, in purine biosynthesis, 330, 331
Inosinic acid pyrophosphorylase, 334
Inosinicase, 344, 345
Insect ecdysones, 311
Insect pheromones, 239
Insulin, effect on pyruvate metabolism, 119
Invertase, 150
Iron, incorporation into porphyrins, 227
Isocaproic acid, 287
Isocholesterol, 252
Isocitrate, 174
Isoleucine, 22
—, biosynthesis, 15
Isoleucovorin, 375
Isomerase, 65, 67, 69, 70
Isopentene units, 249
Isopentenyl diphosphate Δ^3-Δ^2-isomerase, 263
Isopentenyl pyrophosphate, 261, 262, 263, 264, 266, 267, 268, 275, 277, 281

—, from mévalonic acid, 260, 262
— incorporation in β-carotene, 279
Isopentenyl pyrophosphate isomerase, 262, 263, 264, 265, 266
Isoprene, 239
Isoprene rule, 255, 277
Isoprene units of squalene, metabolic precursors, 257
Isoprenoids, biosynthesis, 255
—, from mevalonic acid, 315
—, in essential oils of plants, 313
—, in insects, 313, 315
Isoprenoid alkaloids, from mevalonic acid, 284
Isoprenoid pathway, 11, 23, 239, 257
Isoprenoid units, 259
—, from mevalonic acid, 260
Isotopes, of biological interest, 12
—, use as tracers, 2, 3, 119
—, use in biosynthesis, 3, 23
Isotope tracers, biosynthetic pathways, 23
Isoxanthopterin, 357, 360
ITP = Inosine triphosphate, 130, 261

Jasus lalandei, 313
Juvenile hormone, 313

β-Ketoacyl-ACP, 183
β-Ketoacyl-ACP reductase, 181, 182, 184, 185
β-Ketoacyl reductase, 180
β-Ketoaldehydes, reaction with glycine, 197, 199
α-Ketobutyric acid, 207
α-Ketoglutarate, 55, 121, 122, 174, 204, 208—211, 218
—, labelled, 212
Ketopentose phosphate, 67, 69
Ketopentose phosphoric acid ester, 64
2-Keto-6-phosphogluconate, 63
Ketophosphohexonate, 63
2-Keto-5-phosphopentonate, 63
Δ^5-Ketosteroid isomerase, 294, 295
Kinetic studies, 97
—, carbohydrate biosynthesis, 98

Knorr reaction, pyrrole synthesis, 213

Labelled lanosterol, 252
Lactic acid, 115, 136
—, conversion to carbon dioxide, 126
—, conversion to glucose, 109
—, oxidation by permanganate, 125
—, reaction with carbon dioxide, 85
^{11}C Lactic acid, 125, 126, 127
Lactic bacteria, 257
Lactobacillus acidophilus, 257
Lactobacillus casei, 76, 362
Lactobacillus delbrueckii, 31
Lactobacillus plantarum, 365
Lactonase, 65
Lactose, biosynthesis from D-glucose, 161, 164
—, from sugar nucleotides, 165
Lactose phosphate, 161, 162, 163
Lactose synthase, 164
Lanosterol, conversion to cholesterol, 252, 253, 273
—, from squalene, 256, 257
—, structure, 250, 252
Lauric acid, 175
L-α-Lecithin, 188
Leucine, 22, 46
—, biosynthesis, 15
—, precursor of porphyrins, 197
[^{15}N, ^{2}H]Leucine, 12, 46
Leuconostoc citrovorum, 373
Leuconostoc mesenteroides, 76
Leucopterin, 357, 359, 360
Leucovorin, 373, 374, 375, 378
Linoleate, CoA ester, 186
Lipids, 171
—, biosynthesis, 186
—, paper chromatography, 91
Lipophilic compounds, paper chromatography, 91
Lithocholic acid, 302
Liver glucose-6-phosphatase, 135
Liver glycogen, 22, 125, 126, 127, 128, 129
Liver mitochondria, 131
Liver phosphorylase, 147

Long chain amino acids, 171
Long-chain fatty acids, from acetate, 173, 174, 175
—, mechanism of formation, 177
—, synthesis from acetyl CoA, 174
Lumiflavin, 357, 358
Lutein, 284
Lycopene, from phytoene, 282, 283
—, incorporation of [^{14}C]isopentenyl pyrophosphate, 279
Lycopersene, 280, 281, 282
Lysine, radioactive, 47, 71
Lysophosphatidic acid, 186, 187

Magnesium, porphyrin, 229, 230
Magnesium protoporphyrin, 229, 230
— IX, 231
Magnesium-protoporphyrin-methyltransferase, 229
Magnesium protoporphyrin monomethylester, 229, 230, 231
Magnesium vinyl phaeoporphyrin a_5, 229
Malic acid, 116, 121, 130, 131, 207
—, distribution of ^{14}C, 92
—, from carbon dioxide, 85
—, labelled, 89
—, paper chromatography, 91
Malic enzyme, 130, 131
Malonic acid, 210, 211
Malonyl-acetyl condensing enzyme, 180
Malonyl-ACP, 180, 181, 183, 184
Malonyl-CoA, 22, 178, 179, 180, 183, 184
—, as intermediate in fatty acid biosynthesis, 175, 177, 178
Malonyl-CoA-ACP transacylase, 181
Malonyl synthase complex, 179
Malonyl transacylase, 180, 184
Maltase, 141
Maltose, 143, 151, 159
—, synthesis from glucose, 141
Maltose glucosyltransferase, 151
Mannitol, 116
Mannose-1-P, 146
Mesoporphyrin, 201

Metabolic pathways, thermodynamic analysis, 30
Metabolic studies, using tracers, 12
Metalloporphyrins, 227
—, biosynthesis, 22
Methaemoglobin, 61
Methane, 111, 113
Methanol, 230
$N^{5,10}$-Methenyl-FH$_4$, 381, 383
5,10-Methenylhydrofolate, 379
5,10-Methenyltetrahydrofolate, 344, 375, 377, 379, 380, 383
5,10-Methenyltetrahydrofolate cyclohydrolase, 378
5,10-Methenyltetrahydrofolate 5-hydrolase (decyclizing), 378
5,10-Methenyltetrahydrofolate: 5'-phosphoribosylglycinamide formyltransferase, 338, 345
Methionine, 230, 232, 369, 379, 380, 383
Methylaminopyrrole, 233, 234
3-Methylcrotonic acid, 259
Methylene blue, 61
5,10-Methylenetetrahydrofolate reductase, 380
Methylenetetrahydrofolic acid, 379, 380, 381, 383
Methylethylmaleimide, 201
3-Methylglutaconic acid, 259
5-Methyltetrahydrofolate, 380
5-Methyltetrahydrofolate: NAD$^+$ oxidoreductase, 380
N^5-Methyltetrahydrofolic acid, 380
Mevaldic acid, 275
Mevalonate: NADP$^+$ oxidoreductase (CoA-acylating), 275
Mevalonic acid, as precursor of cholesterol, 274
—, biosynthesis from acetic acid, 273
—, conversion to geranylgeraniol, 276
—, — to isoprenoid alkaloids, 284
—, — to isoprenoid quinones, 284
—, — to pyrophosphates, 260, 262
—, — to rubber, 275

—, — to squalene, 260
—, — to sterols and steroid hormones, 287
—, deuterium-labelled, 17
—, in alarm substances, 315
—, incorporation into β-carotene, 279
—, in the biosynthesis of squalene, 16, 17
—, in the isoprenoid pathway, 257
—, relation to 3-hydroxy-3-methylglutaric acid, 259
Mevalonic kinase, 261, 262
Mevalonic lactone, 257
Microbial mutants, in the analysis of metabolic pathways, 15
Micrococcus aerogenes, 375
Micrococcus lysodeikticus, 131, 281
Microsomal elongation system, 183
Mitochondrial elongation system, 183
Mono-acylglycerol phosphate, 186
Monobenzoylphosphate, 55
Monoglycerides, association with bile salts, 300
Monoglyceride path, 189
Monosaccharides, conversion to complex saccharides, 141
Mucor hiemalis, 279
Muscle phosphorylase, 151
Mutants, in microorganisms, 2
Myristic acid, 175
Myxinol disulphate, 309

NAD = Nicotinamide adenine dinucleotide, 42, 130, 136, 174, 294
NADH, 42, 130, 136, 173, 174, 183, 260, 372, 380
NADH$_2$, 136
NADP = Nicotinamide adenine dinucleotide phosphate, 42, 63, 64, 65, 68, 72, 130, 185
—, in synthesis of hexose phosphate, 102, 103
NADP-linked triose phosphate dehydrogenase, 102
NADPH, 21, 42, 64, 68, 72, 77, 83, 85, 100, 130, 174, 175, 177,

178, 179, 180, 183, 185, 266, 268, 272, 275, 294, 371, 372
—, synthesis by pentose phosphate cycle, 61
NADPH-dependent reductive phase, 97
Nachmansohn's reaction, 38, 39
Nakanishi's benzoate chirality rule, 270
Neisseria meningitidis, 151
Nerolidol, 266
Nerolidyl pyrophosphate, 266
Neurospora crassa, 15, 164, 241, 281, 343
Neurosporene, 282, 283
Nicotinamide adenine dinucleotide, see under NAD
Nicotinamide adenine dinucleotide phosphate, see under NADP
Nitrogen, radio-isotopes, 13
—, ^{13}N, 85
—, ^{15}N, 12, 325
Nitrogenous fractions, in pea seeds, 53
2-Nonaprenyl-6-methoxy-3-methyl-1,4-benzoquinone, 285, 286
Norite eluate factor, 362
Nucleic acids, 49, 77, 325
—, biosynthesis, 72, 329, 348
Nucleic acid purines, 369
—, biosynthesis in the pigeon, 328
—, — in yeast, 328
Nucleoproteins, 325
Nucleoside diphosphate, 130
Nucleoside phosphorylase reaction, 331
Nucleoside pyrophosphate, 143
Nucleoside triphosphate, 130
5′-Nucleotidase, 41, 348
Nucleotides, biosynthesis from purine and pyrimidine bases, 330, 331
—, synthesis, 77
Nutrition and biosynthesis, 17
Nutrition studies, 109

Octanoate hydroxamate, 175
Octanoyl-CoA, 183

2-Octaprenyl-6-methoxy-3-methyl-1,4-benzoquinone, 285, 286
Oestradiol-17β, 294, 295, 296, 297, 298
17β-Oestradiol dehydrogenase, 300
Oestriol, biosynthesis pathway, 298, 299
Oestrogens, 289, 297
—, biosynthesis, 298
Oestrone, 294, 295, 296, 297, 298
Oleic acid, 175
Oligosaccharides, 159
—, biosynthesis from monosaccharides, 141
—, glycosidic linkages, 157
Ornithine, 20
—, cycle, 15, 18, 19, 20
Orotic acid, 334
Orotidylic acid, 334
Osazones, 123
Otala lactea, 352
Ovalbumin, biosynthesis by the hen's oviduct, 51
Ovarian hormones, 296
Oxalate, labelled, 115
Oxaloacetic acid, 133, 136, 207, 330, 331
—, as precursor of phosphoenolpyruvate, 133
—, biosynthesis, 119
—, conversion to citric acid, 112
—, — to phosphoenolpyruvate, 133
—, from carbon oxide, 85
—, from pyruvic acid, 116, 119, 121, 130, 131, 133
β-Oxidation, 9, 22, 171, 173, 303
—, enzymes, 22, 173, 174, 179
Oxidocyclase, 257
2,3-Oxidosqualene, 256, 257
2,3-Oxidosqualene lanosterol-cyclase, 256, 257
2,3-Oxidosqualene mutase (cyclizing, lanosterol-forming), 257
3-Oxoacyl-ACP, 185
α-Oxoglutarate, biosynthesis, 117, 119
—, in synthesis of protoporphyrin, 215, 216

3-Oxohexanoyl-ACP, 185
Oxygen, ^{18}O, 12
—, radio-isotopes, 13

PABA = *Para*-aminobenzoate, see under *p*-Aminobenzoic acid, 383
Palmitate, from acetyl-CoA, 178
Palmitic acid, 11, 175, 179, 181, 183
Palmitoyl-CoA, 188
Palmityl-ACP, 181
Pantothenic acid, 37, 38, 57, 250
Papain, 49
Paper chromatography, 17, 91, 96, 98
Pectin, from sugar nucleotides, 165
Pentose, in biosynthetic reactions, 77
Pentose enediol, 63
Pentose phosphate, 69, 89, 144
—, conversion to hexose phosphate, 67, 101
Pentose-5-phosphate, 63, 71, 72
Pentose phosphate cycle, 21, 61, 72, 77, 102, 103, 163
Pentose phosphate pathway, 75—77
Peptides, dynamic equilibrium with amino acids, 52
Peptide bonds, 21, 45
—, synthesis, 54
—, — in homogenates, 55
Peptide linkages, in amino acid replacement, 46, 47
Peptide synthesis, inhibition by oxidation, 54
Peptide transfers, 49
Permeability barriers, 20
Petromyzonol, 309
Phenols, carboxylation, 81
Phenylalanine, radioactive, 47
Phenylbenzoylphosphate, 57
o-Phenylenediamine, 329
Phenylhydrazine, 217
Phenyl phosphate, 134
Pheromones, 239, 315
Phosphatases, 134, 150
Phosphate cycle, 31
Phosphate dihydroxyacetone, 99
Phosphate esters, 260

—, paper chromatography, 91
Phosphatidate phosphatase, 186, 188
L-α-Phosphatidate phosphohydrolase, 186
L-α-Phosphatidic acid, 186, 189
—, biosynthesis, 187, 188
Phosphatidyl choline, 187, 188
Phosphatidyl ethanolamine, 187, 188
Phosphobilinogen synthase, 220, 221
Phosphocreatine, 30, 33
Phospho-dihydroxyacetone, 100, 104
Phosphoenolpyruvate, 92, 111
—, from oxaloacetate, 133
—, from pyruvate, 130, 133
—, in the initiation of gluconeogenesis, 129
Phosphoenolpyruvate carboxykinase, 129, 130, 133
Phosphofructokinase, 134
Phosphoglucomutase, 161, 165
6-Phosphogluconate, 63, 64, 68
Phosphogluconate dehydrogenase, 63, 65, 68, 69
6-Phospho-D-gluconate: NADP$^+$ 2-oxidoreductase (decarboxylating), 67
6-Phosphogluconate carboxylase, 67
6-Phosphoglucono-δ-lactone, 64, 65
Phosphoglucose isomerase, 72
3-Phosphoglyceraldehyde, 99, 100, 101, 104
2-Phosphoglycerate, 136
3-Phosphoglycerate, 136
3-Phospho-D-glycerate carboxylyase (dimerising), 96
Phosphoglyceric acid, 76, 87, 91, 92, 93, 94, 331, 339
—, as an early intermediate, 85
—, distribution of ^{14}C, 92
—, enzymatic formation from ribulose diphosphate, 96
—, formation with cell-free enzyme preparations, 94

—, labelled from $^{14}CO_2$, 94, 95
3-Phosphoglyceric acid, 89, 96, 97, 103
—, conversion to fructose-6-P, 99
—, reduction to triose phosphate, 97, 98
Phosphoglycerides, 187
Phosphoglycolic acid, 92
Phosphoisomerase, 96
Phosphoketopentose epimerase, 104
5-Phosphomevalonic acid, 261, 262, 263, 264
Phosphomevalonic kinase, 261, 262
Phosphomonoesterase, 161
4'-Phosphopanthoteine prosthetic group, 181
Phosphopentoisomerase, 69, 104
Phosphopentokinase, 95, 104
Phosphopentonate, 63
Phosphopyruvate, 92, 128, 135, 136
5-Phospho-α-D-ribofuranose, 335
Phosphoriboisomerase, 67, 69, 95, 96, 97
5-Phospho-α-D-ribose-1-diphosphate, 335, 337
Phosphoribose epimerase, 67, 69
5-Phospho-β-D-ribosylamine, 336—338
5-Phosphoribosylamine: glycine ligase (ADP-forming), 338
5-Phosphoribosylamine: pyrophosphate phosphoribosyltransferase (glutamate-amidating), 338
5'-Phosphoribosyl-5-aminoimidazole, 339, 341, 342, 343
5-Phosphoribosyl-5-amino-4-imidazole-carboxamide, 342, 343
5'-Phosphoribosyl-5-aminoimidazole-4-carboxylate, 342, 343
5'-Phosphoribosyl-5-amino-4-imidazole-carboxylate carboxy-lyase, 343
Phosphoribosylaminoimidazole synthetase, 339, 341
5'-Phosphoribosyl-4-carboxy-4-amino-imidazole: L-aspartate ligase (ADP-forming), 343

5'-Phosphoribosyl-5-formamido-imidazole-4-carboxamide, 344
5'-Phosphoribosyl-N-formylglycinamide, 337, 339
5'-Phosphoribosylformylglycinamide: L-glutamine amido-ligase (ADP-forming), 341
Phosphoribosylformylglycinamidine synthetase, 339, 341
5'-Phosphoribosylformyl-glycinamidine cycloligase (ADP-forming), 341
5-Phosphoribosylglycinamide, 337
Phosphoribosylglycinamide formyltransferase, 337, 338, 344, 379, 380
Phosphoribosylglycinamide synthetase, 337, 338
5-Phosphoribosyl-1-pyrophosphate, see also under PRPP, 333, 334, 349
5-Phosphoribosylpyrophosphate amidotransferase, 338
5'-Phosphoribosyl-4-(N-succino-carboxamide)-5-aminoimidazole, 342
Phosphoribulokinase, 95, 96, 97
Phosphoribulose isomerase, 69
Phosphoric esters, 153
Phosphorolysis, 95
Phosphorylases 141, 143, 144, 146, 159
—, activity, 147
Phosphorylase theory, 150, 151, 156, 159, 160
—, glycogen and starch biosynthesis, 144
Phosphorylated pyruvic acid, 87
Phosphorylating enzymes, 145
Phosphorylation, 36, 150
4-Phosphotetronate, 63
Phosphotransacetylase, 38
Phosphotriose isomerase, 100, 104
Photosynthesis, 22, 81, 83, 111
—, application of kinetic studies, 98
—, carbon dioxide assimilation, 96
—, formation of sugar, 83, 153
Photosynthetic bacteria, 111

Photosynthetic carbon cycle, 103
Photosynthetic cycle of carbon re-
    duction, 81
Photosynthetic   phosphorylation,
    103, 105
Photosynthetic plants, 81
Phycomyces blakesleeanus, 243,
    279
Phytoene, 280, 281, 282, 283
—, biosynthesis, 277
—, — in carrot root preparations,
    282
—, conversion to lycopene, 282,
    283
Phytofluene, 282, 283
Phytol, 230
—, in chlorophylls, 239
Phytosterols, 284
Pieris brassicae, 360
Pigments, of butterflies, 357
Plakalbumin, 51
Plant terpenes, biosynthesis, 276
Polyglutamates, of pteroic acid, 381
Polyisoprenoid side chain, 284
Polypeptide chain elongation in
    protein synthesis, 51
Polysaccharides, 143
—, biosynthesis from monosaccha-
    rides, 141
—, enzymatic synthesis in vitro, 147
—, formation from glucose-1-phos-
    phate, 159
—, glycosidic linkages, 157
—, synthesis in vitro by phosphoryl-
    ase, 148
Polyterpenes, 247
Porphine, 193, 196
Porphobilinogen, 213, 214, 216,
    217
—, conversion into porphyrins, 225
—, — to tetrapyrroles, 221, 228
—, — into uroporphyrin I, 225
—, distribution of δ-carbon of gly-
    cine in, 208
—, from δ-amino-laevulinic acid,
    220
—, in acute porphyria, 213
—, precursor of vitamin B_{12}, 232

Porphobilinogen   ammonia-lyase
    (polymerizing), 225
Porphobilinogen deaminase, 225
Porphobilinogen synthase, 220, 221
Porphyrins, biosynthesis, 22, 193,
    203, 204, 245
— formation, relationship with tri-
    carboxylic acid cycle, 207, 209
—, fully reduced derivatives, 223
—, incorporation of iron, 227
—, metabolic sources, 196
—, role of aminolaevulinic acid, 219
—, structure, 196, 223
Porphyrins I and III, 213
Porphyrin IX, numbering of atoms,
    206, 207
Porphyrinogens, 223
Potato phosphorylase, 151
PRA = Phosphoribosylamine, 336,
    338
Pregn-5-ene-3,20-dione, 291, 294,
    295
Pregnenolone, biosynthesis in the
    corpus luteum, 297
—, — in the ovary, 296
—, — in the placenta, 297
—, — in the testis, 294
—, conversion to cortisol and C_{21}
    steroids, 291
—, — to progesterone, 289
—, — to progesterone by testis tis-
    sue, 294, 295
—, enzymes in the conversion to
    progesterone, 300
—, from cholesterol, 287
—, in the adrenal cortex, 289
Pregnenolone[^{3}H], 297
Prenols, 264
Prenyl pyrophosphates, 264
Prenyltransferases, 264, 265
Presqualene, 265
Presqualene alcohol, 269, 270
Presqualene pyrophosphate, 269
—, biosynthesis from farnesyl pyro-
    phosphate, 270, 271
—, conversion to squalene, 272
Primary bile acids, 301
Prodigiosin, biosynthesis, 22, 233,
    234

Progesterone, 291, 292, 295
—, from pregnenolone, 289, 300
—, placental production from acetate, 298
Progesterone[^{14}C], 297
Proline, precursor of porphyrins, 196, 197
Propionibacterium stermanii, 233
Propionic acid, 205, 206, 207, 208, 241
—, bacteria, 113, 116, 119
—, carboxylation, 124
—, from fermentation of glycerol, 112
—, from glycerol, 115
—, labelled, 115
Propionyl-CoA, 303, 305, 306
Proteins, acyl carrier, 180
—, biosynthesis, 45, 51
—, — in cells, 54
—, chemistry, 14
—, conversion to fats, 171
—, determination, 14
—, incorporation by substitution, 45
—, metabolism, 47
—, origin of glucose, 109
—, replacement reactions, 45
—, — during synthesis, 49
—, synthesis, peptide hypothesis, 51, 52, 53
—, —, template hypothesis, 51, 52, 53
Proteolytic enzymes, 54
Protochlorophyllide, 231
Protohaem ferrolyase, 229
Protoplasm, 1, 2
—, distribution of labelled carbons from glycine, 202, 203
Protoporphyrins, 196, 207, 208, 209, 216, 217, 227
—, biosynthesis, 205
—, from δ-amino-laevulinic acid, 217
—, from α-carbon of glycine, 204
—, from labelled aminolaevulinic acid, 218, 219
—, labelling from radioactive succin-

ate, 210, 211
—, pyrrole units, 206, 207
—, synthesis from glycine and oxoglutarate, 215
Protoporphyrin IX, 196, 202, 206, 208, 209, 219, 221, 223, 225, 227, 228, 229, 231
Protoporphyrin-monomethyl ester, 229
Protoporphyrinogen, 227
Protoporphyrinogen IX, 228
PRPP = 5-Phosphoribosyl-1-pyrophosphate, 333
—, conversion to FGAR, 336, 337
— in formation of phosphoribosylamine, 338
—, origins, 334, 335
—, reaction with hypoxanthine, 334
Pseudomonas saccharophila, 143, 150
Pteridines, 357, 358, 363
—, biosynthesis, 359
—, from guanine, 359
—, from purines, 359, 360
—, phosphorylated, 365
Pterins, 357, 358
Pterin pigments, 357
Pteroic acid, 363
—, polyglutamates, 381
Pteroin proteins, with coenzymes from tetrahydrofolic acid, 368
Pteroprotein, 379
Pteroprotein enzymes, 368, 369
Pteroylglutamic acid, (see also Folic acid) 362, 363
Pteroylheptaglutamate, 373
Pteroyltriglutamic acid, 360
Purines, 358, 369
—, as constituents of nucleic acids, 325
—, as precursors of pteridines, 359, 360
—, biosynthesis, 23, 325, 350, 351, 352
—, —, extensions on the pathway, 357
—, in biosynthesis of nucleotides, 330

412

SUBJECT INDEX

—, of cellular nucleic acids, 325
Purine catabolism, 352
Purine nucleotides, 348
—, interconversions, 345
Purine rings, biosynthesis through ribotide intermediates, 333
—, precursors of the atoms, 332
Purinolytic enzymes, 352
Pyrethrin I, 270
Pyridine nucleotides, 185, 371
Pyridoxalphosphate, 383
Pyrimidines, 325, 327, 359
—, in biosynthesis of nucleotides, 330
Pyrimidine nucleotides, 348
Pyrophosphomevalonate decarboxylase, 262, 263
5-Pyrophosphomevalonic acid, 261, 262, 263
Pyrophospho-3-methylbut-3-ene-1-ol, 262
Pyrophosphorylase, 158, 161
Pyrrolle-2-carboxylic acid, 204
Pyrrole rings, from glycine, 205
Pyrrole units, precursor, 207
Pyrroles, 193, 196, 205, 206, 208
—, biosynthesis, 197, 209
—, dicarboxylic, 214
—, synthesis, 213
Pyrrolidonecarboxylic acid, 197
Pyruvate, 112, 116, 117, 124, 128, 135, 136, 204, 207, 241, 331
—, biosynthesis and metabolism, 117, 119
—, conversion to glucose, 109
—, conversion to oxaloacetate, 131
—, — to phosphoenolpyruvate, 113
—, oxidation, 33
—, reaction with carbon dioxide, 85, 119, 121
Pyruvate:carbon dioxide ligase (ADP-forming), 131
Pyruvate carboxylase, 131, 133
Pyruvate kinase, 111, 130, 131
Pyruvate oxidase system, 371

Q-Enzyme, 151
Q$_{10}$, 286

Rabbit reticulocyte, 52
Radio-isotopes, in biological studies, 12, 13
Raffinose, from sugar nucleotides, 165
Ranol, 311
Δ^4-5α- and 5β-Reductases, 300
Reductive carbohydrate cycle, 103, 104, 105
Regenerative phase, 101
Reversible zymohydrolysis, 141, 171
Rhamnose, 116
Rhizopterin, 372, 373
Rhizopus nigricans, 372
Rhodopseudomonas spheroides, 218, 225, 227, 229, 230, 231
Rhodopterin, 357
Rhodospirillium rubrum, 38, 218, 285, 286
4-Ribitylamino-5-amino-2,6-dioxo-tetrahydropyrimidine, 359
Riboflavin, 357, 358
—, biosynthesis, 357, 358, 359
Riboflavin synthetase, 358
Ribonucleotides, 330
5'-Ribonucleotide phosphohydrolase, 348
Ribose, 21, 93, 101, 336, 360
Ribose-1,5-diphosphate, 334
Ribose-1-phosphate, 143, 331, 348
Ribose-5-phosphate, 63, 64, 67—70, 72, 73, 95, 335, 336, 360
—, conversion to 3-phosphoglyceric acid, 97
—, reaction with ATP, 334
D-Ribose-5-phosphate ketolisomerase, 68, 69
Ribosephosphate isomerase, 68, 69
Ribosephosphate pyrophosphokinase, 334, 335
Ribulose, 92, 93, 101
Ribulose diphosphate (RuDP), 93, 94, 95, 96, 101, 103, 104
[^{14}C]Ribulose diphosphate, 94
Ribulose diphosphate carboxylase, 96, 97, 102, 105

Ribulose monophosphate, 103
—, phosphorylation by ATP to ribu-
    lose diphosphate, 96
Ribulose-5-phosphate (Ru-5-P), 61,
    63—65, 67—69, 75, 93, 95, 104
Ribulose phosphate-3-epimerase, 68,
    69
R-5-P, see under Ribose-5-phos-
    phate
Rubber, degradation, 275
—, incorporation of mevalonic acid,
    275
RuDP, see under Ribulose diphos-
    phate
Ru-5-P, see under Ribulose-5-phos-
    phate
Ru-6-P, 72
Ru-5-P kinase, 102

Saccharides, biosynthesis, 22
Saccharide formation, by reversible
    zymohydrolysis, 141
Salicylic acid, 81
Scenedesmus, 91, 93
Scymnol, 302, 306
Sedoheptulose, 92, 93
Sedoheptulose-1,7-biphosphate,
    103, 104, 134
Sedoheptulose monophosphate, 93,
    94
Sedoheptulose-7-phosphate, 70, 71,
    73, 101, 103, 104
Sedum, 92
Serine, 203, 338, 339, 340, 379
—, biosynthesis, 369, 371
—, conversion to formate, 328
—, from glycine, 199, 383
—, in biosynthesis of uric acid, 328
Serratia marcescens, 233
Sesquiterpenes,   biosynthesis
    through isoprene pathway, 240
—, from farnesol, 276
Silver adenylate, 39
L-Sorbose, 151
Squalene, 7, 279
—, as cholesterol precursor, 247,
    249

—, biological synthesis from acetate,
    249
—, biosynthesis, 16, 17, 284
—, — from mevalonic acid, 259,
    260
—, — through isoprene pathway,
    240
—, conversion to lanosterol, 256,
    257
—, cyclisation, 250, 251, 252, 253,
    255, 257
—, degradation, 260
—, from farnesyl pyrophosphates,
    264, 265
—, from presqualene pyrophosphate,
    272
—, natural occurrence, 249
—, synthesis from mevalonic acid,
    265, 266, 267
—, — mechanism from farnesyl py-
    rophosphate, 268, 269
Squalene epoxidase, 256
Squalene hydrogen-donor: oxygen
    oxidoreductase   (2,3-epoxidiz-
    ing), 257
Squalene hydroxylase, 256
Squalene isoprene units, metabolic
    precursors, 257
Squalene   mono-oxygenase   (2,3-
    epoxidizing), 256
Squalene oxidocyclase, 256
Staphylococcus, 119
Staphylococcus aureus, 155, 282
Starches, action of branching en-
    zymes, 151
Starch, biosynthesis, 144, 156
—, — using UDPG, 160
—, from sugar nucleotides, 165
—, structure, 148, 149
—, synthesis by phosphorylase, 148
Stearate, CoA ester, 186
Stearic acid, 183
Steroids, 239
—, C_{18}, 297
—, C_{19}, 294
Steroid alcohols, in bile, 301
Steroid hormones, 239
—, biosynthesis, 289

—, enzymes in the biosynthesis, 300
—, from cholesterol, 287
Sterols, 239
—, biosynthesis, 11, 240, 241
—, — from mevalonic acid, 259
—, — from squalene, 250, 251, 252, 253, 255
—, structure, 7
Streptococcus faecalis, 362, 372
Streptococcus lactis, 372
Succinamido-acetic acid, 216
δ-Succinamido-laevulinic acid, 216
Succinic acid, 116, 207, 208, 216, 217, 241
—, $^{14}CO_2$ fixation, 86
—, carboxy-labelled, 210, 211
—, formation of aminolaevulinic acid, 212
—, formation by propionic acid bacteria, 22
—, from fermentation of glycerol, 112
—, in synthesis of protoporphyrin, 215, 216
—, labelled, 89, 210, 211
—, methylene-labelled, 210, 211
—, paper chromatography, 91
Succino-AICRP, 342, 343, 345
Succino-AICRP synthetase, 342, 343
Succinocarboxamide ribonucleotide, 343
Succinyl-CoA, 208, 212, 214, 218
—, conversion to porphobilinogen, 220
Succinyl-CoA: glycine C-succinyltransferase (decarboxylating), 218
Succinylcoenzyme complex, 208, 209
Sucrose, 89, 143
—, biosynthesis, 150, 156, 158
—, —, phosphorylase theory, 150
—, from sugar nucleotides, 165
—, synthesis from fructose and glucose, 150
—, — from inverted sugar, 150
—, — using UDPG, 157, 158

Sucrose glycosyltransferase, 143, 151
Sucrose phosphate, 162, 165
—, synthesis using UDPG, 157, 158
Sucrose phosphorylase, 150, 158
Sugars, 87
—, biosynthesis, 83, 153
—, ^{14}C carbon dioxide fixation, 86
—, paper chromatography, 91
—, photosynthesis, 103
Sugar esters, 143
Sugar nucleotides, 141, 151
—, conversion to complex saccharides, 164, 165
—, function in the interconversions of sugars, 153
—, role in transglycosylation, 156, 157
Sugar nucleotide transglycosylases, 143
Sugar phosphates, 141, 153, 165
Sugar polymers, biosynthesis, 156
Sulphanilamides, 369
—, acetylation, 35, 36
Sulphonamides, inhibition of bacterial growth, 368, 369
Surviving reductant, 86
Synthetase reaction, 160

Tartronic acid, 325, 350
Taurine, 301
—, conjugates with bile acids, 307, 309, 311
Taurodeoxycholic acid, 311
Teropterin, 360, 373
Terpenes, 239
—, biosynthesis, 276
Terpene alcohols, 264
Terpenoid quinones, 284
Testosterone, 289, 294, 295
—, biosynthetic pathway to oestrone and oestradiol, 296, 297, 298
17β (Testosterone)-dehydrogenase, 294, 295
Tetracetyl-tetrapropionylporphyrins, 196
Tetrahydrofolic acid = FH_4, 23, 336, 338, 344, 357, 377, 383

—, biosynthesis, 367
—, derivatives as coenzymes, 368
—, enzymatic production, 371, 372
—, from folic acid, 371, 372
Tetrahydropteroyldiglutamic acid, 381
Tetrahydropteroyltriglutamic acid, 381
3α,7α,12α,26-Tetrahydroxy-5β-cholestane, 303, 305
Tetrahymena vorax, 212
Tetramethyl-tetrapropionylporphines, 196
Tetrapeptides, 51
Tetrapyrroles, 231
—, biosynthesis, 193
—, from porphobilinogen, 221, 228
Tetrapyrrole pigments, 230, 231
Tetrose-4-phosphate, 63
Thiamine diphosphate, 71
Thiobacillus denitrificans, 105
Threonine, 22, 230
Thymine, 165, 383
Tocopherols, 239
Torpedo marmorata, 33
Tracer isotopes, 2, 3, 23
Transacylases, 181
Transaldolase, 71, 72, 73, 101, 102
Transaldolase reaction, 162, 163
Transamidation, 49
Transferases, 143, 151, 158
Transformylases, 380, 381
Transglycosidase, 143
Tranglycosylases, 143, 165
Transglycosylation, 141, 156
Transketolase, 69, 70, 72, 73, 101, 102, 104
—, mechanism of action, 70, 71
Transpeptidation, reactions, 45, 49, 51
Trehalose, from sugar nucleotides, 165
Trehalose phosphate, 153, 156
—, synthesis using UDPG, 157
Triacylglycerol, 188
2,4,5-Triamino-6-hydroxypyrimidine, 360
Tricarboxylic acid cycle, 20, 117,

204, 209, 210, 211, 212
—, relation with porphyrin formation, 207, 209
Triglycerides, biosynthesis, 22, 186
—, hydrolysis, 300
—, monoglyceride path, 189
3α,7α,12α-Trihydroxy-5β-cholestan-26-al, 305
3α,7α,12α-Trihydroxy-5β-cholestane, 301, 305
3α,7α,12α-Trihydroxy-5β-cholestan-26-oic acid, 305
3α,7α,12α-Trihydroxycoprostanic acid, 303, 307, 310
3α,7α,12α-Trihydroxycoprostanyl-CoA, 307
4,4′,14-Trimethylcholestadienol, 253
4,4′,14-Trimethylcholestane, 252
Trioses, conversion to hexoses, 101, 102
Trioseisomerase reaction, 163
Triose phosphates, 67, 75, 92, 98, 101, 103
—, from phosphoglyceric acid, 97, 98
Triose phosphate dehydrogenase, 100, 104
Triose phosphate isomerase, 76, 99, 100, 163
Triose phosphate mutase, 100
Tripyrrolic red pigments, 233
Triterpenes, biosynthesis through isoprene pathway, 240
—, structure, 250, 252
—, tetracyclic 255
Tryptophan, 22
—, precursor of porphyrins, 196
[15N]Tyrosine, 12

Ubichromenol-50, 286
Ubiquinones, 239
—, from p-hydroxybenzoic acid, 284, 285
Ubiquinone-8, 286
Ubiquinone-9, 286
Ubiquinone-10, 286
Ubiquinone-50, 286

UDP = Uridine diphosphate, 136, 155, 157, 158, 159, 164
UDP-N-acetylmuramic acid, 155
UDP-N-acetylmuramic peptides, 155
UDPAG = Uridine diphosphate-N-acetyl-D-glucosamine, 164
—, in chitin biosynthesis, 164
UDPG = Uridine diphosphate glucose, 153, 154, 155, 165
—, as glucose donor, 157, 158, 159, 160
—, as glycosyl donor, 157, 158
—, in biosynthesis of glycogen, 159
—, — of starch, 160
—, in synthesis of cellulose in microorganisms, 158
—, — of sucrose, 157, 158
—, — of trehalose phosphate, 157
UDP-galactose, 155, 156
—, as galactose donor, 161
[^{14}C]UDP-D-galactose, 163, 164
UDP-D-galactose 4-epimerase, 161
UDP-D-galacturonic acid, 156
UDP-glucose, 155, 156, 157, 161
UDP-D-glucose—glycogen glucosyltransferase, 160
[^{14}C]UDP-D-glucose, 163, 164
UDP-D-glucuronic acid, 156, 157
UDP-glycosides, 143
Uranium isotopes, 12
Urate oxidase, 350
Urate: oxygen oxidoreductase, 350
Urea, 325, 348, 350
—, cycle, 21
—, formation from ammonia, 19
—, from carbon dioxide, 124
—, in biosynthesis of uric acid, 327, 328, 329
—, synthesis from carbon dioxide, 111
—, — from citrulline, 20
—, — through arginase, 19
[^{14}C]Urea, 20, 351
Urease, 350, 351
Ureogenesis, 18, 19, 352
Uric acid, 325, 369
—, from amino acids, 352
—, degradation by enzymes, 350

—, formation in birds, 328
—, from AICA, 333
—, from inosine phosphate, 348, 349
—, in uricotelic animals, 352
349
—, source of carbon atoms, 327, 328
—, ureide carbon source, 327
Uric acid formation, Wiener's theory, 350, 351
Uricase, 350
Uricolysis enzymes, 350
Urico oxhydrase, 350
Uridine, 154, 165
Uridine diphosphate, see also under UDP, 155
Uridine diphosphate-N-acetyl-D-glucosamine, 164
Uridine diphosphate-D-galactose, 165
Uridine diphosphate glucose, see also under UDPG, 136, 137, 153, 154, 155, 165
Uridine diphosphate nucleotides, 155
Uridine diphosphate sugars, 155, 156
Uridine diphospho-galactose-4-epimerase, 156
Uridine 5'-phosphate, 154
Uridine triphosphate, 165
Urobilinogen, 213
Uroporphyrins, 216, 217, 225, 227
Uroporphyrin I, 196, 225
Uroporphyrin I synthase, 227, 228
Uroporphyrin III, 196, 208, 221, 223
Uroporphyrin III cosynthase isomerase, 228
Uroporphyrinogen, 223
Uroporphyrinogen I, 225, 228
Uroporphyrinogen III, 225
—, from porphobilinogen, 227, 228
Uroporphyrinogen-III carboxylyase, 227
Uroporphyrinogen III co-synthase, 225, 227

Uroporphyrinogen decarboxylase, 227
Uroporphyrinogen isomerase, 225
Uroporphyrinogen I synthase, 225, 227, 228
UTP, 136, 155, 261

Valine, 51
—, biosynthesis, 15
—, labelled, 52
2-Vinylphaeoporphyrin a_5 monomethylester, 231
Vinyl phosphate, 94
Vital creation, 18
Vitamin B_2, 357, 358
Vitamin B_{12}, 231, 232
—, from cobyrinic acid, 233
Vitamin B_c, 362

Warburg's theory of oxygen activation, 61
Water metabolism, 3
Wiener's theory, uric acid formation, 350, 351
Wood-Werkman reaction, 119, 130, 131

Xanthic bases, 325
Xanthine, 349

Xanthine oxidase, 330, 349
Xanthine: oxygen oxidoreductase, 350
Xanthophylls, 282, 284
Xanthopterin, 357, 360
Xanthosine, 347
Xanthosine phosphate, 348
—, synthesis, 345
Xanthosine-5'-phosphate: glutamine amido-ligase (AMP-forming), 348
Xanthylic acid, 345
Xenopus, 359, 360
XMP = Xanthosine phosphate, 347
D-Xyloketose, 151
Xylulose, 93
D-Xylulose-5-phosphate, 67, 69, 70, 71, 73, 104
4-D-Xylulose-5-phosphate, 68
D-Xylulose-5-phosphate 3-epimerase, 69

Yeast aceto-CoA-kinase, 41

Zymase, 14
Zymohydrolysis, 22
—, reversible, 141, 165, 171

Name Index

Abderhalden, E., 197
Abraham, S., 75, 251
Abrams, R., 328, 345, 347, 348
Abramsky, T., 218, 219
Ach, K.L., 184
Ackermann, W.W., 333, 369
Ackroyd, H., 325
Adamson, L.F., 259
Agnew, W.S., 272
Agranoff, B.W., 261, 263, 264
Aiyar, A.S., 284
Albert, A., 359
Alberts, A.W., 180, 181, 182
Aldrich, P.E., 257
Alexandry, A.K., 19
Allen, M.B., 105
Allman, D.W., 174, 179, 180
Altman, K.I., 199, 203, 229
Amdur, B.H., 260
Anderson, D.G., 279, 281
Anderson, P.E., 155
Anderson, R.C., 372
Anfinsen, C.B., 51, 52
Angier, R.B., 362, 363
Anton, A.H., 371, 375
Archer, B.L., 275
Arigoni, D., 255
Arion, W.J., 135
Arnon, D.I., 105
Aronoff, S., 98
Arreguin, B., 243, 247
Arth, G.E., 372
Aschoff, L., 7
Ashmore, J., 159

Ashwell, G., 69
Aubert, J.P., 105
Audley, B.G., 275
Austrian, R., 156
Avery, O.T., 24
Avigan, J., 273
Axelrod, B., 69, 72, 76, 95
Axelrod, J., 156, 157
Ayrey, G., 275

Babad, H., 164, 165
Bach, S.J., 19
Bachawat, B.K., 259
Bachelard, G., 1
Bachmann, B.J., 221
Bacon, J.S.D., 161
Bacq, Z.M., 24
Baggett, B., 295
Baguena, R., 25
Baker, C.F., 253
Bakerman, H.A., 377
Balderas, L., 295
Baldwin, E., 350, 351
Bamann, E., 14
Bandurski, R.S., 131
Baranowski, T., 144, 186
Barber, G.A., 158, 165
Bard, R.C., 76
Bardos, T.H., 372, 374
Barger, F.L., 374
Barker, H.A., 38, 93, 113, 143, 171, 173, 174, 328
Barnard, D., 275
Barnes, F.W., 325, 327

Barratt, R.W., 241
Barron, E.S.G., 37
Bartholomäus, E., 223
Bartley, W., 131
Bartran, K., 277
Bassham, J.A., 91, 92, 94, 96, 98, 101, 103, 104
Baugh, C.M., 360, 362, 368
Baumann, C.A., 373
Baxter, C.F., 162
Beadle, G.W., 15
Bean, R.C., 165
Beeler, D., 282
Beinert, H., 38, 173
Beisbarth, H., 232
Beisenhertz, G., 100
Belic, I., 277
Bendich, A., 371
Benedict, C.R., 265
Bennett, L.L., 369
Bennett-Clark, T.A., 92
Benson, A., 86, 87, 89—96, 98, 101, 103, 105
Bentley, M., 345, 347, 348
Bentley, R., 203, 284
Berg, B.N., 11, 301
Berg, P., 39—41, 369
Bergmann, M., 48, 49, 54
Bergström, S., 303, 311
Berman, M., 37, 38
Bernard, C., 17
Bernard, K., 185
Bernfeld, P., 143, 148, 149, 155, 156
Bernhauer, K., 232
Bernheimer, H.P., 156
Berséus, O., 303
Berthelot, M., 17
Binkley, S.B., 362
Birch, A.J., 276
Birchall, K., 299
Bird, O.D., 362, 381
Black, A.L., 162
Black, J., 1
Black, S., 38
Blackley, R.L., 379
Blinc, M., 148
Bloch, E., 287

Bloch, K., 53, 171, 180, 185, 196, 204, 241, 242, 243, 245, 247, 249, 250, 251, 252, 253, 257, 260, 261, 262, 263, 264, 265, 266, 267, 273, 281, 301, 327
Bloom, B., 75
Bloom E.S., 362, 371, 373, 381
Bloomfield, D.K., 185
Bogorad, L., 221, 223, 224, 225, 227, 233
Bohlman, F. 277
Bohonos, N., 373
Bond, T.J., 372
Bonner, D., 15, 241
Bonner, J., 38, 243, 247, 275
Bonsignore, A., 65, 71
Booij, H.L., 227
Boothe, J.H., 362
Borek, E., 241, 243
Borgström, B., 300
Borsook, H., 12, 20, 25, 45, 49, 51, 52, 54, 55
Boschetti, A., 281
Boulter, D., 276
Bourne, E.J., 151
Boyd, G.S., 250, 303
Boyer, P.D., 41, 49, 50
Bradley, R.M., 178
Bradlow, H.L., 251
Brady, R.O., 173, 174, 177, 178, 243, 287
Braganca, B., 156
Braganza, B., de, 41
Braithwaite, G.D., 279
Bratton, A.C., 340
Braunstein, A.E., 57
Bray, R.C., 232, 350
Bremer, J., 311
Brenner-Holzach, O., 360
Breslow, R., 71
Bressler, R., 179
Bricteux-Grégoire, S., 351
Bridgwater, R.J., 303, 311
Briggs, D.R., 160
Brodie, A.F., 65
Brodie, J.D., 179
Bronk, J.R., 19, 20
Brooke, M.S., 347

Broquist, H.P., 371
Brown, A., 89
Brown, A.H., 105
Brown, D.H., 164
Brown, E.G., 217
Brown, G.M., 360, 362, 363, 365,
  366, 368, 381
Brown, J.L., 189
Brown, M.A., 162
Brown, R.A., 362, 381
Brown, R.H., 294
Brownell, K.A., 289
Brummond, D.O., 95, 96
Bublitz, C., 186, 274
Buchanan, J.G., 158
Buchanan, J.M., 16, 25, 127, 128,
  129, 243, 326, 327, 330, 331,
  332, 333, 334, 336, 338, 339,
  340, 341, 343, 344, 345, 348,
  349, 351, 379, 380
Bucher, N.R.L., 249, 252
Bücher, T., 100
Büchner, E., 14
Buel, G.C., 186
Bufton, A.W.P., 203
Bunge, G., 54
Bunyan, J., 286
Burian, R., 349
Burkard, J., 150
Burma, D.P., 158
Burnham, B.F., 229, 231, 232
Burchenal, J.H., 371
Burton, R.M., 164
Butenandt, A., 311
Byrne, W.L., 134

Cabib, E., 156, 157
Cahill, G.F., Jr., 159
Calkins, D.G., 373, 381
Calvin, M., 86, 87, 88, 89, 91, 92,
  93, 94, 95, 96, 98, 101, 103,
  104, 105
Campbell, C.J., 381
Campbell, J.W., 352
Campbell, P.N., 53
Capindale, J.B., 105
Caputto, R., 153, 154, 161
Carbone, J.V., 157

Cardini, C.E., 153, 154, 157, 158,
  159, 160, 165
Carey, J.B., Jr., 303
Carson, S.F., 113, 115, 116
Carter, C.E., 343, 345
Casarett, G.W., 199, 203
Cathou, R.E., 380
Cathro, D.M., 299
Cerny, V., 313
Chaikoff, I.L., 75, 81, 122, 243,
  251
Chan, P.C., 365
Chandler, B.V., 279
Channon, H.J., 7, 247, 248, 249
Chantrenne, H., 25, 55, 56, 57
Chapman, D.D., 243
Chaykin, S., 260, 261, 262—264
Cheldelin, V.H., 75
Chen, R.W., 243
Chichester, C.O., 279, 281, 282, 284
Childs, Jr., C.R., 267
Chiriboga, J., 158
Chou, T.C., 38
Christian, W., 61, 63, 65
Christiansen, K., 184
Claes, H., 282, 284
Clark, B., 189
Clarke, H.T., 3, 4, 5, 6, 9, 12
Clayton, R.B., 252, 257
Cleland, W.W., 265
Coates, C.W., 33
Cobey, A., 259
Cockbain, E.G., 275
Cohen, L.H., 343, 345
Cohen, P.P., 20, 55, 57
Cohen, S.S., 63, 64
Cohn, M., 143
Colas, A., 298
Coleman, D.L., 225
Collie, J.N., 4
Colowick, S.P., 114, 145, 146, 147
Conant, J.B., 125, 128
Condon, G.P., 287, 298
Conn, J.B., 372
Consden, R., 91
Cookson, G.H., 213, 216, 217
Coon, M.J., 259
Cooper, J.R., 70

Corcoran, J.W., 232
Corey, E.J., 256, 257
Cori, C.F., 144, 145, 146, 147, 148, 149, 151, 159
Cori, G.T., 144, 145, 146, 147, 148, 149, 151
Cori, O., 64
Cornforth, Sir J.W., 243, 244, 245, 249, 255, 259, 264, 265, 266, 267, 269, 273, 275
Cornforth, Lady R., 244, 258, 264, 266, 267, 269
Cornudella, L., 25
Coryell, C.D., 31
Cosulich, D.B., 363
Cox, R.T., 33
Cramer, R.D., 127, 128
Crane, R.K., 135
Cross, M.J., 369
Cutolo, E., 155

Daft, F.S., 377
Dagley, S., 187, 220, 273
Dahm, K.H., 313
Dahms, G., 360
Dakin, H.D., 109
Dale, Sir H.H., 24, 33
Dancis, J., 287, 298
Danielsson, H., 256, 267, 275, 301, 303, 304, 307, 311
Dauben, W.G., 245, 251
Daves, G.D., Jr., 284, 285, 286
Davidson, A.G., 273
Davidson, H.M., 161
Davies, B.H., 281
Davis, E.M., 289
Davis, P., 277
Dean, G.A., 289
Dean, P.D.G., 257
Deasy, C.L., 51
de Fekete, M.A.R., 160, 165
De Flora, A., 65
De La Haba, G., 70, 71, 101
Della Rosa, R.J., 229
Delluva, A.M., 327, 351
De Moss, R.D., 76
Dennis, D.T., 276
Deubig, M., 184

de Waard, A., 261, 262—264
Dialameh, G.H., 284
Dickens, F., 25, 62, 63, 65, 68, 72
Dietschy, J.M., 300
Dijksterhuis, E.J., 3
Dikanowa, A., 150
Dillard, R., 368
Di Marco, A., 232
Dimter, A., 249
Diplock, A.T., 286
Dische, Z., 67
Disraely, M.N., 365
Dituri, F., 260
Donninger, C., 267, 282
Dorfman, R.I., 287, 289, 293, 297
Doudoroff, M., 143, 150, 151
Dowben, R.M., 287
Dresel, E.I.B., 216, 217, 220, 221, 223, 225
Drysdale, G.R., 369
Dubnoff, J.W., 20, 54, 55
Dubos, R.J., 24
Duchâteau, G., 350
Dulit, E., 273
Dunham, J.L., 105
Durr, I.F., 274, 275
Dutton, G.J., 156, 157

Eakin, R.E., 333, 369
Edmond, J., 269, 270, 271, 272
Edsall, J.T., 46, 49, 112, 113, 121, 124
Eggerer, H., 261, 262, 263—265
Eggleston, L.V., 119
Eik-Nes, K.B., 289, 292—294, 297, 298
Einarsson, K., 301, 303
Eisenberg, M.A., 38
Eisner, T., 315
Elbein, A.D., 165
Elliott, W.H., 311
Embden, G., 76
Emmett, A.D., 362
Engel, L.L., 295
Epstein, W.W., 269, 270
Erwin, M.J., 343, 379
Eschenmoser, A., 255
Eugster, C.H., 277

Evans, E.A., Jr., 11, 117—119, 121, 122, 124, 125, 126, 128
Ewart, M.H., 160
Eyring, H., 105

Fager, E.W., 89, 94
Falcone, A.B., 41
Falk, J.E., 216, 217, 220, 221, 223, 225
Fanshier, D., 165
Feingold, D.S., 165
Feldberg, W., 24, 37, 38
Ferguson Jr., J.J., 274
Fernandes, J.T., 334
Fieser, L.F., 294, 297, 298
Fieser, M., 294, 297, 298
Fildès, P., 125
Fink, E., 197
Fink, K., 91
Fink, R.M., 91
Fischer, E., 16
Fischer, H., 193, 197, 213, 223
Fischer, H.O.L., 16, 193, 197, 213, 223
Fischer, R.B., 19, 20
Fish, C.A., 253
Fish, W.A., 273
Fiske, C., 362
Fitelson, J., 249
Fitting, C., 151
Flaks, J.G., 338, 343, 345, 379
Fles, D.A., 277
Florapearl, F.D., 259
Florkin, M., 350, 351
Flynn, R.M., 38
Folin, O., 12, 47
Folkes, J.P., 47, 49
Folkers, K., 232, 257, 258, 284, 285, 286, 372
Forsyth, C.C., 299
Foster, J.W., 113, 115
Fox, C.L., Jr., 333, 369
Franck, J., 105
Fraenkel-Conrat, H., 49, 54
Frantz, Jr., I.D., 249, 273
Fred, E.B., 112
Freull, F., 173, 183
Friedkin, M., 381
Friedrich, W., 232, 233

Fries, N., 241
Friis, P., 284, 285
Fruton, J.S., 16, 21, 31, 49, 50, 213
Fryer, R.I., 276
Fujita, Y., 240
Fuld, M., 149
Fuller, R.C., 95, 96, 105
Futterman, S., 371

Gabrilove, J.L., 293
Gaffron, H., 89
Galbraith, M.N., 313
Gale, E.F., 47, 49
Gan, M.V., 184
Gander, J.E., 161, 162
Ganguli, N.C., 155
Ganguly, J., 177, 178
Garrett, A.B., 87
Gaskell, W.H., 1
Gasser, H.S., 33
Gautschi, F., 273
Gehring, L.B., 345, 347
Gentner, W., 128
Gest, H., 93
Getzendaner, M.E., 333, 369
Gibbons, G.C., 150
Gibbs, M., 76, 100, 101, 105
Gibbs, M.H., 259
Gibson, D.M., 171, 174, 175, 178
Gibson, K.D., 212, 218, 221, 229, 230
Giles, N.H., 343
Gilfillan, J.L., 257
Gillespie, R., 162, 163
Ginsburg, V., 153, 155, 165
Gladstone, G.P., 125
Glaser, L., 158, 164
Glass, B., 53
Glock, G.E., 67
Gloor, U., 303, 311
Glover, J., 273
Glucksohn, S., 7
Gold, P.H., 284
Goldman, P., 180, 188
Goldthwait, D.A., 335, 336, 371
Gomori, G., 134
Goodale, T.C., 91

Goodman, DeW.S., 264, 266, 267, 269
Goodwin, T.W., 25, 273, 278, 279, 281, 282
Gordon, A.H., 91
Gordon, M., 333, 369
Gore, I.Y., 245, 259, 260
Gorzkowski, B., 351, 352
Gosselin, L., 260
Gould, R.G., 260
Graebe, J.E., 276
Granick, S., 221, 222, 223, 225, 227, 229, 231
Grant, G.A., 153, 161
Grath, H., 350
Gray, C.H., 203, 213
Grazi, E., 65, 77
Green, D., 38, 171, 174, 178, 179, 180
Green, J., 286
Greenberg, D.M., 259, 375, 378
Greenberg, G.R., 328, 330, 331, 333, 335, 336, 338, 339, 345, 371, 374, 375, 379
Gregory, R.A., 24
Greull, G., 70
Griffin, M.J., 365, 381
Grinstein, M., 197, 203
Grisolia, S., 20, 25
Grob, E.C., 243, 279, 281
Grodsky, G.M., 157
Grunberg-Manago, M., 38
Guillory, R.J., 159
Gunsalus, I.C., 76
Gurin, S., 172, 173, 174, 177, 243, 259, 260, 289, 303
Gut, M., 289
György, P., 357

Haagen-Smith, A.J., 51, 276
Haas, V.A., 91, 98
Hall, A.G., 92, 98
Hall, P.F., 289
Hammarsten, E., 328
Hammerstein, J., 297
Hampshire, F., 311
Hanes, C.S., 144, 148
Hanke, M., 125
Hansen, R.G., 162, 163

Happ, G.M., 315
Hardenbrook, H., 162, 163
Harlan, W.R., 183
Harris, A.Z., 94, 101, 103
Harris, S.A., 372
Hart, P., 186
Hartman, F.A., 289
Hartman, S.C., 331, 332, 336, 338, 340, 344, 345
Hartman, W.E., 289
Hartmann, G.R., 16, 25
Haslewood, G.A.D., 25, 301, 306, 307, 308
Hassid, W.Z., 81, 83, 84, 93, 122, 123, 142, 143, 148, 149, 150, 151, 156—158, 163, 164, 165
Hastings, A.B., 25, 119, 120, 121, 124, 125, 126, 127, 128, 129, 159
Hatch, M.D., 345
Hatefi, Y., 377
Haworth, W.N., 148, 149, 151
Hayano, M., 20, 289, 297
Heath, H., 225, 227
Hechter, O., 287, 289, 292, 293
Hehre, E.J., 143
Heidermanns, C., 351
Heinle, R.W., 372
Heinrich, M.R., 328
Heinrichs, W.L., 298
Hele, P., 38, 175
Helfant, M., 77
Helfenstein, A., 247
Heller, J., 351, 352
Helmreich, E., 160
Helmreich, M.L., 289, 294
Hemingway, A., 115, 116
Hemming, F.W., 281
Hench, P.S., 292
Henning, U., 260, 261, 262, 263, 264, 265, 274, 275
Hensen, H., 250, 252
Herrington, K.A., 349
Herriott, R.M., 14
Hevesy, G., 4, 9
Heyworth, R., 161
Hiatt, H.H., 71, 135
Hickman, J., 69
Hift, H., 38

Higgins, G.M.C., 275
Hill, A.C., 141
Hill, R., 105
Hill, R.L., 181
Hilz, H., 39
Himes, R.H., 381
Hirsch, H.E., 92
Hirsch, P.F., 75
Hirst, E.L., 148, 149
Hoare, D.S., 225, 227
Hocks, P., 311
Hodgkin, A.L., 24
Hoffman, C.H., 257
Hoffman-Ostenhof, O., 14
Hoffmeister, H., 313
Hogan, A.G., 362
Holland, B.R., 369
Hollmann, S., 72
Holloway, P.W., 183, 265
Holman, R.T., 184
Hopkins, F.G., 1, 325, 357
Hoppe-Seyler, F., 193
Hora, J., 313
Horecker, B.L., 25, 63, 65, 66, 67,
    69, 70, 71, 77, 93, 94, 95, 96,
    100, 102, 103, 105, 134, 135
Horn, D.H.S., 313
Horowitz, M.G., 161
Hotta, S., 251
Hough, L., 93, 100
Hsu, R.Y., 179
Huang, R.L., 243, 247
Hubbard, N., 227
Hübscher, G., 189
Huennekens, F.M., 29, 363, 371,
    372, 373, 374, 375, 377, 379,
    381, 382
Huff, J.W., 257, 259
Humphrey, J.H., 47
Hunter, G.D., 245
Hurwitz, J., 69, 70, 95, 96
Hutchings, B.L., 362, 373
Hutchinson, D.J., 371
Hutton, H.R.B., 303
Huxley, A.F., 24

Ikeda, K., 232
Illingworth, B., 149, 159

Inhoffer, H.H., 277
Inone, S., 227
Isherwood, F.A., 148
Isselbacher, K.J., 155, 157, 189

Jackson, R., 368
Jacob, M.I., 174
Jacobsen, R.P., 289, 292
Jaenicke, L., 25, 338, 365, 374,
    375, 376, 383
Jakoby, W.B., 95, 96
Jang, R., 69, 95
Janney, N.W., 109
Jeanloz, R.W., 289, 292
Jeger, O., 250, 252, 255
Jezewska, M.M., 351, 352
Jobbling, J.W., 7
Joffe, S., 162, 163
John, H.M., 36, 37
Johnson, M.J., 112, 113, 131, 155
Johnston, J.M., 189
Jones, D., 281
Jones, J.K.N., 93, 100
Jones, M.E., 38, 39
Jones, M.J., 369
Jones, O.T.G., 230, 231
Julian, G.R., 63

Kaczka, E.A., 372
Kagawa, Y., 65
Kalckar, H.M., 31, 130, 144, 155,
    156, 164, 165, 331
Kalnitksy, G., 131
Kaltenbach, J.P., 131
Kamen, M.D., 13, 25, 81, 82, 83,
    85, 86, 113, 115, 116, 117, 121,
    122, 123, 124, 126, 127, 197,
    203
Kandutsch, A.A., 273, 281
Kaplan, N.O., 38, 150
Karlson, P., 311, 313, 314
Karlsson, J.L., 328
Karrer, P., 247, 277
Kasal, A., 313
Kast, L., 7
Kaufman, S., 209
Kawada, M., 65
Kawachi, T., 275
Kawaguchi, S., 92

Kay, L.D., 94, 101, 103, 337
Kazuno, T., 307
Keech, D.B., 121, 137
Kekwick, R.G.O., 275
Keighley, G.L., 12, 51
Kendall, E.C., 289, 292
Kennard, O., 213, 216
Kennedy, E., 186, 187, 188
Keresztesy, J.C., 372, 373
Kessel, I., 260, 262—265
Kett, J., 265
Khorana, H.G., 334
Kiessling, W., 147
Kikuchi, G., 218
Kimizuka, T., 249
King, D.S., 313
King, T.E., 75
Kinoshita, J.H., 77
Kirchner-Kühn, I., 351
Kirschner, K., 281
Kirtley, M.E., 275
Kisliuk, R., 379
Kistiakowsky, G., 125, 127
Kittinger, G.W., 161, 164
Kiyasu, J.Y., 186, 188
Kleiber, M., 162
Klemperer, F.W., 127, 128
Klenow, H., 69, 70, 101, 334
Klevstrand, R., 92
Klybas, V., 70
Knappe, J., 274, 275
Knauss, H.J., 274
Knoop, F., 11, 19
Knopf, L., 109
Koeppe, O.J., 41
Koft, B.W., 365
Kohler, A.R., 371
Kohler, R.E., 7, 9, 11, 12, 25
Koike, S., 227
Korey, S.R., 38, 41
Korn, E.D., 333, 348
Kornberg, A., 25, 130, 334, 349
Kornberg, H.L., 71
Kosobutskaya, L.M., 231
Kosterlitz, H.W., 153
Krampitz, L.O., 70, 131
Krasnovsky, A.A., 231
Krebs, H.A., 18, 19, 25, 109, 111,
    112, 113, 117, 119, 121, 122,

124, 125, 128, 130, 135, 136,
    331
Kriukova, N., 150
Krogh, A., 4
Kroppling-Rueff, L., 274
Krumbieck, C.L., 362
Kubowitz, F., 61
Kühn, R., 357
Kumar, A., 218
Kumin, S., 209, 210, 211, 219
Kunitz, M., 14
Kurahashi, K., 130, 137
Kurssanov, A., 150
Kuster, E.G.F., 193

Labbe, R.F., 227
Labler, L., 313
Labows, J., 315
Lack, L., 212
Lafaye, J., 371
Lagerkvist, U., 331, 345, 346, 347,
    348
Laidman, D.L., 286
Lampen, J.O., 93, 369
Lands, W.E.M., 186
Langdon, R.G., 249
Lanman, G., 295
Lanman, J.T., 299
Lansford, M., 374
Lardy, H.A., 130, 133, 369
Larner, J., 149, 155
Lasater, M.B., 289, 294
Lascelles, J., 212, 363, 369
Laver, W.G., 212, 217, 218
Lavoisier, A., 17
Law, J., 260, 261, 262, 264
Lawrence, E., 12, 13, 81, 122
Leathes, J.B., 11
Lebedev, A.F., 63, 111, 150
Leder, I.G., 70, 71
Leder, J.G., 101
Lee, T.W., 352
Legge, J.W., 204, 207
Leloir, L.F., 25, 152, 153, 154,
    155, 156, 157, 158, 159, 160,
    165
Lemberg, R., 204, 207
Lennarz, W.J., 180
Le Roy, G.V., 289

Leuthardt, F., 360
Levenberg, B., 331, 332, 336, 338, 339, 340, 341, 345
Levin, E., 281
Levin, E.Y., 225
Levitz, M., 287, 298
Levy, H.R., 261, 289, 292
Li, C.H., 15
Lieben, F., 193
Lieberman, I., 334, 345, 349
Lieberman, S., 289, 298
Lifson, N., 76, 129
Light, R.J., 180
Lin, I., 331
Lin, T.S., 165
Lincoln, R.E., 282, 283
Lindberg, M.C., 297
Lindstedt, S., 303
Linn, B.O., 257
Lipmann, F., 22, 30, 31, 32, 35, 36, 37, 38, 39, 41, 51, 55, 57, 63, 64, 65, 85, 131, 173
Lipton, M.A., 37
Litt, C.F., 151
Little, H.N., 241, 243, 247
Littlefield, J.W., 209
Lockwood, W.H., 225
Loeb, J., 30
London, E.S., 19
Lorah, C.L., 273
Lorber, V., 76, 129
Lord, K.E., 257
Lowy, P.H., 51
Luchsinger, W.W., 41
Luick, J.R., 162
Lukens, L.N., 341, 343, 345, 349
Lukton, A., 279
Lumry, R., 105
Luppis, B., 134
Lusk, G., 109
Lynch, V., 92
Lynen, F., 39, 173, 178, 179, 180, 183, 260, 261, 262, 263, 264, 265, 274, 275, 281
Lynn, W.S., Jr., 289, 294

McArthur, C.G., 289
McBee, R.H., 151
McCane, R.A., 24

McCann, M.P., 365
McFall, E., 348
McFarlane, A.S., 47
McGarrahan, K., 249
McGeown, M.G., 162
McGilvery, R.W., 55, 57, 134
McHale, D., 286
McKenna, R.M.B., 249
McKeon, J.F., 233
McSweeney, G.P., 275
Machado, A.L., 33, 35, 37
MacKinney, G., 279, 282
Magasanik, B., 345, 347, 348
Magnus-Levy, A., 171
Mahler, H.R., 350
Maier-Leibnitz, W., 128
Majerus, P.W., 180, 181, 182
Maley, G.F., 358
Malpress, F.H., 161, 162
Mandel, A.R., 109
Maner, F.D., 298
Mangiarotti, G., 65
Mangum, J.H., 380
Mani, N., 300
Mann, P.J.G., 33
Mann, T., 37, 38
Manners, D.J., 159
Mapson, L.W., 156
Markley, K., 261
Marks, G.S., 233
Marks, P.A., 71
Marrian, G.F., 298
Marshall, E.K., Jr., 340
Martin, A.J.P., 91
Mason, M., 4
Mason, N.R., 289
Massini, P., 93
Masters, R.E., 199, 203
Masurat, T., 77
Matoulkova, V., 377
Mauzerall, D.C., 223, 225, 227, 229
Maxwell, E.S., 156, 164
May, M., 374
Mayaudon, J., 96
Mayor, F., 25
Mehler, A.H., 103, 130
Meinke, W.W., 369
Meinwald, J., 315

Meister, A., 58
Melnick, I., 340
Mendelsohn, D., 301
Mendicino, J., 131
Mercer, E.I., 273, 281
Merola, A.J., 365
Meyer, K.H., 149, 150
Meyer-Arendt, E., 100
Meyerhof, O., 29, 35, 36, 76, 125, 128
Middleton, E.J., 313
Mijovic, M.W., 250, 252
Milas, N.A., 277
Milhaud, G.M., 105
Miller, A., 377
Miller, C.S., 330, 333
Miller, I.M., 232
Miller, R.W., 343
Millerd, A., 38
Millet, J., 105
Mills, G.T., 156, 164
Mitchell, F.L., 299
Mitchell, H.K., 362
Mitchell, J.H., Jr., 369
Mittelman, N., 153
Mobberley, M.L., 273
Mohrhauer, H., 184
Mokrasch, L.C., 134
Moldave, K., 58
Molnar, D.A., 365, 368
Mommaerts, W.H.F.M., 159
Moore, C.G., 275
Moore, C.V., 197, 203
Moore, H.W., 286
Morrison, A.B., 161
Mortimer, D.C., 158
Morton, R.A., 281, 286
Möslein, E.M., 261, 263, 264
Mowat, J.H., 362
Mozingo, P.L., 372
Moyed, H.S., 347, 348
Muir, H.M., 52, 199, 201–203, 205, 208, 213, 215
Muller, O., 232
Munch-Petersen, A., 155, 156
Murray, A.W., 311
Murray, H.C., 293
Myrbäck, K., 14

Nachmansohn, D., 24, 25, 33, 34, 35, 36, 37, 38, 41
Nahinsky, P., 116
Nakayama, T.O.M., 279, 281, 282, 284
Nedswedski, S.W., 19
Needham, J., 351
Neiderhiser, D.H., 373
Nelson, N.J., 343
Nelson, N.M., 151
Nencki, M., 193
Nesbett, F.B., 129
Neuberg, C., 150
Neuberger, A., 22, 25, 52, 199, 200, 202, 203, 205, 208, 212, 213, 215, 216–218, 221, 229, 230, 245
Neufeld, E.F., 165
Neumann, E., 24
Ngan, H.I., 272
Nichol, C.A., 371, 372, 275, 279
Nicholas, H.J., 239, 240, 276
Nicholson, D.E., 187, 220, 273
Nier, A.O., 115, 116
Nimmo, C.C., 276
Nimmo-Smith, R.H., 363
Noonan, T.R., 199, 203
Nord, F.F., 14
Nordal, A., 92
Nordlie, R.C., 133, 135
Northrop, J.H., 14
Novelli, G.D., 38
Nutegeren, D.H., 184

Ochoa, S., 25, 38, 95, 96, 105, 130
O'Dell, B.L., 362, 371, 373, 381
Ogston, A.G., 20
Ohlmeyer, P., 128
Okuda, K., 303
Olby, R., 24
Olsen, R.K., 286
Olson, R.E., 284
Oparin, A., 150
Oro, J., 25
Orström, A., 331
Orström, M., 331
Ortiz de Montellano, P.R., 256, 257
Osborn, M.J., 363, 371, 372, 373, 374, 375, 377, 379, 381
Ottke, R.C., 241

Pääbo, K., 303
Paladini, A.C., 153, 154
Pardee, A.B., 14
Park, J.T., 155
Park, R.B., 275
Parnas, J.K.L. von, 144, 145
Parson, W.W., 284, 286
Partington, C.W.H., 343
Paterson, J.Y.F., 286
Pauling, L., 31
Paulus, H., 281
Peabody, R.A., 335, 336
Pearson, C.M., 159
Peat, S., 151
Peeters, G.H., 77
Peeters, G.J., 162
Pennock, J.F., 281, 286
Perrone, J.C., 52
Peters, J.M., 375, 378
Petersen, W.E., 161, 162
Peterson, D.H., 293
Peterson, W.H., 112, 113, 362
Peyer, U., 7
Pfeffer, W., 105
Pfiffner, J.J., 289, 362, 371, 373,
    381
Phelps, A.S., 113
Philips, A.H., 260, 261, 262—264
Piloty, O., 193
Pincus, G., 287, 289, 292, 293
Plaut, G.W.E., 358, 369
Plentl, A.A., 327
Plotz, J., 289
Pogell, B.M., 134
Polley, H.F., 292
Pommer, H., 277
Pontremoli, S., 65, 71, 77, 134
Popjàk, G., 25, 174, 175, 177, 243,
    245, 246, 249, 250, 255, 259,
    260, 261, 264, 265, 266, 267,
    269, 270, 271, 272, 273, 275,
    282
Poretti, G.G., 243
Porter, J.W., 174, 179, 262, 265,
    274, 279, 281, 282, 283
Prandini, B.D., 71
Preedy, J.R.K., 298
Pricer, W.E., Jr., 375, 377

Pridham, J.B., 165
Proner, M., 92
Pugh, E.L., 180
Purrmann, R., 357
Putman, E.W., 151, 165

Quastel, J.H., 33
Quayle, R., 95, 96

Raacke, I.D., 52, 53
Rabinowitch, E.I., 89, 105
Rabinowitz, J.C., 375, 377, 381
Rabinowitz, J.L., 259, 260, 287,
    294, 303
Racker, E., 67, 70, 71, 72, 74, 76,
    96, 100, 101, 102, 103, 105
Radin, N.S., 201, 202, 204, 208
Ramakrishnan, C.V., 38
Raman, P.B., 293
Ramsay, Sir W., 4
Ramsey, V.G., 284
Rappoport, D.A., 93
Ratner, S., 12, 46
Ravel, J.M., 374
Recondo, E., 165
Reich, H., 289, 294
Reichert, E., 173
Reichstein, X.X., 289
Reiser, R., 186
Reithel, F.J., 63, 161, 164
Remy, C.N., 333, 334, 348, 349
Remy, W.T., 334, 349
Reynolds, J.J., 360, 362
Reynolds, J.W., 299
Rice, B.F., 297
Rice, C.N., 116
Richardson, G.M., 125
Richert, D.A., 212
Richey, D.P., 368
Rickes, E.L., 372
Rietz, P., 232
Rilling, H., 260, 264, 265, 266,
    267, 268, 269, 270
Rilling, H.C., 287
Rimington, C., 203, 213, 216, 217,
    221, 225, 226, 227
Ringelmann, E., 274, 275
Ringer, A.J., 109

Rittenberg, D., 5, 9, 10—12, 46, 81, 171, 196, 197, 201, 202, 203, 204, 208, 240, 241, 243, 301, 328
Rizki, M.T., 233
Robinson, R., 240, 247, 250, 251
Robinson, W.G., 259
Roeller, H., 313
Roepke, R.R., 369
Rogers, L.L., 372
Röse, H., 223
Rose, I.A., 38, 130
Rosenberg, J.L., 89
Rosenblum, C., 232
Rosenkrantz, H., 297
Rosenthal, S., 380
Rousselot, L., 11
Ruben, S., 13, 81, 83, 84, 85, 86, 103, 113, 115, 116, 117, 122, 123, 124, 126, 127
Rudney, H., 259, 274, 275, 284, 286
Rummert, G., 277
Rumpf, J.A., 303
Rushton, W.A.H., 24
Russell, A.E., 273
Russell, C.S., 214, 216, 217, 218, 219
Russey, W.E., 256
Ruzicka, L., 240, 250, 252, 254, 255, 257, 276, 277
Ryan, K.J., 297, 298, 311
Ryhage, R., 266, 267, 269

Saba, N., 289
Sable, H.Z., 67
Sachs, J., 105
Sachs, P., 213
Saffan, B.D., 298
Saffran, M., 334
Sakami, W., 243, 328, 369, 379
Salokangas, A., 287
Salomon, H.S., 300
Salomon, J., 199, 203
Salomon, K., 229
Samec, M., 148
Samuels, L.T., 287, 288, 289, 292, 293, 294, 297, 298

Samuelsson, B., 303
Sanadi, D.R., 209
Sandek, W., 232
Sano, S., 227
Santer, U.V., 233
Sauberlich, H.E., 373
Sauer, F., 180
Savard, K., 289, 297
Sawyer, E., 218
Scarano, E., 334
Schachter, D., 57
Schambye, P., 77, 162
Schelble, W.J., 279
Schenker, V., 289, 292
Schlenk, F., 67
Schliep, H.J., 360
Schmiedeberg, O., 54
Schmidt, G., 144, 146, 147
Schmidt, R., 221
Schoenheimer, R., 2, 3, 8, 9, 11, 12, 45, 46, 47, 81, 240, 241, 325, 327
Schopfer, W.H., 243
Schreyegg, H., 249
Schroepfer, G.J., 273
Schulman, M.P., 211, 221, 330, 331, 333
Schwartz, S., 232
Schwenk, E., 253
Schwimmer, S., 14
Scott, D.B.M., 63, 64
Scott, J.J., 215, 216, 217, 221
Scrimgeour, K.G., 380
Searaydarian, K., 159
Seegmiller, J.E., 63, 65, 69, 95
Selye, H., 289
Semb, J., 362, 363
Senior, J.R., 189, 300
Seubert, W., 173, 183
Shack, J., 14
Shah, D.H., 265
Sharma, D.C., 293
Shaw, E., 360, 362
Shemin, D., 22, 25, 197, 198, 201, 202, 203, 204, 205, 206, 207, 208, 209, 210, 211, 212, 213, 214, 215, 216, 217, 218, 219, 221, 232, 245, 328

Shimazono, N., 65
Shimizu, K., 289
Shiota, T., 362, 363, 365, 368, 370, 381
Shive, W., 333, 369, 372, 374
Shunk, C.H., 257
Siddall, J.B., 313
Siegel, I., 371
Siekewitz, P., 328
Sih, C.J., 151
Silverman, M., 338, 373, 374, 377
Siminovitch, D., 160
Simmonds, S., 31
Simms, E.S., 334, 349
Simon, H., 359, 360
Simpson, K., 279
Siperstein, M.D., 300, 311
Siu, P., 162
Skeggs, H.R., 257
Skipper, H.E., 369
Slàma, K., 313
Slaunwhite, W.R., 289, 294
Slavik, K., 377
Sloane, N.H., 373
Slocumb, C.H., 292
Slotin, L., 121, 122, 124, 126, 128
Smallman, E., 261
Smedley-MacLean, R.D., 171
Smith, E.E.B., 155, 156, 164
Smith, J.L., 286
Smith, O.W., 297
Smith, P., 371
Smith, R.A., 218
Smith, S.W., 186
Smyrniotis, P.Z., 63, 65, 69, 70, 71, 93, 94, 95, 101
Snell, E.E., 362
Snoke, J.E., 53
Sobel, H., 249
Solomon, A.K., 127, 128
Solomon, S., 289, 298
Soloway, A.H., 251
Sonderhoff, R., 240
Sonne, J.C., 327, 330, 331, 351
Sörbo, B., 274
Sörm, F., 313
Southwick, P.L., 372
Sowden, J.C., 93

Spalla, C., 232
Spikes, J.D., 105
Springer, M.C., 284
Srere, P.A., 70
Stadtman, E.R., 38, 171, 173, 174
Stadtman, F.H., 162
Stansly, P.G., 173
Staple, E., 289, 301, 303
Steinberg, D., 51, 52, 273
Stekol, J.A., 369
Stephenson, M., 1, 21
Stepka, W., 91
Stern, A., 223
Stern, J.R., 38
Stetten, D., 75
Stetten, M.R., 75, 333, 369
Stiles, P.G., 109
Still, J.L., 76
Stocking, C.R., 160
Stoffel, W., 184
Stockstad, E.L.R., 362, 363, 373
Stokes, J.L., 372
Stokes, W.M., 273
Stone, D., 289
Stone, N.E., 53
Stone, R.W., 76
Storey, I.D.E., 156, 157
Stotz, E.H., 25
Strickland, K.P., 177, 178
Strominger, J.L., 156
Stuckwish, C.G., 115
Stumpf, P., 180
Stumpf, P.K., 69, 70, 93, 100
SubbaRow, Y., 362, 363, 373
Suld, H.M., 303
Sumiya, C., 227
Summer, J.B., 14
Sutherland, E.W., 159
Sutherland, G.L., 374
Suyter, M., 186
Suzue, G., 282
Suzuki, I., 70
Swanson, M.A., 148
Sweeley, C.C., 313
Swingle, W.W., 289
Switzer, R.L., 335
Synge, R.L.M., 91
Szent-Györgyi, A., 89

Tabor, H., 377, 378
Taggart, J.V., 57
Tait, G.H., 25, 229, 230
Takemura, K.H., 245
Takiguchi, H., 65
Talalay, P., 294
Talamo, B., 180, 181
Talmage, P., 218
Tanabe, Y., 227
Tatum, E.L., 15, 241
Täufel, K., 249
Tavormina, P.A., 259
Tchen, T.T., 251, 256, 257, 260,
    261—264, 267, 275, 311
Tener, G.M., 334
Tennebaum, M., 33
Tesar, C., 197
Tewfik, S., 93, 100
Thaler, K., 249
Thimann, K.V., 85
Thomas, H., 240
Thomas, K., 5
Thompson, J.D., 298
Thompson, R.H.S., 14
Thomson, J.A., 313
Thomson, P.J., 276
Tietz, A., 174, 175, 177
Titchener, E.B., 175, 178
Todd, D., 253
Tolbert, N.E., 92
Toomey, R.E., 182
Trams, E.G., 178
Traniello, S., 134
Trebst, A.V., 105
Trenner, N.R., 372
Tristram, G.R., 249
Tros, B.M., 313
Trowell, A.O., 19
Trucco, R.E., 153, 154, 155, 161
Trudinger, P.A., 105
Tsujimoto, H.Y., 105
Tsujimoto, M., 249
Turner, D.H., 155
Turner, J.F., 155

Udenfriend, S., 212, 217
Upper, C.D., 276
Urey, H.C., 3, 12

Utter, M.F., 25, 76, 89, 96, 130,
    131, 132, 133, 137

Vagelos, P.R., 180, 181, 182, 188
Vanaman, T.C., 181
Van Baalen, J., 174
Vandenbelt, J.M., 371
Van de Wiele, R., 289, 298
Van Niel, C.B., 21, 113, 123
Van Tamelen, E.E., 257
Varma, T.N.R., 279
Vennesland, B., 125, 127, 128, 129
Vieira, E., 360
Villar-Palasi, C., 155
Villemez, C.L., 165
Vishniac, N., 96, 105
Vogel, H., 233
Vogelmann, H., 232
Vogt, M., 293
Von Bülow-Köster, J., 185
Von Korpp, R.W., 38
von Muralt, A., 243
Voser, W., 250, 252

Wacker, H., 359, 360, 362
Waelsch, H., 36, 37, 371, 377
Wagner, F., 232, 233
Wagner, H., 185
Wagner-Jauregg, T., 357
Waite, M., 178
Wakil, S.J., 25, 174, 175, 176, 177,
    178, 179, 180, 181, 182, 183,
    184
Waldenström, J., 213
Waldschmidt, M., 359, 360
Waldvogel, M.J., 67
Wallach, O., 239
Waller, C.W., 362, 363
Wang, C.H., 75
Wang, V.S., 294
Warburg, O., 61, 63, 65
Warms, J.V.B., 259,
Warren, L., 338, 345, 379
Wasilejko, H.C., 333, 348
Wasserman, H.H., 233
Wasson, G., 179
Watkins, W.M., 163, 164, 165
Watson, C.J., 232

Watt, W.B., 357
Weaver, W., 3, 4
Weidenhagen, R., 14
Weigl, J.W., 98
Weisman, R.A., 365, 368
Weiss, K., 371
Weiss, S., 371
Weiss, S.B., 186, 187, 188
Weissbach, A., 94, 95, 96
Welch, A.D., 371, 372, 379
Wells, L.W., 279, 282
Wells, W.W., 273
Werbin, H., 287
Werkman, C.H., 22, 76, 110, 111, 112, 115, 116, 119, 125, 131
Wertheim, M., 149
West, C.A., 276
Westall, R.G., 213
Weygand, F., 357, 359, 360
Whatley, F.R., 105
Wheatley, V.R., 249
Whiteley, H.R., 29, 375, 379
Wiechert, R., 311
Wieland, O., 186
Wiener, H., 325, 350, 351
Wiener, M., 297
Wiggins, R.A., 298
Wilkinson, J.F., 153
Willett, J.D., 257
Williams, R.H.J., 282
Williams, R.J., 362
Williams, V., 269, 270, 271, 272
Williams, W.J., 331
Williamson, I.P., 181
Williamson, D.H., 169
Williamson, S., 19
Willis, J.L., 276
Willstätter, R., 193, 194
Wilson, A.T., 94, 101, 103
Wilson, D.W., 328
Windaus, A., 7

Wintersteiner, 289
Wittenberg, J., 201, 204, 205, 206, 207, 208, 209, 214, 221
Witting, L.A., 262, 273
Wolf, D.E., 257, 372
Wolf, G., 350
Wolfe, R.G., 63
Wong, S.M., 269, 270, 271, 272
Wood, H.G., 18, 22, 25, 76, 77, 89, 111, 112, 113, 114—116, 121, 123, 124, 125, 129, 131, 134, 162, 163
Wood, R.G., 76
Woods, D.D., 21, 113, 363, 364, 368, 369
Woodward, R.B., 250, 251, 252, 253
Wormall, A., 249
Woronick, C.L., 131
Wright, D.L., 257
Wriston, J.C., 212
Wu, H., 197
Wuersch, J., 243, 247
Wyngarden, L., 377, 378

Yamaha, T., 157
Yamamoto, H., 279, 284
Yamamoto, S., 257
Yanari, S., 53
Yefimochkina, E.F., 57
Yokoyama, H., 279
Yoshida, A., 63

Zabin, I., 241, 243
Zaffaroni, A., 287, 292
Zakrzewski, S.F., 371, 372, 375
Zaleski, J., 193
Zerweck, W., 223
Ziegler, J.A., 130
Ziegler-Günder, I., 359, 360
Zottu, S., 159